THE RADIATION CHEMISTRY
OF MACROMOLECULES

VOLUME I

CONTRIBUTORS

Malcolm Dole

Péter Hedvig

K. C. Humpherys

Ronald M. Keyser

Leo Mandelkern

Roger H. Partridge

Osamu Saito

Kozo Tsuji

Ffrancon Williams

H. J. Wintle

The Radiation Chemistry of Macromolecules

Edited by   *Malcolm Dole*

Department of Chemistry
Baylor University
Waco, Texas

VOLUME I

 1972

ACADEMIC PRESS   *New York and London*

CHEMISTRY

ACADEMIC PRESS, INC.
111 Fifth Avenue, New York, New York 10003

United Kingdom Edition published by
ACADEMIC PRESS, INC. (LONDON) LTD.
24/28 Oval Road, London NW1

LIBRARY OF CONGRESS CATALOG CARD NUMBER: 71-182668

PRINTED IN THE UNITED STATES OF AMERICA

Contents

FUNDAMENTAL PROCESSES AND THEORY

Chapter 1 Introduction

Malcolm Dole

Chapter 2 Early Processes in Radiation Chemistry and the Reactions of Intermediates

Ffrancon Williams

v

Contents

EXPERIMENTAL TECHNIQUES AND APPLICATIONS TO POLYETHYLENE

* Section II, D was contributed by K. C. Humpherys.

Contributors

Numbers in parentheses indicate the pages on which the authors' contributions begin.

MALCOLM DOLE, Department of Chemistry, Baylor University, Waco, Texas (3, 265, 335)

PÉTER HEDVIG, Research Institute for Plastics, Budapest, Hungary (55, 73, 127)

K. C. HUMPHERYS, Nuclear Science and Instrumentation Labs, EG & G, Goleta, California (274)

RONALD M. KEYSER, Union Carbide Corporation, Nuclear Division, Oak Ridge, Tennessee (145)

LEO MANDELKERN, Department of Chemistry and Institute of Molecular Biophysics, Florida State University, Tallahassee, Florida (287)

ROGER H. PARTRIDGE, Division of Materials Applications, National Physical Laboratory, Teddington, Middlesex, England (25, 193)

OSAMU SAITO, Department of Physics, Chuo University, Tokyo, Japan (223)

KOZO TSUJI, Central Research Laboratory, Sumitomo Chemical Company, Ltd., Osaka, Japan (145)

FFRANCON WILLIAMS, Department of Chemistry, University of Tennessee, Knoxville, Tennessee (7, 145)

H. J. WINTLE, Department of Physics, Queens University, Kingston, Ontario, Canada (93, 109)

Preface

Ten and twelve years have now elapsed since the publication of the classic works on the radiation chemistry of high polymers by Chapiro and by Charlesby, respectively, and since then an enormous amount of material in this field has been published, both of theoretical and experimental interest. Furthermore, many valuable industrial applications have been made of the high-energy irradiation of polymers in this intervening period. Accordingly, a comprehensive review of the radiation chemistry of macromolecules should now be of considerable interest and benefit to those scientists, both academic and industrial, who are working in this field. We believe that many of the contributions to this two-volume treatise will prove to be of significance to radiation chemistry in general and that scientists working with small molecules may find certain chapters worthy of their attention.

Theoretical treatment of the radiation chemistry of macromolecules is concentrated for the most part in the first volume of this treatise. Chapter 2 by Professor Ffrancon Williams deals with fundamental processes of radiation effects in general, while the following chapter by Dr. R. H. Partridge discusses basic ideas concerning energy transfer. Chapter 4 by Dr. Hedvig gives the theory of free radicals, and it is not until we come to his Chapter 5 that the subject matter, this time on the description of molecular mobilities in polymers, is devoted exclusively to polymers. Professor H. J. Wintle's Chapters 6 and 7 on reactions and electrical conductivity in the solid state are similarly general in approach, but polymers are also discussed. Chapter 8 by Dr. Hedvig gives more detailed information on electrical conductivity induced in polymers by radiation. Chapter 9 by Keyser, Tsuji, and Williams and Chapter 10 by Partridge describe the trapping of electrons in polymers and thermal luminescence phenomena resulting therefrom. Inasmuch as the radiation cross-linking of polyethylene and of other polymers is of considerable industrial as well as of theoretical interest, Chapter 11 on statistical theories of cross-linking by Professor O. Saito is most necessary to round out the theoretical treatment. Chapter 12 on experimental techniques is followed by one giving the first detailed description of the radiaton chemistry of a single polymer, that of polyethylene. Professor Mandelkern's Chapter 13 includes a discussion of the complicated morphology of polyethylene as well as of cross-linking, vinyl group decay, and G-values. Finally, this volume ends with a short chapter on free radicals in irradiated polyethylene.

The cooperation of the authors and of the staff of Academic Press in writing and assembling this volume is greatly appreciated. Baylor University generously supplied considerable secretarial help. Grateful acknowledgment is expressed to the U.S. Atomic Energy Commission for its support over a number of years of fundamental research in the radiation chemistry of polymers. My research has also benefited from income from the chair in Chemistry at Baylor University endowed by a gift from The Robert A. Welch Foundation.

Malcolm Dole

Contents of Volume II

FUNDAMENTAL PROCESSES AND THEORY

1

Introduction

Malcolm Dole

Department of Chemistry, Baylor University, Waco, Texas

I. Importance of Macromolecular Radiation Chemistry

Long-chain molecules are unique in so far as their response to high-energy radiations is concerned because a few radiologically produced cross-links can alter profoundly the physical and mechanical properties of the irradiated sample. In a typical polyethylene, for example, of weight average molecular weight of 10^5, weight to number average molecular weight ratio of 9, and assuming no chain scission, only one cross-link per 14,400 monomer units is necessary (Saito *et al.*, 1967) to attain the point of incipient insolubility or gelation. Conversely, a few chain scissions would also have a significant effect, far greater in proportion to their number per monomer unit in a long-chain molecule than in a low molecular weight analog. Inasmuch as the molecular weight distribution in a synthetic polymer influences the degree of gelation as a function of dose, the latter data can be used in a reverse fashion to draw inferences concerning the molecular weight distribution in the polymer.

Most macromolecules are solid at room temperature, unless the molecular weight is quite low, and very often free radicals can be trapped at room temperature and their reactions studied on a convenient time scale. In other words, pulse radiolysis experiments are not entirely necessary to deduce the reactions of the major free radical species. Perhaps pulse radiolysis studies might reveal exciting new information about "very early effects" (Burton, 1970), especially as regards electronically excited or ionized groups in the solid, but, as yet, such experiments have not been carried out. Silverman and Nielsen (1968) performed pulse radiolysis studies on molten polyethylene in order to detect conjugated dienes, and Lerner (1969) studied the ESR spectra of oriented poly(tetrafluoroethylene) subjected to 20-nsec electron beam pulses, but in neither case were "early effects" investigated.

Synthetic polymers can usually be prepared in the form of thin films which are very convenient for infrared and ultraviolet studies. The thickness of the films can be varied to produce the optimum optical absorption. Furthermore, the use of thin films promotes the rapid attainment of gaseous equilibrium between gases in the ambient atmosphere and dissolved in the solid. In the case of studies such as those (Waterman and Dole, 1970) involving the catalytic influence of molecular hydrogen on free radical conversion reactions, the solubility equilibrium is attained so rapidly that the latter is not the rate-controlling process.

The radiation improvement of solid polymers is the basis for an ever growing list of industrial products. In solid state physics one talks of "radiation damage," but very often in the radiation chemistry of macromolecules, we can speak of "radiation improvement." Such improvement is usually based on the formation of insoluble, highly cross-linked three-dimensional network structures. In the case of synthetic polymers the formation of covalent intermolecular bonds, i.e., of cross-links, can be done at room temperature after the solid has been shaped into some industrially important form, such as the tubing or film, for example. The production of such a covalent link in the solid at room temperature is a really remarkable reaction.

II. Relationship between the Radiation Chemistry of Low and High Molecular Weight Compounds

There are, of course, many similar radiological reactions in macromolecules and in low molecular weight analogs. The author (1950) noticed this at once when irradiating a low-density polyethylene in the Argonne Heavy Water Pile in 1948. Table I lists the composition of the gas produced by this irradiation as well as the composition of the gases obtained by Breger (1948) during the irradiation of hexadecane with alpha particles. The comparison given in Table I is striking. Other examples could be given.

The differences in the radiation chemistry of macromolecules as compared to low molecular weight compounds are chiefly the pronounced effect on macromolecular properties of cross-linking and chain scission mentioned

TABLE I

ANALYSIS OF GAS LIBERATED ON IRRADIATION (PERCENT COMPOSITION)

	Polyethylene (pile irradiation)	Hexadecane (alpha particles)
H_2	96	95.5
CH_4	0.4	0.9
C_2H_6	0.9	1.0
C_3H_8	0.2	1.0
$n\text{-}C_4H_{10}$	0.7	1.2
$i\text{-}C_4H_{10}$	0.3	
C_3H_6	0.2	
C_4H_8	0.2	
N_2	1.1	
	100.0	99.6

above, the fact that the radiation chemistry of small molecules has been devoted mostly, although not exclusively, to the gaseous or liquid state, the greater importance of cage effects in the radiation chemistry of solid macromolecules as compared to the liquid state cage effects, and the relatively greater ease of studying free radical reactions in the solid polymers. As an example of the latter, the catalytic effect of molecular hydrogen in converting alkyl radicals to allyl was observed (Waterman and Dole, 1970) in solid polyethylene at room temperature, whereas this effect had not hitherto been observed in the radiation chemistry of low molecular weight hydrocarbons.

III. Comparison of Photolysis and Radiolysis

Although the study of the photolysis of synthetic high polymers is important from the standpoint of illuminating the effect of sunlight in promoting polymer degradation, the photolysis of high polymers is not an industrial method for the beneficiation of polymers, as far as this author is aware. Although the cross-linking of polyethylene has been accomplished (Oster, 1960) by means of ultraviolet light, photosensitizers have to be added and the penetration power of the ultraviolet light does not compare to that of the highly energetic and ionizing gamma rays. In some cases uv light depolymerizes polymers—polystyrene (Calvert and Pitts, 1966), for example.

Radiation chemistry has been defined (Dole, 1965) "as that body of science that encompasses chemical effects produced by ionizing radiations." Most of the studies on macromolecules have been carried out using ^{60}Co-gamma rays or high-speed electrons. Radiation from atomic piles was used in the early days by Charlesby (1952) and the author (1950), but it is not much used today. Ionizing radiation can produce electronic excitation as well as ions and may also produce ions and free radicals in excited states. Thus, it is important that possible chemical effects due to electronically excited groups as well as to ionized groups be borne in mind.

REFERENCES

Breger, I. A. (1948). *J. Phys. Colloid Chem.* **52**, 551.

Burton, M. (1970). *In* "Symposium on Very Early Effects" (W. P. Helman, A. Mozumder, and A. Ross, eds.). Buenos Aires, Argentina.

Calvert, J. G., and Pitts, J. N., Jr. (1966). "Photochemistry," p. 505. Wiley, New York.

Charlesby, A. (1952). *Proc. Roy. Soc. (London)* **215A**, 187.

Dole, M. (1950). *Rep. Symp. IV. Chem. Phys. Radiat. Dosimet.* p. 134. Army Chem. Center, Maryland.

Dole, M. (1965). *In* "Crystalline Olefin Polymers" (R. A. V. Raff and K. W. Doak, eds.), Vol. II, p. 846. Wiley (Interscience), New York.

Lerner, N. R. (1969). *J. Chem. Phys.* **50**, 2902.

Oster, G. (1960). *Proc. Conf. Large Radiat. Sources Ind., Warsaw, 1959* **1**, 321.

Saito, O., Kang, H. Y., and Dole, M. (1967). *J. Chem. Phys.* **46**, 3607.

Silverman, J., and Nielsen, S. O. (1968). *Polymer Preprints, Amer. Chem. Soc.* **9**, 296.

Waterman, D. C., and Dole, M. (1970). *J. Phys. Chem.* **74**, 1913.

2

Early Processes in Radiation Chemistry and the Reactions of Intermediates

Ffrancon Williams

Department of Chemistry, University of Tennessee, Knoxville, Tennessee

I. Introduction

Progress in radiation chemistry during the last decade has resulted in a more complete inventory of the elementary physical and chemical processes which can contribute to the mechanisms of radiation effects in molecular systems (Ausloos, 1968). However, it is only in model systems that these fundamental processes have been clearly demonstrated, and their relative importance in any real system is difficult to predict. For many materials of considerable practical importance, and especially polymers, relatively few elementary reactions have been elucidated by experiment, and frequently one can only speculate about the nature of the processes that are responsible for the overall chemistry.

It is a formidable task to unravel the entire spectrum of chemical reactions induced by high-energy radiation in any particular system, and the problems are comparable to those encountered in other applied fields of chemical kinetics. Some idea of this overall complexity can be gained from the fact that some 20 elementary reactions are used to describe the chemistry of the primary species in the radiolysis of oxygenated water (Thomas, 1969). Nevertheless, by careful experimental design it is frequently possible to study certain characteristic reactions without hindrance from other processes, and many advances have followed from this line of attack. The ability to detect and identify transient intermediates is clearly of central importance, and most experimental research is now concentrated in this area. Indeed, radiation chemistry has made a number of novel contributions to chemical kinetics by virtue of the ease with which certain important intermediates can be generated. After a brief account of the particulars of energy absorption, this chapter will attempt to give a general description of the main classes of intermediates which are formed in radiation chemistry.

II. General Features of Energy Absorption and Track Structure Produced by Charged Particles

The principles governing the absorption of high-energy radiation are well established (Whyte, 1959), and an excellent account (Hart and Platzman, 1961) has been given of the underlying physical theory. For most high-energy radiations of practical interest, the bulk of the energy is eventually deposited in the medium through the release of fast electrons. It is the interaction of these high-energy electrons with atoms or molecules that provides the commonest mode of excitation in radiation chemistry.

The motion of a fast electron through condensed matter is punctuated by a copious number of well-separated excitation events, each of which may constitute either a single or a group of individual molecular processes (see below). In this way the particle is slowed down and its energy is progressively transferred to the medium. The incremental energy loss with distance is called the linear energy transfer (LET) and this is a function of the velocity

v and numerical charge z of the ionizing particle according to the Bethe formula (Fano, 1963),

$$-dE/dx = (4\pi e^4 z^2 NZ/mv^2) \ln (2mv^2/L)$$

where NZ refers to the number of electrons per unit volume (1 cm^3 for cgs units) of the absorber, e and m are the charge and mass of the electron, and L is the mean excitation potential of the absorber, which for all practical purposes can be set equal to its ionization potential. This relation is valid for heavy particles and electrons below relativistic energies, i.e., below $mc^2/2$ or 0.255 MeV for electrons. For an electron of energy E (in electron volts) traveling through polyethylene, the LET is given in units of electron volts per angstrom by the numerical relation $(980/E) \log 0.2E$. Thus, the LET is slightly greater than 0.01 eV/Å when $E = 0.25$ MeV and increases to 2.3 eV/Å at $E = 1$ keV.

Turning now to the effect of the high-energy particle on the medium, a consequence of the increase in LET with decreasing energy is that the spacing between consecutive excitations along the main track decreases from a few thousand angstroms at an electron energy of ≈ 1 MeV to the order of 100 Å at 5 keV. Below 1 keV, the successive excitation processes that are produced in the wake of the electron's path begin to merge together so that in pictorial terms the distribution changes from a "string of beads" above 1 keV to a continuous ribbon at lower energy.

Hitherto, excitation has been used as a general term to denote the electronic interaction with the medium. As long as the kinetic energy of the electron is well above the ionization potential, there is always a high probability that molecular ionization will occur in each of these excitation events, resulting in the production of a secondary electron. This has important consequences with regard to the detailed track structure, for the secondary electron will frequently possess enough energy to bring about further ionization on its own account. Where the energy imparted to the secondary electron is below 100 eV, a compact group of ions and excited molecules commonly referred to as a spur is formed along the main track. Electrons ejected with higher energies up to 5 keV produce short secondary tracks of their own where the successive spurs are now more closely spaced than in the adjacent main track. Much less frequently, an electron is knocked out of a molecule with sufficient energy to produce a distinctive long track, and this is called a delta ray.

Mozumder and Magee (1966) have discussed these geometrical effects in considerable detail. They classify the spatial distribution of sequential events, including both ionization and excitation, into three regions comprised of isolated spurs, short tracks, and blobs, according to the mean separation distance between the individual spurs. Isolated spurs are considered as noninteracting entities separated in space over the crucial period of diffusive expansion during which time most of the nonhomogeneous reactions occur in the individual spur. Short tracks are defined as regions in which consecutively formed spurs begin to overlap. In the blob, the spurs merge together to form a larger continuous region of high ionization density. For recoil electrons from ^{60}Co-γ rays, the fractions of the total energy dissipated in isolated spurs, short tracks, and blobs are computed to be 0.60, 0.25, and 0.15, respectively (Mozumder, 1969).

When the above description is put into suitable quantitative form, it provides the basis for the well-known model of track reactions in radiation chemistry which has been developed principally by Magee and his co-workers (Samuel and Magee, 1953; Kuppermann, 1961). It has long been recognized that important kinetic consequences follow from the initial formation of a nonhomogeneous distribution of reactive species such as free radicals. Accordingly, the basic problem is to calculate the extent to which the various possible bimolecular reactions take place in the localized regions of high concentration before spatial homogeneity of the reacting species is attained by diffusive transport. Most of the advanced theoretical treatments of non-homogeneous kinetics have been concerned with reactions in aqueous systems (Kuppermann, 1961). For liquids of high dielectric constant, ions can be included within the scope of the general treatment developed for neutral radicals whenever dielectric relaxation is faster than translational diffusion, which is usually the case. On the other hand, the problem of geminate ion recombination in nonpolar liquids must be treated differently if the mean separation between ions is less than about 280 Å, because the strong Coulomb interaction between the ions will perturb the mutual diffusive motion in these circumstances (Williams, 1964). These questions are considered further below.

III. Excitation by Charged Particles

Although the marked spatial inhomogeneity of energy deposition by charged particles serves to differentiate radiation chemistry from high-energy (vacuum-ultraviolet) photochemistry (Ausloos, 1966), there is believed to be little or no intrinsic difference in the nature of the excitations induced by energy losses from fast charged particles and those brought about by the absorption of high-energy photons (15–20 eV) in single encounters. Only in the case of low-energy electrons is it necessary to consider specifically the excitation of optically forbidden transitions (Kuppermann and Raff, 1963), and in most circumstances the contribution of such processes in radiation chemistry is thought to be minor as compared to optically allowed transitions.

Platzman (1962) has emphasized that the interaction of fast charged particles with molecules is formally equivalent to the absorption of photons in the energy range of 10–30 eV wherein lies most of the oscillator strength for the electronic transitions of valence electrons. Accordingly, there is no basic distinction between far ultraviolet photochemistry and radiation chemistry, and the term "ionizing radiation" can be applied to both. Summarizing, the very high energy of charged particles is of theoretical significance in radiation chemistry only insofar as it affects the spatial distribution of the excitation events. Where "high-energy chemistry" is used synonymously with radiation chemistry, it should be made clear that the "high energy" refers to ≈ 20-eV excitations and *not* to the high energy of the incident particle.

From a practical standpoint, the action of high-energy radiation in bringing about the formation of highly excited states of molecules assumes particular importance because the energetic particles or quanta are able to penetrate deeply into condensed matter. In material of density 1 g cm^{-3}, the range of 1.25-MeV electrons is 0.5 cm and a beam of 1.25-MeV photons is attenuated by less than 5% in traveling 1 cm. Hence, X rays or γ rays can be used to produce high-energy electrons more or less uniformly throughout a large body of material, and the simplicity of this irradiation technique is suitable for much routine work when there is no restriction on the time required to attain the required dose. It should also be mentioned that, for photons and electrons of incident energies around 1–2 MeV, the details of energy absorption are only modified in a very minor way by differences in the molecular structure of the irradiated object. Consequently, all molecules are affected by high-energy radiation in contrast to the highly selective absorption of visible and ultraviolet radiation.

Hitherto, we have not considered the actual nature of the electronic states excited directly by high-energy radiation except to point out that these excitation energies are considered to be comparable to the lowest ionization potential. Platzman (1962, 1967) has calculated the excitation spectrum to be expected for a number of small molecules, including methane. The yield of excited states is proportional to the ratio of the oscillator strength at the transition energy to that energy, and this function reaches a maximum at the lowest ionization potential of the absorber, as shown for methane in Fig. 1. Excitation above the ionization potential leads not only to direct ionization but also to the formation of superexcited states. These latter states may undergo either a delayed ionization or some type of internal conversion process whereby ionization is avoided through degradation of the excess electronic energy. In a qualitative way, the existence of superexcited states helps to explain the well-established fact that the average expenditure of energy required to produce an ion pair greatly exceeds the ionization potential.

Thus, it is found that ionization yields in radiation chemistry generally do not exceed 4 events/100 eV, so about half the total energy transferred from charged particles is used to form excited states, either of ions or of neutral molecules. Some progress has been made in characterizing the chemical reactivity of superexcited molecules in the gas phase (Ausloos and Lias, 1968), but at present we have little understanding of these precursor states in the condensed phase.

The preceding discussion of excitation by fast charged particles is based upon the optical approximation which assumes that only optically allowed transitions are excited. For electrons of energy greater than about 100 eV, this assumption is considered to be valid. However, the degradation spectrum

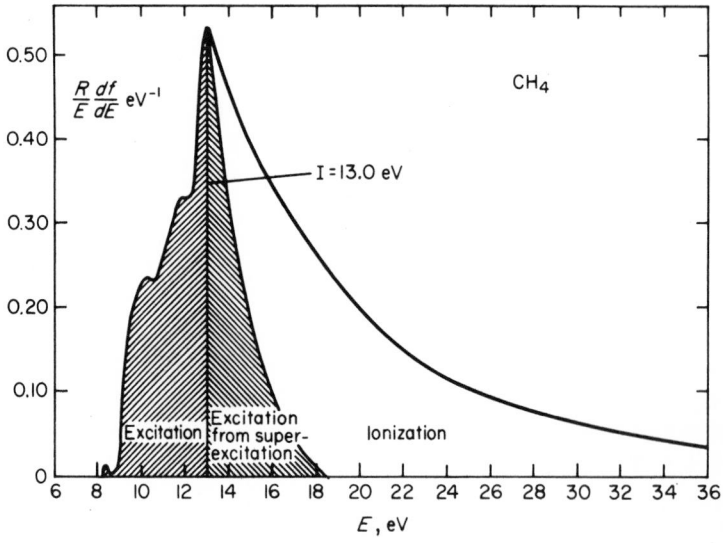

Fig. 1. Excitation spectrum of methane, with inclusion of the superexcited states. The ordinate is proportional to the number of excited states of energy E generated by fast charged particles. R is the Rydberg energy (13.60 eV), and f is the oscillator strength. [Reproduced with permission from R. L. Platzman, *The Vortex* **23,** 372 (1962).]

(Spencer and Fano, 1954) of electron energies includes a large proportion of slow electrons below 100 eV, so the possibility of optically forbidden transitions cannot be excluded. Unfortunately, it is hard to obtain convincing evidence for such processes in radiation chemistry because in most systems there are several reaction paths which can account for an observed intermediate. Of course, this lack of uniqueness is even more applicable to attempts to trace the origin of reaction products. Thus, it appears that definitive evidence for a contribution from optically forbidden transitions would require the detection of specific excited states under macroscopic conditions, a task that is most difficult to achieve.

IV. Fates of Low-Energy Electrons

As already described, molecular ionization is one of the characteristic processes of radiation chemistry. Therefore, a knowledge of the interactions of low-energy electrons with molecules is complementary and assumes prime importance. This branch of radiation research has developed rapidly in recent years, and many original contributions have emerged.

In this discussion, low-energy electrons are those electrons whose energies have been reduced below the lowest electronic transition energy in the system. These electrons are further moderated to thermal energies through the excitation of vibrational and rotational transitions in molecules. This energy-transfer process may proceed through the virtual formation of negative ions (Chen and Magee, 1962), but many of these states are not usually regarded as chemical intermediates because their lifetimes are extremely short, generally on the order of a few vibrational or rotational periods ($\approx 10^{-12}$ sec). A theoretical description of these temporary negative-ion states of simple molecules has been given in considerable detail (Chen, 1969).

There is now abundant evidence that permanent electron attachment takes place to many molecular species in radiation chemistry. Therefore, this process assumes equal importance with that of ionization in many systems. The term "permanent" is used here to mean that the negative-ion product is stable with respect to electron detachment, so as to differentiate it from the temporary negative ions mentioned earlier. Actually there are long-lived temporary negative ions such as SF_6^- which, at least on the basis of their gas phase properties, appear to be intermediate between these two classes (Compton and Huebner, 1970).

Almost all the detailed information which is relevant to radiation chemistry about the energetics of direct electron-capture processes comes from low-energy electron spectroscopy in the gas phase. A thorough review of this field has been published (Compton and Huebner, 1970). It should be noted that both homogeneous and heterogeneous electron-transfer processes have been studied extensively in solution electrochemistry, and this work provides data on the electron affinities of molecules, but in these reactions the electron is rarely considered as a separated entity (Marcus, 1968). For electron-attachment processes in the radiation chemistry of the condensed phase, it is uncertain whether the energy dependences observed in electron beam experiments with gases are at all applicable. This uncertainty springs from the lack of any detailed understanding of the radiation chemical mechanism of charge separation in liquids and solids.

In spite of these reservations, some attention needs to be focused on the possible significance of resonance and thermal electron-capture processes in radiation chemistry. For resonance capture, the cross section is a maximum at some epithermal electron energy whereas thermal capture involves a large cross section for thermalized electrons. Therefore, the efficiency of resonance capture in the radiation chemistry of gases would seem to depend on the moderation of the electrons and, specifically, on the increments of energy loss sustained by the electrons in passing through the critical region of resonance capture. Unfortunately, the electron-degradation spectrum is generally unknown in this low-energy region of interest (below 5 eV), so *a priori* estimates are impossible. Judging from the results of radiation chemical experiments in gases, liquids, and solids, most of the effective electron scavengers are molecules which are known to undergo thermal electron capture. At present there is little evidence to suggest that resonance capture of electrons with several electron volts of energy contributes significantly to radiation chemical processes.

Molecules which undergo thermal electron capture invariably possess low-lying orbitals to accommodate the additional electron (Whelan, 1969). These include typical aromatic molecules such as biphenyl and most compounds having functional groups such as

$$\diagdown_{\diagup}C{=}O, \quad \text{and} \ {-}C{\equiv}N$$

with π antibonding orbitals. The optical and ESR spectra of many radical anions have been thoroughly characterized by chemical studies, so their identification in radiation chemical investigations is generally straightforward.

Dissociative electron attachment usually results in the formation of a stable diamagnetic negative ion. A large number of halogen compounds function as good electron scavengers in this class, well-known examples being CH_3Br and CCl_4. The lowest unoccupied orbital in these compounds is σ-antibonding, and it is likely that electron capture is followed almost immediately (within a few bond vibrations) by dissociation to a radical–ion pair formed on a repulsive potential energy surface. Recently, weakly interacting methyl radical–bromide ion pairs have been observed directly by ESR following dissociative electron capture by methyl bromide in a crystalline matrix (Sprague and Williams, 1971).

There is no evidence to suggest that thermal electron attachment to molecules such as water, ammonia, and the saturated hydrocarbons results either in dissociative electron capture or in the formation of a monomer radical anion. However, electrons released on γ irradiation or photoionization are found to be stabilized in the liquid and glassy states of many of these compounds, and it is customary to regard these excess electrons as being localized in physical traps. There has been much speculation about the nature of the trapping sites, but it is generally agreed that they are inter-molecular in origin. The most popular theoretical model (Jortner, 1964) is based on electron localization in cavities or voids which are associated with structural imperfections in liquids and disordered solids. In this approach (Copeland *et al.*, 1970) the polarization of the surrounding medium by the trapped electron is described in terms of bulk properties such as dielectric constant and surface tension. By adjustment of the cavity radius, this model gives a value for the optical transition energy ($h\nu_{max}$) of the hydrated electron in agreement with the experiment result. On the other hand, the continuum model does not readily explain the observation that the absorption spectrum of the trapped electron in ice resembles that of the hydrated electron (Taub and Eiben, 1968), and this particular result favors the tetrahedral vacancy description (Natori and Watanabe, 1966) which emphasizes a well-defined local structure of the hydration shell.

According to both of these models for the solvated electron, the binding energy of the electron originates from the polarization of the surrounding molecules. Historically, these descriptions were attractive because solvated electrons appeared to be produced only in polar solvents. However, the discovery of trapped electrons in nonpolar glasses (for a review, see Eiben, 1970) raises some awkward problems, and the emphasis on solvent polarity as a stabilizing factor may be misplaced.

Recent work by the writer and his colleagues (Kerr and Williams, 1971) has provided evidence that the structure of certain radical anions involves the population of supramolecular orbitals built from molecular antibonding orbitals of two separated molecules. An important feature here is that the lowest supramolecular orbital is bonding between the molecules so that the excess electron is stabilized relative to its occupation of a single molecular orbital. Extrapolation of this concept to explain the stability of solvated and trapped electrons is premature, but the general idea of the excess electron residing in a supramolecular orbital embracing several molecules is not unreasonable. One might add that excess electrons in liquids and solids can hardly avoid encountering the accessible molecular orbitals during the process of thermalization, so the mechanism of stabilization could even proceed through a transient negative ion. Moreover, this mechanism could explain the fact that solvent-trapped electrons can be produced in saturated hydro-carbons by γ irradiation even though chemical methods based on the use of alkali metals or other electron donors are ineffective.

To conclude this section on electron trapping, mention should be made of the exact physical theory (Fox and Turner, 1967) which predicts bound states of electrons in the field of a stationary electric dipole possessing a dipole moment greater than 1.625 D. At present there is no convincing experimental evidence to support this interaction. Moreover, no realistic theory has been presented to show how the electron–dipole binding energy is modified by the constituent electrons of the dipolar molecule (Byers-Brown and Roberts, 1967).

Figure 2 summarizes the various processes leading to charge separation and excited state formation in the radiation chemistry of liquids and solids.

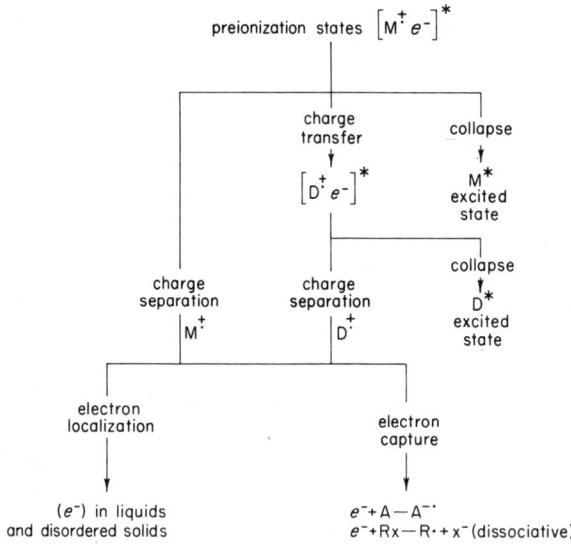

Fig. 2. Schematic representation of processes leading to charge separation in the condensed state.

In a solid, the preionization or superexcited states can be regarded essentially as exciton states such that the positive ion–electron pair is mobile. Consequently, charge transfer may take place efficiently to a suitable molecule D of lower ionization potential although this electron donor is only present in very low concentration. Thus, it is conceivable that positive charge transfer is an extremely fast process which precedes charge separation. Reliable information is lacking about the distances of positive ion–electron separation achieved in the condensed state, but most estimates are in the range 50–100 Å. These distances should not necessarily be identified with the migration distance of the mobile or "quasifree" electron before trapping, particularly if the motion of the preionized state resembles that of an exciton.

In writing about the very early stages of radiolysis, one becomes aware of the limitations involved in a purely deductive approach. Although generalities can be stated, theory cannot predict the specific molecular processes which follow excitation, ionization, and electron attachment with any degree of confidence. Even where information about these elementary steps is available from gas phase experiments, it is hazardous to assume these findings can be applied indiscriminately to liquid and solid systems. This is not to deny the usefulness of theory and experimentation as they apply to isolated molecules, but the problems in the condensed phases are less tractable and an inductive approach from experiment is generally more useful.

V. Kinetics of Reactive Species in the Condensed Phase

A comprehensive description of a radiation chemical system involves the identification of the reactive species and elucidation of the reaction paths through kinetic studies. We now turn to examine some of the general kinetic problems encountered in the radiolysis of the condensed phase. As mentioned earlier, a nonuniform distribution of intermediates is laid down by the radiation. Some of the species will react in the spur while the remainder will undergo diffusive separation to achieve a uniform spatial distribution and so contribute to the steady state population. Whereas homogeneous kinetics can be used to describe the subsequent reactivity of the uniformly distributed species, the spur reaction does not obey a simple kinetic formulation. This problem has been discussed in considerable detail (Williams, 1968, 1969) for radicals and ions in mobile liquids, and a brief summary will suffice here. For neutral reactive species, the spur recombination time τ_s is of the order of $r^2/6D$, where r is the distance through which the spur expands during the major portion of this nonhomogeneous recombination and D is the sum of the diffusion coefficients for the reactants. In the corresponding case of ionic reactions, the characteristic time τ_g of geminate recombination is given (Williams, 1964) by the formula $r^3/3Dr_c$, where r is now the distance of initial separation between the ions, D again represents the diffusion coefficients, and r_c is the distance at which the Coulombic potential energy between the ions e^2/sr_c equals kT, the thermal energy.

First, let us consider the situation in mobile liquids where D assumes a typical value of 2.5×10^{-5} cm^2 sec^{-1}. Magee and his co-workers (Samuel and Magee, 1953; Ganguly and Magee, 1956) have shown mathematically that the spur reaction between neutral species occurs almost entirely as the spur radius parameter r expands diffusively from an initial value of ≈ 10 Å to about 50 Å. This corresponds to τ_s of 1.6×10^{-9} sec. Even allowing for the relatively small amount of recombination which takes place as r increases from 50 to 500 Å, we can set an upper limit for τ_s of 10^{-7} sec for the nonhomogeneous reaction. For the case of ionic species, the average distance r of initial separation (≈ 60 Å) being less than r_c means that the Coulomb interaction will be the dominant factor governing recombination, so to a good approximation we can regard the time of geminate recombination as the time required for spur contraction. In nonpolar liquids (dielectric constant $\varepsilon = 2$) at 300°K, r_c is 280 Å, so τ_g becomes 1.0×10^{-9} sec for $r = 60$ Å. The probability of geminate recombination is given by the Onsager (1938) expression, $1 - \exp(-r_c/r)$, and this diminishes with increasing r. At $r = r_c$, the recombination probability is 0.633 and τ_g becomes 10^{-7} sec. Thus, if the initial distribution of separation distances between the ions is located almost entirely at r values less than 100 Å, most of the ions undergo geminate recombination in times shorter than 10^{-7} sec.

From the preceding analysis, we surmise that spur processes involving diffusion-controlled reactions of either radicals or ions are virtually complete in mobile liquids within 10^{-7} sec. On the other hand, the reactive species which undergo homogeneous reactions according to steady state kinetics have much longer lifetimes. At a dose rate of 1 Mrad/hr in nonpolar liquids, it has been calculated (Williams, 1968) that the steady state lifetimes τ_{ss} of both ions and radicals are on the order of 10^{-2} sec. The value of τ_{ss} is calculated from the relation $(1/R_i k_t)^{1/2}$ where R_i is the rate of formation of reactive species in the steady state and k_t is the diffusion-controlled rate constant for recombination. As implied by this expression, the lifetime varies inversely as the square root of the dose rate. The reason that τ_{ss} values for ions and radicals are approximately the same in nonpolar systems comes about because in the ionic reaction the lower R_i (based on free or separated ions) is offset by a larger value of k_t.

To summarize the situation for mobile liquids, the nonhomogeneous combination of radicals and of ions occurs in each case with times which are less than 10^{-7} sec, and the mean lifetime of these spur processes is probably less than 10^{-9} sec. For nonpolar systems, more than 90% of the ionic recombination occurs in the spur, whereas the radical recombination occurs mainly, but not exclusively, by reaction in the steady state. Branched hydrocarbons appear to provide an exception to this rule since the fraction of ions entering the steady state is considerably larger than that for linear alkanes (Schmidt and Allen, 1969, 1970). Lastly, the yield of free or separated ions increases with the polarity of the solvent (Freeman and Fayadh, 1965), and the distribution between spur and homogeneous ion-recombination processes approaches that which obtains for neutral radicals.

The preceding description is corroborated by three independent sources of experimental data. Originally, evidence was sought through the use of scavengers for the occurrence of ionic as well as free radical reactions in nonpolar liquids. Table I lists a number of positive ion and electron scavengers which have been employed in the radiolysis of nonpolar liquids. The scavenger technique is essentially a titration method, and the vital kinetic information is contained in the plot of yield versus scavenger concentration. It can easily be shown that reactive species with lifetimes greater than τ_s are readily scavenged at concentrations $[S]$ greater than $1/k_s\tau_s$, where k_s is the appropriate rate constant for the scavenging reaction. Thus, in order to scavenge reactive species with lifetimes in the range of 10^{-9} to 10^{-10} sec, the product $k_s[S]$ should approach 10^{10} sec^{-1} which corresponds to a diffusion-controlled rate constant ($\approx 10^{10}$ liter mole^{-1} sec^{-1}) and 1 M scavenger concentration. A significant difference is observed in the concentration dependence of scavenging for positive ions and electrons, and it appears that electron mobilities are greater than those of positive ions in mobile liquids by more than a factor of 10 (Rzad et al., 1969a, 1969b). Other experimental evidence which supports the above picture comes from studies of the electrical conductivity of liquids during irradiation (Schmidt and Allen, 1969, 1970) and pulse-radiolysis experiments (Thomas et al., 1968; Capellos and Allen, 1969) with suitable electron scavengers.

VI.   Processes in Rigid Media, Including Macromolecules

We conclude this chapter by drawing attention to some particular features of radiation chemical processes in rigid media. Depending on the nature of the process, the results may be similar or substantially different from those described for mobile liquids. This aspect is certainly relevant to studies of macromolecules in the bulk state.

As far as we know, the most immediate electronic processes of ionization, excitation, and electron capture are not influenced to any significant extent by the viscosity of the medium. The one important experimental fact which stands out in this connection is that charge separation over 50–100 Å can take place in rigid media just as in liquid systems (Eiben, 1970), provided a mechanism is made available for electron capture or localization in the matrix. The qualification in the last statement is not trivial because, in general, the chemical and perhaps even the morphological structure of a polymer will determine whether such a mechanism exists. For example, acrylate polymers contain the carboxylate group which, in small molecules, is known to attach electrons readily (Whelan, 1969), whereas in polyethylenes made free from additives, the only available mechanism of charge separation is through

TABLE I

POSITIVE ION AND ELECTRON SCAVENGERS IN THE RADIOLYSIS OF NONPOLAR LIQUIDS

Scavenger	Type	Reaction	Product	Reference
Ammonia-d_3	+	$RH^{+\cdot} + ND_3 \rightarrow R\cdot + ND_3H^+$	HD	F. Williams, *J. Amer. Chem. Soc.* **86**, 3954 (1964)
Ethanol-d	+	$RH^{+\cdot} + C_2H_5OD \rightarrow R\cdot + C_2H_5ODH^+$	HD	J. W. Buchanan and F. Williams, *J. Chem. Phys.* **44**, 4377 (1966)
Cyclopropane-d_6	+	$C_nH_{2n+2}^{+\cdot} + c\text{-}C_3D_6 \rightarrow C_nH_{2n}^{+\cdot} + CD_2HCD_2CD_2H$	Propane-d_6	P. Ausloos, A. A. Scala, and S. G. Lias, *J. Amer. Chem. Soc.* **88**, 5701 (1966)
Nitrous oxide	−	$N_2O + e^- \rightarrow N_2 + O^-$	N_2	G. Scholes and M. Simic, *Nature* **204**, 1187 (1964)
Methyl bromide	−	$CH_3Br + e^- \rightarrow CH_3\cdot + Br^-$	CH_4	J. M. Warman, K.-D. Asmus, and R. H. Schuler, *Advan. Chem. Ser.* **82**, 25 (1968)
Biphenyl	−	$\phi_2 + e^- \rightarrow \phi_2^{-\cdot}$	$\phi_2^{-\cdot}$	J. K. Thomas, K. Johnson, T. Klippert, and R. Lowers, *J. Chem. Phys.* **48**, 1608 (1968)

electron localization and stabilization in the matrix (Keyser and Williams, 1969).

Most of the evidence for rapid positive-charge migration comes from experiments in rigid media (Hamill, 1968), and it would seem reasonable to suppose that this process should be particularly favored by intramolecular transport in macromolecules. Unfortunately, really definitive experiments are lacking in this area. The increasing availability of "tailor-made" polymers incorporating specific groups should help to stimulate further work on this important problem.

Turning now to the processes that follow excitation and ionization, those involving molecular dissociation and reaction without mass transport do not seem to be greatly influenced either by the rigidity of the medium or molecular size. Admittedly this conclusion is largely based on analogies in the gross results of radiation damage. A good illustration is provided by the relatively large yield of C—C bond scission which occurs both in branched hydrocarbons of low molecular weight and in polymers containing quaternary-substituted carbon atoms in the main chain (e.g., polyisobutylene, polymethylmethacrylate), a result which has been interpreted in terms of ionic dissociation (Williams, 1961). This general mechanism is supported by a recent study on the radiolysis of neopentane in the liquid phase (Collin and Ausloos, 1971).

All processes which depend on atomic (hydrogen atom) or molecular diffusion are profoundly affected by the viscosity of the matrix. Thus, free radical reactions in irradiated polyethylene at room temperature (Waterman and Dole, 1970) take place on a time scale more than a million times longer than that which applies for the liquid systems discussed earlier. More surprisingly, trapped electrons undergo a thermal decay which seems to be affected by viscosity as instanced by their greater stability in 3-methylhexane as compared to 3-methylpentane at $77°K$ (Lin *et al.*, 1968), a finding which correlates with a lower glass-transition temperature for the latter compound.

In the limiting case of extremely slow diffusion, the reactive species are frozen in the solid and their concentration builds up as the irradiation proceeds. Under these conditions, a limit to the free radical concentration is not attained until very high values are achieved, typically 10^{-2} to 10^{-3} M instead of $\approx 10^{-8}$ M as in liquid systems at conventional dose rates. Clearly, the conventional steady state description for liquids cannot be applied to solids. One simple consequence of this restricted diffusion is that free radical scavengers will be much less effective in solid than in liquid systems. Moreover, this conclusion applies also to the results of warm-up experiments carried out on irradiated solids. Since the initial radical concentration under these conditions is very much larger than the steady state value in a liquid, the probability of radical–radical reactions is greatly enhanced relative to that of radical scavenging.

In the very wide range of viscosities between the mobile liquid and the rigid solid as exemplified by a molten polymer, one expects to find a smooth gradation between the extremes of kinetic behavior which have been described. Smaller diffusion coefficients D will tend to blur the distinction between the steady state reaction and the spur reaction since the lifetimes of reactive species participating in these two kinetic processes depend on D^{-1} and $D^{-1/2}$, respectively. This follows from our previous discussion of diffusion-controlled recombination where $\tau_s = r^2/6D$ and $\tau_{ss} = (1/k_t R_i)^{1/2} = (1/4\pi\sigma D R_i)^{1/2}$ if we adopt the Smoluchowski expression $k_t = 4\pi\sigma D$, where σ is a reaction distance (Williams, 1968). Essentially the steady state concentration $(R_i/k_t)^{1/2}$ increases with a decrease in D resulting in a reduction of the average spacing between the uniformly distributed species, whereas of course, the spacing in the spur remains unaffected.

It is evident from the preceding discussion that the final chemical changes induced in irradiated polymers may well be affected by the postirradiation treatment of the polymer. In future studies, more emphasis should be given to experiments involving the characterization of specific processes in temporal sequence. Also, there is a particular need for definitive work on charge and energy migration in the radiation chemistry of macromolecules.

REFERENCES

Ausloos, P. (1966). *Ann. Rev. Phys. Chem.* **17**, 205.
Ausloos, P., ed. (1968). "Fundamental Processes in Radiation Chemistry." Wiley, New York.
Ausloos, P., and Lias, S. G. (1968). *Ber. Bunsenges. Phys. Chem.* **72**, 187.
Byers-Brown, W., and Roberts, R. E. (1967). *J. Chem. Phys.* **46**, 2006.
Capellos, C., and Allen, A. O. (1969). *J. Phys. Chem.* **73**, 3264.
Chen, J. C. Y. (1969). *In* "Advances in Radiation Chemistry" (M. Burton and J. L. Magee, eds.), Vol. 1, p. 245. Wiley, New York.
Chen, J. C. Y., and Magee, J. L. (1962). *J. Chem. Phys.* **36**, 1407.
Collin, G. J., and Ausloos, P. (1971). *J. Amer. Chem. Soc.* **93**, 1336.
Compton, R. N., and Huebner, R. H. (1970). *In* "Advances in Radiation Chemistry" (M. Burton and J. L. Magee, eds.), Vol. 2, p. 419. Wiley, New York.
Copeland, D. A., Kestner, N. R., and Jortner, J. (1970). *J. Chem. Phys.* **53**, 1189.
Eiben, K. (1970). *Angew. Chem. Int. Ed.* **9**, 619.
Fano, U. (1963). *Ann. Rev. Nucl. Sci.* **13**, 1.
Fox, K., and Turner, J. E. (1966). *J. Chem. Phys.* **45**, 1142.
Freeman, G. R., and Fayadh, J. M. (1965). *J. Chem. Phys.* **43**, 86.
Ganguly, A. K., and Magee, J. L. (1956). *J. Chem. Phys.* **25**, 129.
Hamill, W. H. (1968). *In* "Radical Ions" (E. T. Kaiser and L. Kevan, eds.), p. 321. Wiley, New York.

Hart, E. J., and Platzman, R. L. (1961). *In* "Mechanism in Radiobiology" (M. Errera and A. Forssberg, eds.), Vol. 1, Chapter 2. Academic Press, New York.

Jortner, J. (1964). *Radiat. Res. Suppl.* **4**, 24.

Kerr, C. M. L., and Williams, F. (1971). *J. Amer. Chem. Soc.* **93**, 2805.

Keyser, R. M., and Williams, F. (1969). *J. Phys. Chem.* **73**, 1623.

Kuppermann, A. (1961). *In* "Actions Chimiques et Biologiques des Radiations" (M. Haissinsky, ed.), p. 85. Masson, Paris.

Kuppermann, A., and Raff, L. M. (1963). *Disc. Faraday Soc.* **35**, 30.

Lin, J., Tsuji, K., and Williams, F. (1968). *J. Amer. Chem. Soc.* **90**, 2766.

Marcus, R. A. (1968). *Disc. Faraday Soc.* **45**, 7.

Mozumder, A. (1969). *In* "Advances in Radiation Chemistry" (M. Burton and J. L. Magee, eds.), Vol. 1, p. 1. Wiley, New York.

Mozumder, A., and Magee, J. L. (1966). *Radiat. Res.* **28**, 203.

Natori, M., and Watanabe, T. (1966). *J. Phys. Soc. Japan* **21**, 1573.

Onsager, L. (1938). *Phys. Rev.* **54**, 554.

Platzman, R. L. (1962). *The Vortex* **23**, 372; *Radiat. Res.* **17**, 419.

Platzman, R. L. (1967). *In* "Radiation Research" (G. Silini, ed.), p. 20. North-Holland Publ., Amsterdam.

Rzad, S. J., Infelta, P. P., Warman, J. M., and Schuler, R. H. (1969a). *J. Chem. Phys.* **50**, 5034.

Rzad, S. J., Hummel, A., and Schuler, R. H. (1969b). *J. Chem. Phys.* **51**, 1369.

Samuel, A. H., and Magee, J. L. (1953). *J. Chem. Phys.* **21**, 1080.

Schmidt, W. F., and Allen, A. O. (1969). *J. Chem. Phys.* **50**, 5037.

Schmidt, W. F., and Allen, A. O. (1970). *J. Chem. Phys.* **52**, 2345.

Spencer, L. V., and Fano, U. (1954). *Phys. Rev.* **93**, 1172.

Sprague, E. D., and Williams, F. (1971). *J. Chem. Phys.* **54**, 5425.

Taub, I., and Eiben, K. (1968). *J. Chem. Phys.* **49**, 2499.

Thomas, J. K. (1969). *In* "Advances in Radiation Chemistry" (M. Burton and J. L. Magee, eds.), Vol. 1, p. 103. Wiley, New York.

Thomas, J. K., Johnson, K., Klippert, T., and Lowers, R. (1968). *J. Chem. Phys.* **48**, 1608.

Waterman, D. C., and Dole, M. (1970). *J. Phys. Chem.* **74**, 1913.

Whelan, D. J. (1969). *Chem. Rev.* **69**, 179.

Whyte, G. N. (1959). "The Principles of Radiation Dosimetry." Wiley, New York.

Williams, F. (1961). *Trans. Faraday Soc.* **57**, 755.

Williams, F. (1964). *J. Amer. Chem. Soc.* **86**, 3954.

Williams, F. (1968). *In* "Fundamental Processes in Radiation Chemistry" (P. Ausloos, ed.), Chapter 8, p. 515. Wiley, New York.

Williams, F. (1969). *In* "Large Radiation Sources for Industrial Processes," p. 247, Publ. No. IAEA-SM-123/18. Int. At. Energy Agency, Vienna

3

Energy Transfer in Polymers

Roger H. Partridge

Division of Materials Applications, National Physical Laboratory, Teddington, Middlesex, England

I. Definitions

The energy expended in many radiation reactions is less than, or roughly equal to, the amount of incident radiation energy absorbed directly by the reactants. But in some cases the extent of reaction is far greater than could be accounted for in this way, indicating that most of the energy absorption has occurred elsewhere and that some of it has then been transferred to the various reaction sites. Examples of this are the well-known "protection" of alkanes by benzene and other additives (see, for instance, Burton and Lipsky, 1957), the scintillation of many organic liquids and solids (Birks, 1964), and the rapid elimination of some forms of unsaturation in polyethylene (Dole *et al.*, 1966).

Such energy transfer, in the broadest sense, can be classified under the headings transport of *mass* (perhaps together with charge and/or excitation); transport of *charge alone*; transport of *excitation alone*.

The first of these covers the normal diffusion through a particular matrix of small reactive chemical species such as hydrogen atoms, free radicals, ions (in ground or excited states), and excited molecules. This should not come under the strict title of energy transfer, but mention of it here is very necessary because it is often difficult to separate experimentally from the other two types of energy transfer. Such a separation is easiest to make in measurements at very low temperatures ($\lesssim 77°$K) because here (especially in polymers, which are well below their glass-transition temperatures) mass transfer is usually confined to migration of hydrogen atoms and protons; intramolecular radical migration may also occur, though this is often essentially intra-molecular hydrogen atom migration.

The term energy transfer is best applied to transport of charge or excitation alone although the specific terms charge transfer and excitation transfer are preferable wherever possible. Unfortunately, for many polymers it is very difficult to distinguish unambiguously between these two processes.

II.   Charge Transfer

Charge transfer is a form of energy transfer in radiation chemistry both because some of the ions eventually produced will be reactive species in their own right (Frankevich, 1966) and because when any of these ions finally recombine with opposite charges electronic excitation energy will be generated which can cause further reaction by direct bond scission or excitation transfer. Theoretical and experimental work on charge generation and transfer in polymers is discussed in detail in later sections on electrical conductivity (Chapters 7 and 8) but some specific points relevant to energy transfer will be noted here.

Charge transfer in polymers is basically electronic (transfer of electrons or positive charges *alone*) or ionic (transfer of protons or larger charged species, already noted under transport of *mass*). Electronic conduction can be further divided into conduction due to the diffusion of electrons which are not localized on particular molecules (as usually found in liquids or gases) and conduction due to positive or negative charges which are localized on particular molecules and can be exchanged between like molecules (or segments of single polymer molecules) without net energy loss (resonant charge transfer). It is found experimentally (for example, Hirsch and Martin, 1969a,b; 1970) that the electrical conductivity of a variety of polymers subjected to a short irradiation pulse consists initially of a "prompt" component that signifies very rapid transfer of a considerable amount of charge over a comparatively short distance (≈ 100 Å) which is terminated by trapping in "shallow" traps. This is followed by a "delayed" component that is very temperature dependent and probably indicates a thermally activated charge-hopping process between the shallow traps until terminated (after ≈ 1 μ) by trapping in deep traps or by recombination. The majority carriers in the delayed component are negative in some polymers, such as polyethyleneterephthalate, and positive in others, such as polyethylene and polystyrene. Thus, it is best to assume that both positive and negative charge transfer is possible in most polymers, with the transfer distance small at low temperatures (below ≈ 0°C in many cases) but much greater at higher temperatures [or for low-temperature irradiations followed by warming to higher temperatures, as shown by glow conductivity measurements (Yahagi and Shinohara, 1966)].

The nature of the positive charge traps is by no means clear. In chain-folded polymers the actual fold areas may act as traps (Hirsch and Martin, 1970). A theoretical study of positive alkane ions by Ovchinnikov (1965) indicates that positive charge transfer will only occur in excited states and that a charge is trapped when it relaxes to the ground state. This is because the lowest energy level in the ground state of a long-chain alkane is one in which the positive charge becomes localized over a few atoms by distortion of the molecular bonds in its vicinity. Extra energy is then needed to return it to a conducting level. The same study predicted that in short-chain alkanes the positive charge will lie in the center of the chain. Frankevich (1966), in a most useful review of ionic reactions in condensed hydrocarbons, has suggested that some positive charge trapping may occur via ion–molecule reactions such as

$$-CH_2{}^+- + -CH_2- \rightarrow -CH_3{}^+- + -\dot{C}H-$$

where the positive charge can not then transfer to another methylene group as the potential barrier would be too high although resonant proton transfer might be possible. Other important potential positive charge traps of radiation chemical interest are foreign molecules, including radiation-produced free radicals (Frankevich and Yakovlev, 1963), that have lower ionization potentials than the polymer molecules. If extensive charge transfer were possible then a relatively high proportion of such foreign molecules could be ionized even if present in low concentration.

The nature of the potential electron traps is somewhat better characterized and includes "cavity" traps, molecules with a positive electron affinity, and free radicals (see Chapters 9 and 10). An important electron-trapping reaction to note is dissociative electron attachment. Here, an electron, which may have migrated a considerable distance and have only thermal energy, attaches to a molecule and in doing so dissociates it into a radical and a radical anion. Such reactions are found particularly in aromatic halides, though also in a few other aromatic molecules with strongly electronegative groups, and have been studied in some detail by Gallivan and Hamill (1965).

The yield of migratory charges produced in a material by ionizing radiation can be obtained moderately well for electrons but not so easily for positive charges. The upper limit for both is the actual ion pair yield. This is usually obtained from gas phase measurements, but the initial yield in condensed phase is probably not much different. For alkanes the G value for ion pair production is about 4 (Meisels, 1964). A lower limit is obtained from electrical conductivity measurements following a radiation pulse although rapid ion recombination often makes this value very small (Freeman and Fayadh, 1965). The best G values for migratory electrons are probably those obtained by addition of small quantities of electron-trapping molecules whose anions have known optical or ESR spectra. The maximum G values obtained in this way are about 1–1.5 for nonpolar materials and 2–3 for polar ones (see review by Roginskii and Kotov, 1967) although hardly any work has been done on polymers. Partridge (1970d) found a maximum biphenyl anion yield of about 1 for polyethylene (although about half of this may have come from ionization of the biphenyl additive) in addition to other electrons still trapped in the polymer itself. The yield of migratory positive charges is much harder to obtain since although in principle additives can be used to trap the positive charges these additives might also be ionized by excitation energy transfer, as described below, and thus the results are often ambiguous. The upper limit of this yield is probably set by the migratory electron yield, since electrons not even captured by a high concentration of electron-trapping molecules will probably have recombined with their parent positive charges before either have significantly migrated.

III. Excitation Energy Transfer. General

Transfer of electronic excitation energy between molecules or molecular groups can occur by a number of different processes which can be classified under six general headings.

A. RADIATIVE TRANSFER

This involves the actual emission of a photon from an excited donor molecule and its absorption by an "acceptor" molecule, thus deexciting the donor and exciting the acceptor. The essential condition for such transfer is that the donor emission spectrum shall overlap the acceptor absorption spectrum, so this is primarily an energy transfer between dissimilar molecules. The efficiency of such transfer depends on the donor emission efficiency, the absorption strength of the acceptor, the concentration of donor and acceptor molecules, and the actual size of the sample (since for a given material this will determine the probability of the photon being absorbed before escaping from the sample). Transfer can occur from the singlet or triplet state of the donor (i.e., as fluorescence or phosphorescence emission) but is only likely to excite a singlet state of the acceptor since transitions to triplet states are usually spin forbidden.

B. EXCHANGE TRANSFER

Transfer by electron-exchange interaction occurs only over short distances (up to ≈ 15 Å) because it requires some overlap between the orbitals of the molecules involved. Such transfer allows triplet as well as singlet states of the acceptor to be produced, but an essential requirement is that the total spin of the system be conserved; for instance, exchange interaction between a triplet donor and a ground state singlet acceptor could give a ground state singlet donor and a triplet acceptor, but not triplet or singlet states of both donor and acceptor. A theoretical treatment of exchange transfer has been given by Dexter (1953) and a useful analysis of experimental data by Inokuti and Hirayama (1965). The very efficient triplet–triplet transfer in aromatic crystals is largely due to exchange transfer (see, for example, Jortner *et al.*, 1965; Rice and Jortner, 1968) although there may be contributions from charge transfer excitons and excimers (see below). Exchange transfer in a liquid occurs largely as diffusive hopping of the excitation from one molecule to another.

C. EXCIMER TRANSFER

An excimer is a transient dimer formed by the combination of an excited (usually aromatic) molecule and a second similar (usually unexcited) molecule. Such a dimer bonds only in the excited state and promptly dissociates on losing its excitation energy. The term "exciplex" has been introduced by Birks (1967) to describe an excited complex of *any* two molecules which is dissociative in the ground state, and an excimer is then just a special case in which the two constituent molecules are identical. Excimer formation has been observed in solids, liquids, and gases. It is usually indicated by the appearance of a broad structureless fluorescence band (due to radiative transitions from the excimer to its dissociative ground state) at rather lower energy than the normal monomer fluorescence band but with no change in the absorption spectrum. Formation occurs as a bimolecular collision process in liquids and gases, so the formation rate is diffusion controlled for a fixed molecular concentration and also increases rapidly with increasing concentration.

Using the notation of Birks *et al.* (1963), the formation of an excimer $D*$ from an excited monomer $M*$ and an unexcited monomer M can be represented as

$$M* + M \rightarrow D* \tag{1}$$

and this will be in competition with the monomer fluorescence emission. The various deexcitation paths of the excimer include

$$D* \rightarrow 2M \tag{2}$$

$$\rightarrow 2M + h\nu \tag{3}$$

$$\rightarrow M + M* \tag{4}$$

and

$$D* + Y \rightarrow 2M + Y* \tag{5}$$

$$D* + Q \rightarrow 2M + Q \tag{6}$$

where Y and Y* are ground and excited singlet states of a fluorescent molecule different to M and Q is a nonluminescent molecule. It can be seen that, when taken in conjunction with (1), processes (2) and (6) represent self-quenching and impurity quenching of the monomer fluorescence, respectively, while (3) represents excimer fluorescence in competition with monomer fluorescence. Processes (4) and (5) represent excitation transfer between similar and dissimilar molecules respectively. Process (4) can be repeated over and over again and is then equivalent to a hopping transfer of excitation energy between the molecules of a solvent, and this may sometimes be terminated by transfer to an additive molecule in process (5). If the solvent is a liquid, then the migration of excitation is by a combination of molecular diffusion and excimer hopping and, thus, occurs at a considerably faster rate than for diffusion alone.

Excimers may also be formed between two excited molecules (usually triplets because of their long lifetime), and dissociation of this excited excimer often yields an excited singlet and a ground state molecule. Such triplet–triplet annihilation processes are well known in aromatic crystals, and a possible mechanism for one case has been suggested by Krishna (1967).

An excellent review of the whole field of aromatic excimers has just been published by Birks (1970).

D. FÖRSTER LONG-RANGE TRANSFER

Förster long-range transfer (also known as inductive resonance or vibrational-relaxation transfer) is caused by a Coulomb interaction between two molecules that are making simultaneous coupled electronic transitions of almost equal energy, one molecule going from an excited state to the ground state while the other goes from the ground to an excited state. This can be considered as an essentially electrostatic interaction between the transition charge densities of the molecules. The transition charge distributions are usually approximated as dipoles, which yields a dipole–dipole interaction between the molecules whose strength is proportional to the inverse cube of the intermolecular distance. If dipole transitions are symmetry forbidden in either molecule, then higher multipoles, or vibrationally induced dipoles, may be significant.

The essential feature of Förster transfer is that the interaction or coupling energy between the two molecules is so small that it is less than the energy width of the individual vibrational bands of the molecules. Because of this small coupling strength, the rate of excitation transfer is relatively slow ($\approx 10^9$ sec^{-1}), and so transfer always occurs from the lowest vibrational level of the appropriate donor excited state. Since transfer is resonant, the energy of the transition induced in the acceptor molecule must precisely match the energy of the transition from the lowest vibrational state of the excited donor to some vibrational level of the donor ground state. In practice this means that, as for radiative transfer, the donor emission spectrum should overlap the acceptor absorption spectrum, and thus, Förster transfer is usually between *dissimilar* molecules. The very weak coupling strength means that there is no detectable difference between the absorption spectrum of the whole system and that of the donor and acceptor molecules separately, but despite this the excitation transfer can occur over distances of up to ≈ 100 Å in a single step in suitable systems.

Förster (1946, 1948) has presented the detailed theory of this type of transfer by both classical and quantum mechanical methods and has also discussed it in several review articles (Förster, 1959, 1960a,b, 1966, 1968). The main results of experimental interest are that the rate of energy transfer is proportional to the inverse sixth power of the intermolecular distance and that the transfer distance depends on the quantum efficiency, but *not* the lifetime, of emission from the donor molecule (in the absence of acceptors). From the latter condition it is clear that Förster transfer can occur from triplet as well as singlet states of the donor, but it will usually produce only singlet excitation of the acceptor (if energetically possible) since the transfer rate also depends on the absorption strength of the acceptor. In Förster transfer the excitation is more or less localized on a particular molecule at a particular time. In a multimolecular system the excitation can hop from one molecule to another by a series of quite uncorrelated transfers.

E. MOLECULAR EXCITON TRANSFER

This is similar to Förster transfer in that it also arises from Coulomb interactions between molecules or molecular units, and indeed, there has been some debate as to whether Förster transfer should be considered an extreme case of molecular exciton transfer (Kasha, 1963; Förster, 1966). The essential difference between the two is that the coupling energy is considerably stronger in the molecular exciton case, with the important consequence that the excitation transfer rate is proportional to the inverse cube of the intermolecular separation rather than the inverse sixth power as in Förster transfer. The molecular exciton model can be best considered in the extreme cases of strong and weak coupling. In all cases it is assumed that electron overlap and electron exchange between the interacting molecules is negligible.

In the weak coupling case the coupling energy is considerably less than the energy width of the whole vibrational envelope of the particular electronic levels associated with the transfer (Simpson and Peterson, 1957) though it is still considerably greater than for Förster transfer. The transfer rate ($\approx 10^{12}$ sec^{-1}) is faster than for Förster transfer, and transfer can occur from the higher vibrational levels of the appropriate donor excited state (after several vibration periods). Thus, transfer between similar molecules is favored although it can also occur between dissimilar molecules if the excited states involved are close in energy. In a lattice of similar molecules the excitation is essentially spread over a number of molecules and moves initially in a coherent manner with a constant group velocity, being then known as an exciton, although lattice vibrations and irregularities tend to destroy the coherence and make propagation purely diffusive (Goad, 1963; Katsuura, 1964). The absorption spectrum of a weak coupling system is generally similar to that of the isolated molecules, but a splitting of some of the individual vibrational bands will occur and will sometimes be observable if the vibrational bands are well resolved.

In the strong coupling case the coupling energy is considerably greater than the energy width of the vibrational envelope of the appropriate electronic transition, with the result that the excitation transfer rate ($\approx 10^{15}$ sec^{-1}) is now much faster than the nuclear vibration rate. The excitation can be considered, in time independent theory, to be spread out over the whole system of (similar) molecules. The vibrational motions of the molecules are so strongly coupled that the whole system vibrates as a single molecule (which means that in the Born–Oppenheimer approximation the total wave function of the system can be written as the product of a single electronic wave function and a single vibrational wave function). In time dependent theory the excitation transfer process can, as for weak coupling, be described in terms of a traveling wave packet or exciton which moves through the system in a coherent manner with constant group velocity. In contrast to most other forms of excitation transfer the absorption spectrum of the system will often differ considerably from that of the isolated molecules of which it is composed. This is because a single energy level in the isolated molecule which is accessible by an optically allowed transition is split into an exciton band in the system. This band is nearly symmetrical in energy about the isolated molecule level, and for a system of N repeating molecular units the exciton band contains N discrete energy levels. The energy width of the exciton band depends on the coupling energy, which in turn depends on the absorption strength of the isolated molecule transition and on the intermolecular separation. Since for strong coupling the absorption strength must be high, this case usually only applies to transfer between singlet states. Often only the top or bottom energy levels of the exciton band will be accessible by an optically allowed transition (depending on the alignment of the molecules in the system), and in this case the absorption spectrum will be strongly blue or red shifted, respectively, as compared with the isolated molecule transition. The strong coupling model can be applied to one-, two-, or three-dimensional arrays of similar molecules, and an excellent description of this at a comparatively simple level has been given by McRea and Kasha (1964) and Hochstrasser and Kasha (1964).

Some useful reviews of the general theoretical and experimental framework of the molecular exciton model, with many references to original work, have been given by Kasha (1963) and by Förster (1959, 1966, 1968). The relationship of molecular excitons to excitons in other very different physical situations has been classified by Kasha (1964), and the relation between exciton bands and conduction bands in molecular systems has also been discussed by Kasha (1959).

F. CHARGE TRANSFER EXCITONS

It is sometimes convenient to consider the molecular excitons discussed above as neutral excitation particles consisting of traveling electron-positive hole pairs that are bound together roughly within the confines of a single molecule or molecular unit (Kasha, 1959); indeed, the molecular excitation process can be pictured as a transfer of an electron to a different molecular orbital, leaving a "positive hole" behind it. However, other excitons can occur in which the electron and hole are on *different* molecules but are bound together by their mutual Coulomb attraction. Such excitons are known as charge transfer (CT) or ion pair excitons in organic systems and as Wannier excitons in inorganic systems. The maximum possible separation of the charge pairs is that at which their Coulomb attraction just outweighs the disruptive effects of thermal collisions, but in practice, CT excitons will usually be of nearest neighbor type with the charges on adjacent molecules.

Charge transfer excitons are ionic species in the sense that some charge separation has occurred, but they cannot contribute directly to a conduction current as the two charges always move together. The energy required to form a CT exciton is usually somewhat greater than for the corresponding neutral exciton (although this is not essential; Merrifield, 1961), due to the greater charge separation, but is considerably less than that required for true ionization (infinite charge separation). Direct optical transitions to a CT state are forbidden because the transition moment is approximately proportional to the overlap integral between the orbitals of the two different charges, and this is very small because the charges are on different molecules (Lyons, 1957). Charge transfer states are thus generally reached by an indirect route such as the recombination (or rather the near recombination) of free positive and negative charges.

The migration of CT excitons, which like neutral excitons does not involve mass transport, is an energy transfer process with some characteristics of both charge and excitation transfer. Thus, the exciton can decay by radiative or nonradiative transitions, giving fluorescence or increased local vibrational energy, respectively, or it may receive sufficient extra thermal energy to split up and so produce two free charges that can contribute to electrical conduction. Annihilation of two CT excitons in a bimolecular collision process is also possible, giving fluorescence or a single true ionization.

Charge transfer excitons have been observed by Pope and his collaborators in some beautiful experiments on anthracene and tetracene crystals (Pope, 1967, 1968). Theoretical studies, including the "mixing" of neutral and CT exciton states, have been made by Lyons (1957), Choi *et al.* (1964), and Merrifield (1961), among others.

IV. Excitation Energy Transfer. Polymers

In doing theoretical calculations on polymers, or indeed on any multiatom systems, a choice invariably has to be made between performing nearly exact calculations on a very idealized system or much less accurate calculations on a more real system. Polymer studies by both these approaches are considered in this section.

A. IDEALIZED POLYMERS

The simplest approach is to treat an "infinite one-dimensional crystal" consisting of a line of identical molecules spaced a regular distance apart. Such a system has been described on the strong coupling exciton model, in the absence of molecular vibrations, by McRae and Kasha (1964). If the individual molecules have only one strong (dipole allowed) transition to a stable singlet excited state and all N molecules have the same orientation, then the polymer will have a band of N discrete excited levels which is symmetrical about the isolated molecule excited level (in the absence of permanent dipoles) and has an energy width (if only nearest neighbor molecular interactions are considered) of four times the transition dipole–dipole interaction energy between two adjacent molecules. The absorption strength of the polymer will be just N times that of a single molecule, but in general, only optical transitions to the highest or lowest level of the exciton band are allowed. (If all the transition dipoles are aligned along the polymer axis, then only transitions to the lowest level are allowed; and if all are aligned perpendicular to this axis, then only transitions to the highest level are allowed.) An important point in relation to energy transfer, however, is that even if a polymer is excited to a higher exciton level of a particular exciton band it will usually fall rapidly via collisional deactivation to the lowest state of that band, and thus, the manner in which excitation transfer occurs will largely be determined by the characteristics of this lowest state; experimental evidence for this comes from the enhancement of phosphorescence emission in dimers or polymers with parallel transition dipoles (McRae and Kasha, 1964; Kasha *et al.*, 1967). Use of time dependent theory (Förster, 1966) shows that the time taken for an exciton to transfer from one molecule to another in such a system is $\approx h/8U$, where U is the coupling energy between two adjacent molecules and h is Planck's constant, and is independent of the distance the exciton has traveled.

Many variations of this simple basic model have been considered in order to obtain results rather closer to the situation in real molecules. The effects of molecular vibrations have been studied on time dependent models by McRae (1961), Goad (1963), Förster (1966), and as noted later, by Magee and co-workers; in general, the vibrations tend to slow down the rate of exciton transfer at larger migration distances and make transfer more similar to a diffusion process. Similar results were obtained, using time independent theory, by Katsuura (1964) for polymers in which the molecules were either regularly spaced but had random orientations or were equally oriented but subject to random variations in position about an average separation position; some more general work on these lines has been done by Takeno (1968).

Calculations on polymers containing one or more foreign molecules in a chain of otherwise identical molecules are of great interest in relation to the role of excitation transfer in radiation chemistry. Katsuura and Inokuti (1963) and Merrifield (1963) found that, under certain specific conditions, the introduction of a foreign molecule creates an extra energy level just below the bottom of the polymer exciton band and another just above the top of the band. The lowest energy level is found to be one in which the excitation is predominantly distributed on and around the foreign molecule itself, and as this is the lowest excited level of the polymer, it is clear that any excitation of the polymer exciton band is likely to revert fairly rapidly to excitation of this level and hence excitation of the foreign molecule itself. This result is even true if the foreign molecule has a higher excitation energy (in the isolated state) than the other polymer molecules, as long as the coupling energy between the foreign molecule and the others is high enough. This therefore constitutes an efficient mechanism for selective excitation of a foreign molecule in a polymer chain, which could well result in observable effects such as ionization, dissociation, or radiative emission. Bierman (1970) has recently obtained an exact time dependent solution of this foreign molecule problem and so was able to compute the actual probability of the foreign molecule being excited, as a function of time. The probability was found to oscillate with time and to depend on parameters such as the difference in energy between the foreign molecule and the other molecules (in the isolated state) and the distance of the initial excitation event from the foreign molecule.

The effect of a single displaced molecule in an otherwise regular chain of molecules was found by Katsuura and Inokuti (1963) and Merrifield (1963) to be rather similar to that of a foreign molecule in that, again, extra energy levels appear just outside the top and bottom edges of the exciton band, and the lowest level has excitation localized in the vicinity of the structural defect.

Charge transfer excitons in a simple polymer were studied by Merrifield (1961), who showed that there would be a series of CT levels (singlets and triplets) of increasing energy, corresponding to increasing charge pair separation, extending up to free charge levels in which electrical conduction can occur. Mixing of the lowest energy charge transfer and neutral exciton bands would occur if these were close in energy.

A most interesting polymer exciton study of a different type was the work of Magee and Funabashi (1961) and Lorquet *et al.* (1962) on a polymer made up of molecules that have *dissociative* excited states in the isolated condition. The actual system considered, a linear chain of hydrogen molecule ions with axes perpendicular to the polymer axis, is certainly idealized but was chosen because it was amenable to reasonably accurate mathematical treatment, and some of the results could well be generalized to more complex real polymers. Time dependent exciton theory was used and the actual motion of the various hydrogen nuclei was calculated as a function of time after one of the molecules had been excited. It was found that as the coupling strength between molecules increased (by decreasing the intermolecular separation) so the probability of molecular dissociation decreased from unity (at weak coupling) down to zero (strong coupling). The complete absence of dissociation in strong (and intermediate) coupling was due to the excitation being passed from one molecule to the next faster than the molecules could dissociate. Conversely, however, it was found that in very short chains of molecules several dissociations could result from a single excitation even in the case of strong coupling. The maximum number of dissociations by a single excitation is limited, on energetic grounds, to less than the ratio of the excitation energy to the molecular dissociation energy.

B. REAL POLYMERS

One of the earliest applications of molecular exciton theory to a real polymer was that of Simpson (1951, 1955) on conjugated polyenes. Although such conjugation might be considered to result in π electron orbitals which are delocalized over the whole length of the chain, Simpson showed that quite reasonable results could be obtained by treating the polyenes according to their classical chemical formulas as containing alternating single and double bonds. The double bonds were treated as isolated molecules on the simple strong coupling exciton model, with no contribution from single bonds. A useful discussion and extension of this work has been given by Murrell (1963) and, more recently, by Tric and Parodi (1967), using CT as well as neutral exciton states. Some work on hydrocarbon polymers containing nonconjugated double bonds is noted later.

A number of synthetic polymers, such as polystyrene, contain aromatic side-chain groups attached to a saturated main chain. As the aromatic groups have excitation energies that are considerably lower than that of the main chain, the lowest energy states of the whole polymer can be approximately represented by considering only interactions between the side groups. This is a very suitable system for use of the molecular exciton model since there is little orbital overlap between the different side-chain groups. The main study of such systems has been by Vala *et al.* (1964, 1965), who did calculations on isolated polystyrene and polyvinylnaphthalene molecules of helical conformation. They found a strong interaction between the singlet states of the aromatic side groups, especially for states accessible by strongly allowed transitions in the isolated aromatic molecules, and this gave rise to strong splitting of the electronic energy levels of the polymer. The size of the splitting and the optical transition strength were also controlled, as usual, by the actual geometrical configuration of the aromatic groups along the polymer chain. Interaction between triplet states was very small since, as noted earlier, these interact by an electron exchange mechanism and this is dependent on appreciable orbital overlap between adjacent molecular units. Mixing of CT exciton states with neutral exciton states was also found to be small, due partly to the small orbital overlap and partly to the much higher energy of even the lowest pure CT state.

The largest amount of theoretical work on real polymers has been directed towards biological polymers, mainly polypeptides and double-stranded polynucleotides, having a helical conformation. Early work on such systems using the simple strong coupling exciton model has been described by, among others, Kasha (1963), Bradley *et al.* (1963), and Tinoco *et al.* (1963). Recent work, using more complex models, includes that of Woody (1968), Miyata and Yomosa (1969), and Rhodes and Chase (1967). These studies aim at evaluating various optical parameters such as optical rotatory dispersion, circular dichroism, and changes in both the energy and intensity of transitions in certain biopolymers caused by changes in their conformation. Reviews of some of this work have been given by Mason (1966), Tinoco (1965), and Deutsche *et al.* (1969).

The possibility of excitation transfer in saturated organic molecules, particularly alkanes and alkane polymers, has sometimes been discounted on the grounds that such molecules have dissociative excited states. However, the recent observations of well-defined vibrational structure in the vacuum ultraviolet absorption and electron impact spectra of ethane (Raymonda and Simpson, 1967; Lombos *et al.*, 1967; Lassettre *et al.*, 1967) and of fluorescence from a wide range of alkanes from pentane to polyethylene (Hirayama *et al.*, 1970) indicate that alkanes do not invariably dissociate immediately when excited, although they may have a considerable probability of doing so at a somewhat later time. Furthermore the calculations of Magee and co-workers on dissociative excitons, noted earlier, show that dissociation of short-chain molecules may be replaced by very efficient excitation transfer in similar but much longer chain molecules. The study of the vacuum ultraviolet absorption spectra of many short-chain alkanes led Raymonda and Simpson (1967) to propose a bond exciton model of alkane excited states. In this it was assumed that each bond, C–C or C–H, can be considered as a quite separate molecular unit, having a ground state and one excited state, which interacts with the other bonds only via a transition dipole–dipole interaction and not via charge exchange. Use of simple strong coupling exciton theory, taking into account the type and geometrical orientation of each bond, then allowed the exciton states of each alkane molecule to be calculated. The excitation energy and absorption strength of hypothetical "isolated" C–H and C–C bonds were obtained by comparison of the methane and ethane spectra with the theoretical results, and these values were then used to predict the spectra of a range of small alkanes. The correspondence between theory and experiment was surprisingly good.

The Raymonda and Simpson bond exciton theory was extended to alkane polymers by Partridge (1968, 1970a), with particular consideration of polyethylene, polypropylene, polyisobutylene, and diamond. Again there was quite good agreement between theory and experiment although, as the polymer absorption spectra are very broad, it was difficult to locate the higher energy transitions with any accuracy. The testing of this alkane exciton model is of considerable radiation chemical interest because it not only predicts the energies, intensities, and polarizations of the various allowed transitions but also predicts the manner in which the excitation energy will be distributed throughout each molecular chain. This latter information leads to some fairly specific predictions on primary radiation reactions in these molecules (Partridge, 1970a) and, in particular, that in linear alkanes the excitation will be essentially divided into two portions, one which is localized in pairs of C–H bonds and the other which is transferred very rapidly from one C–C bond to another (in $\approx 10^{-16}$ sec) along the main chain. The C–H bond pair

excitation should lead to scission of one or both bonds, ejecting H or H_2.
The C–C bond excitons, predicted to have an average energy ≈ 8 eV, will
normally travel along a single polymer chain but may sometimes transfer to
another chain or to foreign molecules on or off the chain. The efficiency of
such transfer should increase steadily with chain length up to some limiting
length (representing the maximum probable exciton travel distance). The
rapid transfer rate between C–C bonds should make dissociation of these
bonds (due to singlet excitation) negligible. In branched alkanes C–C bond
excitons should also occur, but they are unlikely to cause scission of side
branches; side-branch scission is predicted to occur only from excitation
deposited initially in these branches or from secondary reactions.

The above scission predictions assume that scission will occur from a
singlet excited state since the bond exciton model only considers singlet
states. Plotnikov (1969a,b) has concluded from theoretical calculations that
the rate of intersystem crossing (from singlet to triplet states) is so high in
aliphatic molecules ($\approx 10^{10}$–10^{11} sec^{-1}) that the triplet state will be reached
before any dissociation can occur and, hence, that any observed dissociations
will occur from the triplet and not the singlet state. However, the alkane
fluorescence experiments of Hirayama and Lipsky (1969) and Hirayama *et al.*
(1970) indicate that the singlet fluorescence occurs in competition with other
singlet nonradiative processes (such as intersystem crossing, dissociation,
and collisional deactivation) that have a total rate constant which varies
nearly linearly with chain length from about 3×10^9 sec^{-1} in *n*-pentane to
2×10^7 sec^{-1} in *n*-hexadecane and presumably to a much lower value still
in polyethylene. These results suggest that intersystem crossing may not be
very significant in long-chain alkanes even if it is important in short-chain
ones. Hirayama and Lipsky suggested that the decrease in rate constant with
increase in chain length might be due to a reduction in molecular distortion
of the excited molecules due to the excitation being spread over a longer
chain length.

The possibility of excitation transfer along a C–C bond chain has also been
discussed by Kaplan and Plotnikov (1967), who did an *ab initio* calculation
on the interaction energy of two adjacent C–C bonds in an alkane chain.
Their value of 1.67 eV for singlet interactions compares quite well with the
empirical value of 2.4 eV from the bond exciton theory (Partridge, 1968)
in view of the approximate nature of the calculation. They also obtained a
value of 0.1 eV for the triplet state interaction of two C–C bonds, which
certainly indicates that triplet excitation transfer could be very efficient if the
triplet states could be excited. The energies of the lowest alkane triplet states
are not known with any certainty although experimental values as low as
3.8 eV have been suggested for *n*-hexane (Hunter *et al.*, 1970). Triplet excitons

on C–H bond pairs have been used by Loeve and Salem (1967) in a theoretical study of the geminal proton–proton coupling constants in alkanes.

Theoretical studies of excitation transfer have also been made on unconjugated diene polymers. Polybutadiene was considered by Lorquet *et al.* (1962) as part of their work on dissociative excitons, in relation to the experimental observation of radiation-induced *cis–trans* isomerism. It was assumed that the excitation interaction was only between double bonds and that, following various theoretical and experimental studies of ethylene, excitation of an isolated double bond would cause its ends to twist through 90° relative to one another. The extent of double-bond rotation, and hence degree of isomerization, in the polymer was calculated as a function of time for different coupling strengths, and it was concluded that a single excitation might be able to give an average of 2.5 isomerizations.

A range of alkene molecules and diene polymers has been studied by Partridge (1972) using the "bond exciton" model and incorporating all bonds including both the π and σ bonds of each double bond. This approach gives a reasonable interpretation of the available spectral data on such molecules and suggests that appreciable excitation transfer may be possible in diene polymers that have all their double bonds in the main chain.

V. Comparison of Charge and Excitation Transfer

In many polymers it may be difficult to distinguish unambiguously between charge and excitation transfer because any acceptor molecules or molecular groups involved in the transfer will usually have both lower ionization potentials and excitation energies than the polymer molecules. Any experimentally detectable differences between the two processes are thus of considerable interest, even though their magnitude will vary from one polymer to another, and some such possibilities are outlined here.

A. DIRECT OBSERVATION

The transfer mechanisms involve migration of charge in one case and of excitation in the other but only the former can be directly observed (as a conduction current). However, excitation transfer can sometimes be observed by its effect on the electronic spectra of the system, such as by an energy shift compared with the monomer spectrum, by the formation of an excimer fluorescence band at high molecular concentrations, or by the depolarization of fluorescence emitted from the system (see, for instance, Förster, 1959). Such evidence does not by itself necessarily link the observed energy transfer process with a particular radiation chemical reaction, but it is a valuable first step.

B. Transfer Dependence on Radiation Energy

The energy required for direct ionization of a molecule is always greater than that for excitation of at least the lower excited states [some excited states may occur above the lowest ionization level, as discussed by Platzman (1962)], so in principle the two forms of transfer may be separated by reducing the energy of the incident radiation until charge formation is eliminated. For polymers containing aromatic or other molecular groups with low-energy excited states such experiments may be fairly easy since visible or near-ultraviolet radiation can be used. However for most aliphatic polymers the main absorption occurs in the vacuum ultraviolet, and so more complex apparatus is required and the radiation dose rate may be low. Further complications are that some charge formation may occur even under normally nonionizing radiation by processes such as double excitation (Charlesby and Partridge, 1965) and exciton–exciton annihilation, although this will usually be characterized by an ion formation rate that is proportional to the square of the irradiation intensity. Detection of any such charge formation may be achieved by direct conductivity measurements or, if the charges can be trapped at low temperatures, by thermoluminescence (see Chapter 10). Some polymer studies have been made in the vacuum ultraviolet, as noted later, but much more work is needed.

C. Transfer Dependence on Temperature

Both charge and excitation transfer could, in principle, occur over large distances, especially in long-chain polymers. However, from the electrical conductivity measurements noted earlier it seems that the charge migration distance may be small in many polymers at low temperature (≈ 100 Å) but increase very considerably at high temperatures ($\approx 1\ \mu$) due to thermally activated trap hopping. Furthermore, charges trapped at low temperature will often migrate at higher temperatures, so radiation chemical reactions in polymers caused by CT should, in general, have a marked temperature dependence due to the variation in migration distance. By contrast excitation transfer in polymers should be little affected by temperature except for excimer and, perhaps, triplet transfer, which depend for their efficiency on the close approach of molecules to one another. However, the secondary reactions following after energy transfer may of course be temperature dependent.

The actual migration time for both processes at low temperature will probably be much too short for direct measurement, but the onset of thermally activated charge hopping at higher temperatures might well yield a CT reaction with an observable time dependence.

D. CHARGE YIELD

If extensive charge or excitation transfer occurs in a polymer then the addition of easily ionizable foreign molecules will usually result in a considerable concentration of additive cations after irradiation. Observation of such cation formation will not by itself favor either mechanism unless the ionization potential of the additive when in the polymer is accurately known and can be compared with the ionization potential of the polymer molecules themselves or with a probable exciton energy. However, a potentially more useful difference follows from the fact that CT essentially constitutes a redistribution of the initial polymer charge between the polymer and additive molecules, while excitation transfer involves formation of additive ions from initial polymer excitation in addition to the charge formed by direct polymer ionization. If the amount of charge formed in the absence of the easily ionizable additive can be estimated, preferably with use of a nonionizing additive to capture all available electrons, then estimation of the total change in charge yield on introduction of the additive may be possible. Any increase in total yield would then indicate the presence of some excitation transfer (Partridge, 1970b).

E. ADDITIVE DISSOCIATION

Actual dissociation of additive molecules is most likely to be caused by excitation transfer although dissociative electron attachment is possible in some molecules while others might be dissociated by positive charge transfer (especially if the donor ion was excited). However, an indication of the dissociation mechanism is possible if the additive can be chosen such that it will dissociate into two fragments which can both be detected experimentally (using pulse radiolysis or low-temperature absorption techniques). If neither fragment reacted further, then observation of a neutral and a positive fragment would favor positive charge transfer, a neutral and a negative fragment would favor dissociative electron attachment, and two neutral (or one positive and one negative) fragments would favor excitation transfer. One possible such additive is hexaphenylethane, since this would be very likely to dissociate at the C–C bond and yield fragments of $\phi_3 C\cdot$, $\phi_3 C^+$ or $\phi_3 C^-$, all of whose optical absorption spectra are known (Leftin, 1960).

VI. Experimental Observation of Energy Transfer in Polymers

This section notes some of the experimental evidence for the occurrence of energy transfer in various polymers. Many such claims have been made, but a number of these, especially in the earlier literature, are now open to considerable doubt. No attempt can be made here to present a complete and critical review of all this work, and consideration is limited to a few cases for which an appreciable amount of data is available.

A.   POLYSTYRENE AND SIMILAR AROMATIC POLYMERS

In polystyrene energy migration by both charge and excitation transfer are possible. Electrical conductivity measurements after a short irradiation pulse indicate that there is a considerable movement of predominantly positive charges over distances of $\approx 0.5 \mu$ at room temperature although the distance is reduced to $\approx 0.03 \mu$ at low temperatures (Hirsch and Martin, 1970). Photoconductivity measurements in the vacuum ultraviolet (Ofran et al., 1968) show that most charge production requires more than 8.8 eV, where the lower energy limit probably represents the ionization potential of the phenyl groups in the polymer.

Excitation transfer can occur in polystyrene by both inter- and intra-molecular processes, and the relative importance of these depends strongly on the physical state of the polymer molecules. In solution at room temperature polystyrene exhibits both normal and excimer fluorescence under near-ultraviolet radiation, where the excimers are formed between neighboring phenyl groups on the *same* molecular chain (Vala et al., 1965); there is competition for the excitation on each phenyl group between normal fluorescence, excimer formation and, probably, intersystem crossing to the triplet state (see, for instance, Kistiakowsky and Parmenter, 1965). In a rigid glass solution at 77°K there is no excimer emission and only normal fluorescence and phosphorescence occur, although these may be associated with efficient triplet transfer and triplet–triplet annihilation as found for polyvinylnaphthalene by Cozzens and Fox (1969a,b). The occurrence of excimer formation only at room temperatures is probably due to the thermally induced motions of the phenyl groups, and of the rest of the chain, intermittently producing conformations of neighboring phenyl group pairs that are suitable for excimer formation.

In solid polystyrene only excimer emission occurs at room temperature [as shown by the spectra of Basile (1964), although the emission was not assigned to excimers at that time]. Under vacuum-ultraviolet irradiation emission spectra due probably to two different types of excimer are observed, and these seem very sensitive to the microstructure of the polymer (Leibowitz and Weinreb, 1966, 1967). The vacuum-ultraviolet absorption spectrum of solid films shows evidence of strong excitation interactions, especially in the high-absorption region around 1900 Å, and these probably involve phenyl groups on different chains as well as on the same chain (Partridge, 1967). The complete absence of normal fluorescence in solid polystyrene indicates either that all phenyl groups are situated in favorable positions for excimer formation or, more likely in view of the absorption spectrum evidence, that efficient exciton transfer occurs between phenyl groups along a chain until encountering a favorable site for excimer formation [as suggested by Klopffer (1969) for polyvinylcarbazole]. It should be noted that exciton transfer will be much slower for the lowest singlet levels (since these involve a forbidden transition at ≈ 2600 Å between the ground and benzene-type $^1B_{2u}$ excited state) than for the upper levels around 1900 Å (corresponding to a strong transition involving the benzene $^1E_{1u}$ state), but it should still occur in the lowest levels with considerable efficiency.

Intermolecular transfer to various aromatic additives has been demonstrated by, among others, Basile (1964) and Leibowitz and Weinreb (1966). This probably involves competition for migrating excitons between the additives and the potentially excimer-forming sites. Triplet transfer is suggested by the work of Alfimov and Shekk (1967).

Excimer formation has also been observed in polyvinylnaphthalene (Vala *et al.*, 1965) and, together with excitation transfer to aromatic additives, in polyvinylcarbazole (Klopffer, 1969). In addition, Cozzens and Fox (1969a,b) have demonstrated triplet transfer and annihilation in polyvinylnaphthalene glass solutions at 77°K, Lashkov and Ermolaev (1967) have observed triplet transfer in polyvinylphthalimide, and David *et al.* (1970) observed triplet transfer from polyvinylbenzophenone to naphthalene.

Several radiation chemical studies have been made of polystyrene with relation to energy transfer. Ho and Siegal (1967, 1969) using pulse radiolysis found that addition of naphthalene or biphenyl gave only triplet states of these molecules (which they ascribed to triplet transfer) and no ions at all, but addition of triphenylamine gave the appropriate cations in high yield ($G \simeq 2.0$). Now the ionization potentials of these three molecules are 8.13, 8.27, and 6.86 eV, respectively, in the gas phase (Watanabe *et al.*, 1962; Terenin and Vilessov, 1964), while that of polystyrene is about 8.8 eV in the solid phase (Ofran *et al.*, 1968); and so if positive charge transfer is widespread, it should have produced cations of all three molecules. However, if the strong polystyrene absorption at ≈ 1900 Å is indeed associated with inter- and intrachain singlet excitons, as indicated earlier, then the ≈ 6.5-eV energy of these might well be sufficient to ionize triphenylamine (as its ionization potential will be even lower in solid phase) but not the other two molecules, as observed.

Harrah (1969) studied radical formation in polysterene under γ irradiation and suggested that excitation transfer along the polymer chains was responsible both for formation of radicals exclusively at chain ends in chains of only ≈ 20 monomer units and also for limiting radical concentrations at high doses in long-chain polystyrenes. It may be significant that on the bond exciton models discussed earlier efficient main-chain C–C bond exciton migration would be expected in polystyrene just as in pure alkane polymers, with the only difference that such migration occurs in the lowest excited singlet state of the alkanes while in polystyrene the side-chain phenyl groups provide still lower excited states. Excitation of end-chain units by C–C bond excitons in polystyrene could occur if the lifetime of these excitons was sufficient to permit appreciable migration before the excitation reverted to the lower excited states.

Wilske and Heusinger (1969b) added pyrene to polystyrene and showed that excitation transfer to the pyrene sharply decreased the radical concentration but had little effect on hydrogen evolution or cross-linking. They concluded that radical formation must occur from the lowest excited singlet states while the other reactions occurred from ionized or highly excited states. Actually, the very lowest polystyrene excited singlet state is essentially a $^1B_{1u}$ benzene-type state on the phenyl groups and this would not be likely to cause bond scission in the phenyls; scission must be due either to a higher excited state (perhaps that corresponding to the strong absorption at ≈ 1900 Å) or to abstraction of a neighboring H atom by the excited phenyl to give a cyclohexadienyl radical, which then reacts further.

Finally, it should be noted that the efficient energy transfer in polystyrene (and polyvinyltoluene) has long been utilized commercially in the production of plastic scintillators for the detection of ionizing radiation. The radiation is absorbed by the polymer and the resulting energy passed to a suitable fluorescent additive (Birks, 1964). The transfer mechanism seems to be predominantly excitation transfer, although some contribution from charge transfer cannot be excluded (see, for instance, Hirayama et al., 1968).

B. POLYETHYLENE AND OTHER ALKANES

Charge transfer in polyethylene is demonstrated by the observation of electrical conductivity during and after irradiation. The charge carriers are predominantly positive species at high temperatures ($\approx 80\,^\circ$C) and can migrate for ≈ 5000 Å (in the direction of an applied electric field) by thermally assisted trap hopping, although migration is probably limited to ≈ 100 Å for any temperature much below room temperature (Hirsch and Martin, 1970).

Excitation transfer from alkanes to other molecules has been discussed for many years but only recently has its existence been established. Laor and Weinreb (1965, 1969) found that solutions of n-hexane or cyclohexane containing small concentrations of 2,5-diphenyloxazole gave appreciable solute fluorescence when exposed to vacuum-ultraviolet radiation that was strongly absorbed (without ionization) by the alkane. They found that cyclohexane was more efficient in transferring excitation than *n*-hexane and that oxygen quenched the excited states of both. Recently, Hirayama and Lipsky (1969) demonstrated excitation transfer from cyclohexane to benzene under vacuum-ultraviolet irradiation in which the cyclohexane fluorescence decreased and the benzene fluorescence increased with increasing benzene concentration. Furthermore, they found that cyclohexane had a considerably greater fluorescence lifetime than *n*-hexane and that oxygen efficiently quenched the fluorescence, both of which agree with the Laor and Weinreb results. (On the bond exciton model of alkanes, described earlier, the greater cyclohexane lifetime could be ascribed to the absence of end effects perturbing the migrating excitons.) Transfer of cyclohexane excitation to benzene under nonionizing vacuum-ultraviolet irradiation was also observed by Holroyd *et al.* (1967, 1968), who found that addition of benzene to cyclohexane caused a much greater reduction in hydrogen evolution than could be accounted for by H atom scavenging.

Intramolecular excitation transfer is indicated by the vacuum-ultraviolet absorption spectra of small alkanes (Raymonda and Simpson, 1967) and alkane polymers (Partridge, 1968) *if* the bond exciton model of alkanes is sufficiently valid. As a test of validity the model was applied to the radiation chemistry of alkane polymers and provided reasonable explanations for the unimportance of main and side-chain scission in polyethylene, the higher scission probability of pairs of side chains, the elimination of main-chain unsaturation, the occurrence of molecular hydrogen elimination, the negligible C–H bond energy migration, and the efficiency of some cross-linking sensitizers (Partridge, 1970a,c). Similar successful application of the bond exciton model to the photolysis of the small alkanes *n*-pentane and trimethylpentane was noted by Holroyd (1969). The further prediction of main-chain excitation transfer in alkanes with an efficiency dependent on chain length was tested by Partridge (1970b), using additives which would ionize or dissociate to give fragments of known absorption spectrum; energy transfer due largely to excitation transfer was indeed found to be very efficient in polyethylene but much less so in small alkanes. The higher efficiency with increasing chain length is probably partly due to the increased exciton migration distance (since the bond exciton model predicts migration to be predominantly intramolecular) and partly to the increased excited state lifetime [which is in turn probably a consequence of the greater excitation delocalization (Hirayama *et al.*, 1970)]. Similar earlier energy transfer studies on 3-methylpentane by Hamill *et al.* (see, for instance, Skelly and Hamill, 1965; Gallivan and

Hamill, 1966) were interpreted solely in terms of charge transfer, but in most cases excitation transfer was not considered. It may be that CT predominates in small alkanes due to a higher intermolecular transfer efficiency, but this remains to be demonstrated.

Various other energy transfer reactions have been observed in the radiolysis of polyethylene and other alkanes, but it is not yet possible to separate these into charge and excitation transfer contributions. The "protection" of alkanes against hydrogen evolution by addition of benzene or other molecules is well known. This is probably due in part to H atom scavenging by benzene since cyclohexadienyl radicals are observed (Sauer and Mani, 1968; Partridge, 1970d). However, this alone cannot account for the appreciable fall in total radical concentration at low temperatures (Molin *et al.*, 1960; Zeman and Heusinger, 1967; Partridge, 1970d) which indicates quenching of the precursors of hydrogen evolution. The vacuum-ultraviolet study of Holroyd *et al.* (1967, 1968) suggests that a part of this quenching under radiolysis may be due to excitation transfer, but equally, there is evidence for CT participation (see, for example, Dyne, 1965), and the relative importance of these processes is yet to be established. Zeman and Heusinger (1967) suggest, from study of alkylbenzenes and use of a model involving random diffusion (rather than coherent migration) of energy, that intra- and intermolecular energy transfer in small alkanes are roughly equally efficient and that intramolecular transfer occurs over only a relatively few methylene groups. (On the bond exciton model this energy transfer refers to C–H bond excitations, which are predicted not to migrate significantly in infinite alkanes although some migration will occur in short chain alkanes due to the mixing of C–H and C–C bond excitations.)

Other energy transfer effects in polyethylene radiolysis include the surprisingly rapid elimination of vinyl, vinylene, and vinylidene unsaturation in the polymer chains. It seems very likely that both excitation and charge transfer may be involved here (see, for example, Dole *et al.*, 1966; Partridge, 1970c), but the reactions are not yet sufficiently well characterized for definite mechanisms to be demonstrated.

A polyethylene excitation transfer reaction of practical interest, although not one associated with its basic alkane structure, is photodegradation. This has been found to be associated with carbonyl impurity groups attached to the polymer chains, where the carbonyls are excited (indirectly) to the triplet state by near-ultraviolet radiation (≈ 3000 Å) and then cause main-chain scission. This reaction can be sensitized or quenched by additives that have triplet states higher or lower, respectively, than those of the carbonyl (see, for example, Heskins and Guillet, 1970). No indication of intramolecular triplet transfer within polyethylene itself has been observed from these or other studies involving similar excitation energies (Lamola *et al.*, 1965), so if any does occur, it will probably involve considerably higher energies.

C. DIENE POLYMERS

Radiolysis of solid 1,4-polyisoprene or 1,4-polybutadiene is found to produce isomerization of the double bonds with initial *G* values as high as 10 and 7.2, respectively (Golub, 1965; Golub and Danon, 1965). This is apparently caused not by double-bond migration but by excitation of the double bonds which causes them to rotate and then to stabilize in a new position. Such isomerization can be induced by near-ultraviolet radiation (not absorbed by the polymer itself) if suitable additives are used whose triplet states are higher than those of the polymer double bonds, thus demonstrating intermolecular excitation transfer (Golub and Stephens, 1966). An energy transfer, presumably intramolecular, is also indicated for isomerization by radiolysis since the electron fraction of the double bonds is too small for all the excitation to be caused by direct energy absorption. Since vacuum-ultraviolet irradiation at 1236 Å (10 eV) produces much less isomerization (Golub and Stephens, 1967), this energy transfer could involve positive charge migration to the double bonds followed by ion recombination, since the ion yield at this irradiation energy is probably considerably less than for much higher energies (Schoen, 1962). Such a process should show a slight temperature dependence in that some electrons are trapped in these polymers at $\sim 77°K$ (see thermoluminescence, Chapter 10), and indeed, some extra isomerization on warming from such temperatures has been observed by Sukhov *et al.* (1967) and by Il'icheva and Slovokhotova (1968). However, the ion yield is not likely to be sufficient to cause all the observed isomerization, unless by a chain reaction such as suggested for elimination of unsaturation in these polymers (Golub, 1965). Furthermore, study of small monoalkenes suggests that in this case all intramolecular transfer is negligible and that the higher isomerization efficiency of the polymers is connected with their long chain lengths (Golub, 1967), quite possibly due to multiple isomerizations by a single excitation as suggested by Lorquet *et al.* (1962) in their study of dissociative excitons.

It is also of interest that polyethylene can apparently act as a sensitizer of isomerization in polybutadiene, and at the same time is "protected" somewhat against hydrogen evolution (Dole and Williams, 1959; Golub, 1962).

D. POLYMETHYLMETHACRYLATE (PMMA)

Considerable study has been made of this polymer and energy transfer has often been invoked to explain the results. Intermolecular triplet excitation transfer seems fairly well established from the work of Fox and Price (1967), who found that degradation (main chain scission) under 2537-Å radiation could be either quenched or sensitized by the use of additives whose triplet states were respectively below or above the probable polymer lowest triplet. (This polymer triplet is likely to be mainly associated with the carbonyl of the side-chain ester group, which provides a link with the work on oxidized polyethylene noted earlier.)

Under ionizing radiation Bagdasar'yan *et al.* (1962) and Milyutinskaya and Bagdasar'yan (1964) found that the reduction in degration by addition of aromatic amines was accompanied by an amine cation yield of as much as 3 in some cases. Borovkova and Bagdasar'yan (1967) ascribed such protection, by these amines and other aromatic molecules, solely to the trapping of migratory polymer positive charges by the additives, thus preventing scission due to electron recombination with a positive polymer ion. They also showed that in the additives used by them the protection efficiency increased steadily with decreasing additive ionization potential. However, they failed to consider excitation transfer to the additives in their reaction scheme, and it may be significant that they could only observe cations from their amine additives. Since their other additives included naphthalene, which has a known cation absorption spectrum, and since ions that are formed in PMMA are usually surprisingly stable even at room temperature (Milyutinskaya and Bagdasar'yan, 1964; Ho and Siegal, 1969), widespread positive CT should have produced other cations. This, and the demonstration of scission by excitation under near-ultraviolet radiation, make it doubtful whether positive charge recombination can really be the only mechanism for chain scission under ionizing radiation. Gardner and Epstein (1961) and Gardner *et al.* (1966) treated their results on protection against degradation in terms of simultaneous charge and excitation transfer, with both having some probability of causing scission. They concluded, without examining the exact mechanism of transfer, that excitation could migrate long distances but had a low probability of causing chain scission while positive charges migrated little but had much more probability of causing scission. This result may help to explain the conclusion of Wilske and Heusinger (1969a) from radical studies that two energy transfer processes occur within a PMMA chain, one involving transfer over more than 1000 monomer units and the other over only two or three such units.

Finally, the pulse radiolysis work of Ho and Siegal (1969) showed that addition of biphenyl to PMMA gives anions with a *G* value of ≈ 1 but no cations and, also, biphenyl triplet states with a yield $\gtrsim 0.7$. The anions were observed only at quite high biphenyl concentrations, suggesting a large number of other electron traps which may be the ester side chains themselves. The triplets were thought to be produced via triplet transfer within the polymer. Addition of naphthalene gave only triplet states but no anions or cations of this molecule.

E. BIOLOGICAL MACROMOLECULES

The application of the exciton model to biological macromolecules and comparison with experimental results is far too large a topic to be discussed here. Some useful reviews have been given by Tinoco (1965), Mason (1966), and Deutsche *et al.* (1969).

REFERENCES

Alfimov, M. V., and Shekk, Yu. B. (1967). *Khim. Vysok. Energii* **1**, 235 (*English transl.: High Energy Chem.* **1**, 201).
Bagdasar'yan, Kh. S., Krongauz, V. A. and Kardash, N. S. (1962). *Dokl. Akad. Nauk. SSSR* **144**, 101 (*English transl.: Proc. Acad. Sci. USSR Chem.* **144**, 379).
Basile, L. J. (1964). *Trans. Faraday Soc.* **60**, 1702.
Bierman, A. (1970). *J. Chem. Phys.* **52**, 4987.
Birks, J. B. (1964). "The Theory and Practice of Scintillation Counting." Pergamon Press, Oxford.
Birks, J. B. (1967). *Nature* **214**, 1187.
Birks, J. B. (1970). *In* "Progress in Reaction Kinetics" (G. Porter, ed.), Vol. 5, p. 181. Pergamon Press, Oxford.
Birks, J. B., Dyson, D. J., and Munro, I. H. (1963). *Proc. Roy. Soc.* **A275**, 575.
Borovkova, L. Ya., and Bagdasar'yan, Kh. S. (1967). *Khim. Vysok. Energii* **1**, 340 (*English transl.: High Energy Chem.* **1**, 295).
Bradley, D. F., Tinoco, I., and Woody, R. W. (1963). *Biopolymers* **1**, 239.
Burton, M., and Lipsky, S. (1957). *J. Phys. Chem.* **61**, 1461.
Charlesby, A., and Partridge, R. H. (1965). *Proc. Roy. Soc.* **283**, 329.
Choi, S., Jortner, J., Rice, S. A., and Silbey, R. (1964). *J. Chem. Phys.* **41**, 3294.
Cozzens, R. F., and Fox, R. B. (1969a). *J. Chem. Phys.* **50**, 1532.
Cozzens, R. F., and Fox, R. B. (1969b). *Macromolecules* **2**, 181.
David, C., Demarteau, W., and Geuskens, G. (1970). *Eur. Polymer J.* **6**, 537.
Deutsche, C. W., Lightner, D. A., Woody, R. W., and Moscowitz, A. (1969). *Ann. Rev. Phys. Chem.* **20**, 407.
Dexter, D. L. (1953). *J. Chem. Phys.* **21**, 836.
Dole, M., and Williams, T. F. (1959). *Discuss. Faraday Soc.* **27**, 74.
Dole, M., Fallgatter, M. B., and Katsuura, K. (1966). *J. Phys. Chem.* **70**, 62.
Dyne, P. J. (1965). *Can. J. Chem.* **43**, 1080.
Förster, Th. (1946). *Naturwissenschaften* **33**, 166.
Förster, Th. (1948). *Ann. Phys.* **2**, 55.
Förster, Th. (1959). *Discuss. Faraday Soc.* **27**, 7.
Förster, Th. (1960a). *Rad. Res. Suppl.* **2**, 326.
Förster, Th. (1960b). *In* "Comparative Effects of Radiation" (M. Burton, I. S. Kirby-Smith, and J. L. Magee, eds.), p. 300. Wiley, New York.

Förster, Th. (1966). *In* "Modern Quantum Chemistry, Istanbul Lectures" (O. Sinanoglu, ed.), Vol. 3, p. 93. Academic Press, New York.

Förster, Th. (1968). *In* "Energetics and Mechanisms in Radiation Biology" (G. O. Phillips, ed.), p. 183. Academic Press, New York.

Fox, R. B., and Price, T. R. (1967). *J. Appl. Polymer Sci.* **11,** 2373.

Frankevich, E. L. (1966). *Usp. Khim.* **35,** 1161 (*English transl.: Russ. Chem. Rev.* **35,** 487).

Frankevich, E. L., and Yakovlev, B. S. (1963). *Zh. Fiz. Khim.* **37,** 1106 (*English transl.: Russ. J. Phys. Chem.* **37,** 581).

Freeman, G. R., and Fayadh, J. M. (1965). *J. Chem. Phys.* **43,** 86.

Gallivan, J. B., and Hamill, W. H. (1965). *Trans. Faraday Soc.* **61,** 1960.

Gallivan, J. B., and Hamill, W. H. (1966). *J. Chem. Phys.* **44,** 2378.

Gardner, D. G., and Epstein, L. M. (1961). *J. Chem. Phys.* **34,** 1653.

Gardner, D. G., Henry, G., and Ward, D. (1966). *In* "Energy Transfer in Radiation Processes" (G. O. Phillips, ed.), p. 37. Elsevier, Amsterdam.

Goad, W. (1963). *J. Chem. Phys.* **38,** 1245.

Golub, M. A. (1962). *J. Phys. Chem.* **66,** 1202.

Golub, M. A. (1965). *J. Phys. Chem.* **69,** 2639.

Golub, M. A. (1967). *In* "Radiation Research" (G. Silini, ed.), p. 339. North-Holland Publ., Amsterdam.

Golub, M. A., and Danon, J. (1965). *Can. J. Chem.* **43,** 2772.

Golub, M. A., and Stephens, C. L. (1966). *J. Polymer Sci. B* **4,** 959.

Golub, M. A., and Stephens, C. L. (1967). *J. Polymer Sci. C* **16,** 765.

Harrah, L. A. (1969). *Mol. Crystallogr. Liquid Crystallogr.* **9,** 197.

Heskins, M., and Guillet, J. E. (1970). *Macromolecules* **3,** 224.

Hirayama, F., and Lipsky, S. (1969). *J. Chem. Phys.* **51,** 3616.

Hirayama, F., Basile, L. J., and Kikuchi, C. (1968). *Mol. Crystallogr.* **4,** 83.

Hirayama, F., Rothman, W., and Lipsky, S. (1970). *Chem. Phys. Lett.* **5,** 296.

Hirsch, J., and Martin, E. (1969a). *Solid State Commun.* **7,** 279.

Hirsch, J., and Martin, E. (1969b). *Solid State Commun.* **7,** 783.

Hirsch, J., and Martin, E. (1970). *J. Noncrystallogr. Solids* **4,** 133.

Ho, S. K., and Siegal, S. (1967). *J. Phys. Chem.* **71,** 4527.

Ho, S. K., and Siegal, S. (1969). *J. Chem. Phys.* **50,** 1142.

Hochstrasser, R., and Kasha, M. (1964). *Photochem. Photobiol.* **3,** 317.

Holroyd, R. A. (1969). *J. Amer. Chem. Soc.* **91,** 2208.

Holroyd, R. A., Yang, J. Y., and Servedio, F. M. (1967). *J. Chem. Phys.* **46,** 4540.

Holroyd, R. A., Yang, J. Y., and Servedio, F. M. (1968). *J. Chem. Phys.* **48,** 1331.

Hunter, L. M., Lewis, D., and Hamill, W. H. (1970). *J. Chem. Phys.* **52,** 1733.

Il'icheva, Z. F., and Slovokhotova, N. A. (1968). *Khim. Vysok. Energii* **2,** 66 (*English transl.: High Energy Chem.* **2,** 56).

Inokuti, M., and Hirayama, F. (1965). *J. Chem. Phys.* **43,** 1978.

Jortner, J., Rice, S. A., and Katz, J. L. (1965). *J. Chem. Phys.* **42,** 309.

Kaplan, I. G., and Plotnikov, V. G. (1967). *Khim. Vysok. Energii* **1,** 507 (*English transl.: High Energy Chem.* **1,** 447).

Kasha, M. (1959). *Rev. Mod. Phys.* **31,** 162.

Kasha, M. (1963). *Radiat. Res.* **20,** 55.

Kasha, M. (1964). *In* "Physical Processes in Radiation Biology" (L. G. Augenstein, R. Mason, and B. Rosenberg, eds.), p. 17. Academic Press, New York.

Kasha, M., Rawls, H. R., and Ashraf El-Bayoumi, M. (1967). *Pure Appl. Chem.* **11,** 371.

Katsuura, K. (1964). *J. Chem. Phys.* **40,** 3527.

Katsuura, K., and Inokuti, M. (1963). *J. Phys. Soc. Japan* **18**, 1486.

Kistiakowsky, G. B., and Parmenter, C. S. (1965). *J. Chem. Phys.* **42**, 2942.

Klopffer, W. (1969). *J. Chem. Phys.* **50**, 2337.

Krishna, V. G. (1967). *J. Chem. Phys.* **46**, 1735.

Lamola, A. A., Leermakers, P. A., Byers, G. W., Hammond, G. S. (1965). *J. Amer. Chem. Soc.* **87**, 2322.

Laor, U., and Weinreb, A. (1965). *J. Chem. Phys.* **43**, 1565.

Laor, U., and Weinreb, A. (1969). *J. Chem. Phys.* **50**, 94.

Lashkov, G. I., and Ermolaev, V. L. (1967). *Opt. Spekt.* **22**, 848 (*English transl.: Opt. Spectrosc.* **22**, 462).

Lassettre, E. N., Skerbele, A., and Dillon, M. A. (1967). *J. Chem. Phys.* **46**, 4536.

Leftin, H. P (1960) *J. Phys. Chem.* **64**, 1714.

Leibowitz, M., and Weinreb, A. (1966). *J. Chem. Phys.* **45**, 3701.

Leibowitz, M., and Weinreb, A. (1967). *J. Chem. Phys.* **46**, 4652.

Loeve, P., and Salem, L. (1967). *In* "Proceedings of the 24th Colloque Ampere" (R. Blinc, ed.), p. 1046. North-Holland Publ., Amsterdam.

Lombos, B. A., Sauvageau, P., and Sandorfy, C. (1967). *J. Mol. Spectrosc.* **24**, 253.

Lorquet, J. C., El Komoss, S. G., and Magee, J. L. (1962). *J. Chem. Phys.* **37**, 1991.

Lyons, L. E. (1957). *J. Chem. Soc.* 5001.

Magee, J. L., and Funabashi, K. (1961). *J. Chem. Phys.* **34**, 1715.

Mason, R. (1966). *In* "Energy Transfer in Radiation Processes" (G. O. Phillips, ed.). Elsevier, Amsterdam.

McRae, E. G. (1961). *Aust. J. Chem.* **14**, 354.

McRae, E. G., and Kasha, M. (1964). *In* "Physical Processes in Radiation Biology" (L. G. Augenstine, R. Mason, and B. Rosenberg, eds.), p. 23. Academic Press, New York.

Meisels, G. G. (1964). *J. Chem. Phys.* **41**, 51.

Merrifield, R. E. (1961). *J. Chem. Phys.* **34**, 1835.

Merrifield, R. E. (1963). *J. Chem. Phys.* **38**, 920.

Milyutinskaya, R. I., and Bagdasar'yan, Kh. S. (1964). *Zh. Fiz. Khim.* **38**, 776 (*English transl.: Russ. J. Phys. Chem.* **38**, 419).

Miyata, T., and Yomosa, S. (1969). *J. Phys. Soc. Japan* **27**, 727.

Molin, Yu. N., Chkheidze, I. I., Petrov, Al. A., Buben, N. Ya., and Voevodskii, V. V. (1960). *Dokl. Akad. Nauk. SSSR* **131**, 125 (*English transl.: Proc. Acad. Sci. USSR Phys. Chem.* **131**, 219).

Murrell, J. N. (1963). "The Theory of the Electronic Spectra of Organic Molecules." Wiley, New York.

Ovchinnikov, A. A. (1965). *Zh. Strukt. Khim.* **6**, 291 (*English transl.: J. Struct. Chem.* **6**, 266).

Ofran, M., Oron, N., and Weinreb, A. (1968). *J. Chem. Phys.* **48**, 4805.

Partridge, R. H. (1967). *J. Chem. Phys.* **47**, 4223.

Partridge, R. H. (1968). *J. Chem. Phys.* **49**, 3656.

Partridge, R. H. (1970a). *J. Chem. Phys.* **52**, 2485.

Partridge, R. H. (1970b). *J. Chem. Phys.* **52**, 2491.

Partridge, R. H. (1970c). *J. Chem. Phys.* **52**, 2501.

Partridge, R. H. (1970d). *J. Chem. Phys.* **52**, 1277.

Partridge, R. H. (1972). *Int. J. Quant. Chem.* **6**, 167.

Platzman, R. L. (1962). *Radiat. Res.* **17**, 419.

Plotnikov, V. G. (1969a). *Opt. Spectrosc.* **26**, 505.

Plotnikov, V. G. (1969b). *Opt. Spectrosc.* **27**, 322.

Pope, M. (1967). *J. Polymer Sci. C* **17**, 233.

Pope, M. (1968). *Mol. Crystals* **4**, 183.

Raymonda, J. W., and Simpson, W. T. (1967). *J. Chem. Phys.* **47**, 430.

Rhodes, W., and Chase, M. (1967). *Rev. Mod. Phys.* **39**, 348.

Rice, S. A., and Jortner, J. (1968). *In* "Physics and Chemistry of the Organic Solid State" (D. Fox, M. M. Labes, and A. Weissberger, eds.), Vol. 3. Wiley, New York.

Roginskii, V. A., and Kotov, B. B. (1967). *Khim. Vysok. Energii* **1**, 291 (*English transl.: High Energy Chem.* **1**, 254).

Sauer, M. C., and Mani, I. (1968). *J. Phys. Chem.* **72**, 3856.

Schoen, R. I. (1962). *J. Chem. Phys.* **37**, 2032.

Simpson, W. T. (1951). *J. Amer. Chem. Soc.* **73**, 5363.

Simpson, W. T. (1955). *J. Amer. Chem. Soc.* **77**, 6164.

Simpson, W. T., and Peterson, D. L. (1957). *J. Chem. Phys.* **26**, 588.

Skelly, D. W., and Hamill, W. H. (1965). *J. Chem. Phys.* **43**, 3497.

Sukhov, F. F., Il'icheva, Z. F., Slovokhotova, N. A., Margolin, D. M., and Terekhov, V. D. (1967). *Khim. Vysok. Energii* **1**, 58 (*English transl.: High Energy Chem.* **1**, 51).

Takeno, S. (1968). *Chem. Phys. Lett.* **2**, 579.

Terenin, A., and Vilessov, F. (1964). *Advan. Photochem.* **2**, 385.

Tinoco, I. (1965). *In* "Molecular Biophysics" (B. Pullman and M. Weissbluth, eds.), p. 269. Academic Press, New York.

Tinoco, I., Woody, R. W., and Bradley, D. F. (1963). *J. Chem. Phys.* **38**, 1317.

Tric, C., and Parodi, O. (1967). *Mol. Phys.* **13**, 1.

Vala, M. T., Silbey, R., Rice, S. A., and Jortner, J. (1964). *J. Chem. Phys.* **41**, 2846.

Vala, M. T., Haebig, J., and Rice, S. A. (1965). *J. Chem. Phys.* **43**, 886.

Watanabe, K., Nakayama, T., and Mottl, J. (1962). *J. Quant. Spectrosc. Radiat. Trans.* **2**, 369.

Wilske, J., and Heusinger, H. (1969a). *Radiochim. Acta* **11**, 187.

Wilske, J., and Heusinger, H. (1969b). *J. Polymer Sci. A-1* **7**, 995.

Woody, R. W. (1968). *J. Chem. Phys.* **49**, 4797.

Yahagi, K., and Shinohara, K. (1966). *J. Appl. Phys.* **37**, 310.

Zeman, A., and Heusinger, H. (1967). *Radiochim. Acta* **8**, 149.

4

Theory of Free Radicals

Péter Hedvig

Research Institute for Plastics, Budapest, Hungary

I. Introduction

Free radicals are extremely important in the radiation chemistry of macro-molecules since they are formed in the initial as well as in the subsequent stages of radiolysis. The process of formation and transformation of radicals can be studied by electron spin resonance and by optical spectroscopy. Many basic problems of radiolysis of macromolecules are connected with the formation, stabilization, and reactions of free radicals.

In considering these problems it is essential to study the electronic structure of radicals. This can be done theoretically by using the methods of quantum chemistry and experimentally by analyzing the electron spin resonance (ESR) spectra. In this section the basic theoretical aspects of free radical structures will be discussed and illustrated by selected examples. It is fortunate that the ESR technique makes it possible to check the theoretical assumptions rather directly.

Our considerations will be centered on the structure of aliphatic radicals which occur more often in radiolysis of macromolecules than aromatic radicals. A more detailed review of the theoretical methods can be found in the book by Memory (1968).

II. The Molecular Orbital Theory of Free Radicals

There are many theoretical approaches to describe the electronic structure of organic molecules and radicals on the basis of quantum theory. Only one of them will be discussed here very briefly, i.e., the molecular orbital (MO) theory.

In the case of a single electron subjected to a potential $V(r)$ the following (Schrödinger) equation holds (see, for example, Pilar, 1968)

$$[-(\hbar^2/2m)\,\nabla^2 - V(r)]\psi = E\psi \qquad (1)$$

where $\hbar = h/2\pi$, h is the Planck constant, m is the mass of the electron, ∇^2 is the Laplacian differential operator, ψ is the wave function, and E is the energy of the electron.

The solution of this differential equation results in a set of eigenfunctions, $\psi_1 \cdots \psi_n$ and a set of energies $E_1 \cdots E_n$, which are the measurable stationary energy states of the system. The eigenfunctions are complex; they have no direct physical meaning. The real $|\psi_i|^2$ values are interpreted as the probability density for an electron to be in a state ψ_i. The probability for an electron to be in a volume element $d\tau$ is thus $|\psi|^2\,d\tau$.

Equation (1) is written in a general form as

$$\hat{H}\psi = E\psi \qquad (2)$$

where $\hat{H}$ is the Hamiltonian operator of the system. It is constructed from the total Hamiltonian function by using the following transformations:

$$r \to r$$

$$p \to (\hbar/i)\,\nabla \qquad (3)$$

where r is the position vector of the electron, p is the momentum vector, $\hbar = h/2\pi$, h is the Planck constant, and i is the imaginary unit.

In conservative force fields the Hamiltonian function is equal to the total energy; thus, the Hamiltonian operator can be easily constructed as the sum of the kinetic, potential, and various interaction energy operators of the system. For an introductory treatment of basic quantum chemistry see Pilar (1968).

The total Hamiltonian operator of an unpaired electron (free radical) in a molecule can be constructed by considering the following energies

(a) The kinetic energy of the electron

$$E_k = (p^2/2m) \rightarrow (\hbar^2/2m) \nabla^2 \tag{4}$$

where m is the mass of the electron.

(b) The Coulombic interaction energy between the electron and the nuclei

$$V_c = \sum_k (eZ_k/r_{ik}) \tag{5}$$

where Z_k is the charge number of the nuclei and r_{ik} is the distance between the electron i and nucleus k.

(c) The Coulombic repulsion energy arising from the interaction with the other electrons is

$$V_{ik} = \sum_{ik} (e^2/r_i r_k) \tag{6}$$

These are the spin-independent interactions; the corresponding Hamiltonian operator is

$$\hat{H}_0 = -\frac{\hbar^2}{2m} \nabla^2 - \sum_k \frac{eZ_k}{r_{ik}} - \sum_{ik} \frac{e^2}{r_i r_k} \tag{7}$$

The main interactions involving the electronic and nuclear spins are as follows.

A. THE SPIN–ORBIT INTERACTION

This is a magnetic dipole interaction between the intrinsic magnetic moment of the electron and the magnetic moment generated by its orbital motion. The corresponding Hamiltonian is

$$\hat{H}_{SO} = (Ze^2/2m^2c^2r^3) \, \hat{L} \cdot \hat{S} \tag{8}$$

where $\hat{L}$ is the orbital angular momentum operator, $\hat{S}$ is the spin momentum operator of the electron, and c is the velocity of light.

B. DIPOLE–DIPOLE INTERACTION BETWEEN THE UNPAIRED ELECTRON AND THE NUCLEI

This interaction depends on the relative orientation of the dipoles; consequently, it can be observed only in single crystals. In solution this interaction is completely averaged out by the tumbling motion of the molecules. In randomly oriented systems, as in polymers, it may be partially averaged out by rotation of groups. It is usually observed as line broadening in the ESR spectra. The Hamiltonian operator of the dipole–dipole hyperfine interaction is

$$H_{dd} = \gamma_e\gamma_n[(1 - 3\cos^2\vartheta)/r^3]\,\hat{I}\hat{S} \tag{9}$$

where $\hat{I}$ and $\hat{S}$ are, respectively, the nuclear and electronic spin operators, γ_n, γ_e are the nuclear and electronic gyromagnetic ratios, r is the distance between the electronic and nuclear moments, and ϑ is the angle between their orientation.

C. FERMI (CONTACT) INTERACTION BETWEEN THE MAGNETIC MOMENTS OF THE UNPAIRED ELECTRON AND THAT OF THE NUCLEUS

This interaction is independent of the orientation of the dipoles. The interaction energy is expressed as

$$E_F = -(8\pi/3)\mu_e\mu_n|\psi(r_n)|^2 \tag{10}$$

where μ_e, μ_n, respectively, are the electronic and nuclear magnetic moments and $\psi(r_n)$ is the wavefunction of the unpaired electron at the nucleus.

The Hamiltonian operator of this interaction is

$$\hat{H}_F = -(8\pi/3)\hat{\mu}_e\hat{\mu}_n\,\delta(r_e - r_n) \tag{11}$$

where $\delta(r_e - r_n)$ is the Dirac function, $\hat{\mu}_e$, $\hat{\mu}_n$ are, respectively, the electronic and nuclear magnetic momentum operators, and r_e, r_n are respectively the radius vectors of the electron and nucleus. When a strong magnetic field is applied to the system in direction z, the Hamiltonian is expressed as

$$\hat{H}_F = (8\pi\hbar^2/3)\gamma_e\gamma_n\hat{I}_z\hat{S}_z\,\delta(r_e - r_n) \tag{12}$$

where γ_e, γ_n are, respectively, the electronic and nuclear gyromagnetic ratios and $\hat{I}_z$, $\hat{S}_z$ are, respectively, the nuclear and electron spin operator components along the direction of the applied magnetic field.

Once the total Hamiltonian is constructed the problem seems relatively simple: One has to construct the Schrödinger equation and solve it for ψ and E. This would result in the stationary energies and the electron probability densities. This, however, in practice cannot be done because of principal mathematical difficulties. For obtaining the eigenfunctions methods of approximation must be used. A usual procedure is to start with a trial function and by using the variational method evaluate the true ψ by successive approximations. The ground state energy of the system is the expectation value of the Hamiltonian operator

$$E_0 = \langle \hat{H} \rangle_{\text{exp}} = \int \psi_0 \hat{H} \psi_0{}^* \, d\tau / \int \psi_0 \psi_0{}^* \, d\tau \tag{13}$$

It has been shown that by substituting any function instead of ψ_0 the value of $\langle \hat{H} \rangle_{\text{exp}}$ will be higher than E_0. Thus, the mathematical problem is to find the function which minimizes expression (13). For this calculation any trial wavefunction can be used. In the molecular orbital theory this function is constructed from the known atomic eigenfunctions in the form of a linear combination

$$\phi = \sum_{i=1}^{N} \lambda_i \varphi_i \tag{14}$$

where λ_i are parameters to be determined by the variational calculation, φ_i are the known atomic eigenfunctions, and N is the number of electrons in the molecule.

It is to be realized that electrons in molecules do not belong to definite atoms; they are delocalized over the molecule. The way of building up the electronic configuration of molecules from atomic orbitals is just an approximation.

Fortunately, the extent of delocalization of an unpaired electron can be quantitatively measured by ESR. This is possible because the ESR hyperfine spectra show what kind and how many nuclei are in contact interaction with the unpaired electron.

III. Spin Density

The probability for an electron to be at a given nucleus is $\psi \, |(r_n)|^2$, where ψ is the wave function of the electron and r_n is the position vector of the nucleus. In the case of an unpaired electron (free radical) the spin states are also to be considered: The electron may be at the position r_n with α-spin $+\frac{1}{2}$ or with β-spin $-\frac{1}{2}$. It is usual to define a spin density operator as

$$\hat{\rho}(r) = 2 \sum_i \delta(r_i - r_n) \hat{S}_{zi} \tag{15}$$

where $\delta(r_i - r_n)$ is the Dirac function and $\hat{S}_{zi}$ is the zth component of the spin operator of the ith electron.

The spin density is defined as the expectation value of the spin density operator with respect to the wavefunction of the unpaired electron

$$\rho = \langle \hat{\rho}(r) \rangle = \int \psi \hat{\rho}(r) \psi^* \, d\tau / \int \psi \psi^* \, d\tau \tag{16}$$

where integration is carried out over the whole molecule and the asterisk means complex conjugate. The Hamiltonian operator of the Fermi interaction expressed in terms of the spin density operator is

$$\hat{H}_F = (8\pi\hbar^2/3)\gamma_e\gamma_n\hat{I}_z\hat{\rho}(r) \tag{17}$$

The Fermi hyperfine interaction energy is the expectation value of $\hat{H}_F$

$$\langle \hat{H}_F \rangle = \int \psi \hat{H}_F \psi^* \, d\tau / \int \psi \psi^* \, d\tau \tag{18}$$

By using the variational method it is possible to calculate the hyperfine energies for particular radical structures. The difference between the subsequent hyperfine energy levels is the hyperfine splitting constant. It is approximately expressed by the McConnell (1956) equation

$$a_i = Q^i \rho(i) \tag{19}$$

where a_i is the hyperfine splitting constant of the nucleus i, which can be measured directly by ESR, $\rho(i)$ is the corresponding spin density, and Q^i is a factor which depends on the electronic structure of the radical. In Table I values of this Q factor are collected for a few radical fragments.

TABLE I

McCONNELL AND KARPLUS–FRAENKEL Q VALUES FOR HYPERFINE SPLITTING BY HYDROGEN, CARBON, AND NITROGEN IN AROMATIC RADICALS[a]

Hydrogen	Carbon	Nitrogen
$Q_{CH}^H = -23.5$	$Q_{CC^1}^C = 13.6 \pm 1$	$Q_{NC}^N = 9.8 \pm 2$
	$Q_{C^1C}^C = -13.6 \pm 1$	$Q_{CN}^N = -2 \pm 2$
$Q_{NH}^H = -38.2 \pm 2$		
	$Q_{CN}^C = 19.0 \pm 1$	$Q_{NH}^N = 14.5 \pm 1$
$Q_{CCH_3}^H = 25 \pm 2$	$Q_{NC}^C = 20.5 \pm 4$	$Q_{ON}^N = 35.8 \pm 3$
	$Q_{CO}^C = 17.7 \pm 1$	
	$Q_{OC}^C = 27.7 \pm 3$	
	$S^C = -12.7$	$S^N = 11.3$

[a] The limits given for the numerical values correspond to those used at calculations of different systems. From Karplus and Fraenkel (1962), Ayscough (1967), Fraenkel (1962), and Das and Fraenkel (1965).

Evidently, spin densities in hydrocarbon radicals can be determined at the protons and at the carbon atoms as well. As ^{12}C has no magnetic moment, the unpaired electron density at these atoms does not appear in the ESR spectra. Carbon-13, however, has a magnetic moment. The corresponding ^{13}C hyperfine splitting can be observed even in the natural abundance of the isotope (Karplus and Fraenkel, 1962). This makes it possible to draw a complete picture of how an unpaired electron is delocalized over the carbon atoms and over protons in a free radical. In aromatic systems this delocalization may involve as many as 30 nuclei resulting in extremely complex ESR spectra. In highly conjugated polymeric solids this delocalization may reach macroscopic magnitudes; the unpaired electron may be in such a delocalized state as conduction electrons are in metals and in semiconductors. For recent reviews on ESR data, see Ayscough (1967), Hedvig and Zentai (1969), and Scheffler and Stegmann (1970).

In the radiation chemistry of polymers aliphatic free radicals are most important. One would think that the radicals in such systems containing only σ bonds are not delocalized at all. Calculations and ESR studies have shown, however, that in hydrocarbon alkyl radicals the unpaired electron is delocalized at least over six to eight C–C bonds. This evidently means that the most common view of σ and π bonds based on simple valence bond theory is only a rough approximation.

It will be shown by the following illustrative examples that in aliphatic groups an extra conjugation, i.e., π character of the bonds, takes place which is not expected from the simple bond theory. This is referred to as hyperconjugation (Dewar, 1962).

IV. Hybrid Orbitals

Carbon is known to exhibit hybrid orbitals in which s-type atomic orbitals are mixed up with p-type ones. In methane, for example carbon exhibits sp^3 hybridization, the molecular orbitals are (see Pilar, 1968)

$$\psi_1 = \tfrac{1}{2}(2s + 2p_x + 2p_y + 2p_z)$$

$$\psi_2 = \tfrac{1}{2}(2s + 2p_x - 2p_y - 2p_z)$$

$$\psi_3 = \tfrac{1}{2}(2s - 2p_x + 2p_y - 2p_z)$$

$$\psi_4 = \tfrac{1}{2}(2s - 2p_x - 2p_y + 2p_z) \tag{20}$$

where s and p_x, p_y, p_z are the atomic orbitals. This represents four tetrahedrally arranged σ bonds with the four hydrogen atoms. Methane indeed exhibits a tetrahedral configuration (Fig. 1).

Another possibility of hybridization is realized in conjugated systems. In these cases the $2s$ electron orbital is combined only with the $2p_x$ and $2p_y$

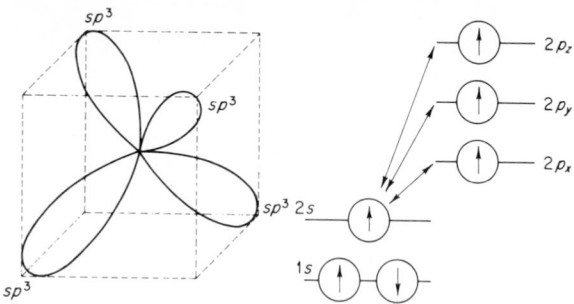

Fig. 1. The sp^3 hybridization of the carbon atom.

orbitals to form sp^2 hybrids. The corresponding orbitals are

$$\psi_1(sp^2) = \frac{1}{\sqrt{3}}(2s) + \sqrt{\frac{2}{3}}(2p_x)$$

$$\psi_2(sp^2) = \frac{1}{\sqrt{3}}(2s) - \frac{1}{\sqrt{6}}(2p_x) + \frac{1}{\sqrt{2}}(2p_y)$$

$$\psi_3(sp^2) = \frac{1}{\sqrt{3}}(2s) - \frac{1}{\sqrt{6}}(2p_x) + \frac{1}{\sqrt{2}}(2p_y)$$

$$\psi_4(p) = 2p_z \tag{21}$$

This configuration is shown in Fig. 2. The three sp^2 orbitals are oriented trigonally; the remaining p_z orbital is perpendicular to their plane.

A third possibility is that the $2s$ electron is combined only with the $2p_x$ electron to form two sp hybrid orbitals; the remaining $2p_y$ and $2p_z$ orbitals are unchanged. The corresponding electron configuration is shown in Fig. 3; it is referred to as sp hybridization.

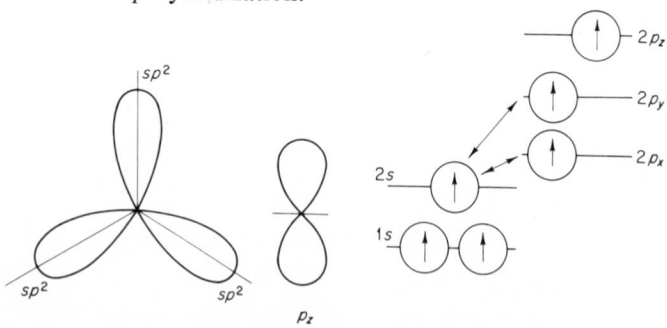

Fig. 2. The sp^2 hybridization of the carbon atom.

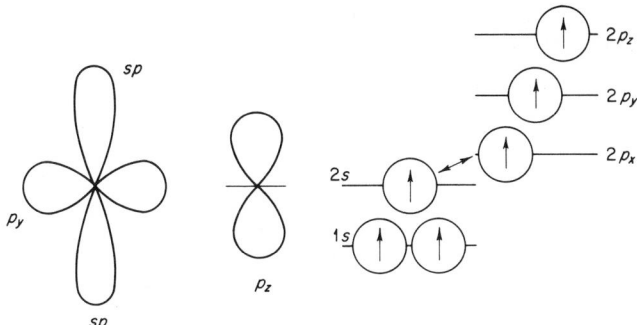

Fig. 3. The sp hybridization of the carbon atom.

The hybridization of carbon is of essential importance in studying organic free radicals. By formation of a radical, for example, by abstracting a hydrogen atom from methane, the hybridization changes from sp^3 to sp^2, i.e., the unpaired electron will be in a p_z orbital. Similarly, by formation of hydrocarbon radicals mentioned in the introduction, the hybridization should be changed which must involve change in the structure.

V. The $>\overset{\bullet}{C}H$ Fragment

As an illustrative example let us consider the electronic structure and spin density in a $>\overset{\bullet}{C}H$ fragment which is connected to an aliphatic or aromatic molecule. Carbon is sp^2 hybridized in this case. One of the sp^2 hybrid orbitals (h) is bound to the hydrogen $1s$ orbital (s), the others to the other part of the molecule; the unpaired electron is at first approximation thought to be localized in the p_z orbital (p). The wave functions of the C–H σ bond are

$$\sigma_B = (1/\sqrt{2})(s + h)$$

$$\sigma_A = (1/\sqrt{2})(s - h) \qquad (22)$$

where s is the atomic orbital of the hydrogen and h is the sp^2 orbital of the carbon.

σ_B is the bonding orbital; σ_A is an antibonding orbital. In the ground state the electron configuration of the $>\overset{\bullet}{C}H$ fragment is $(\sigma_B)^2(\pi)^1$, as shown in Fig. 4. An excited state is formed when one of the electrons in the σ_B bonding level is promoted to the antibonding level resulting in a configuration $(\sigma_B)^1(\sigma_A)^1(\pi)^1$. The ground state is described by a wavefunction ϕ_0, the excited states by wavefunctions φ_1, φ_2, and φ_3. As the ground state is

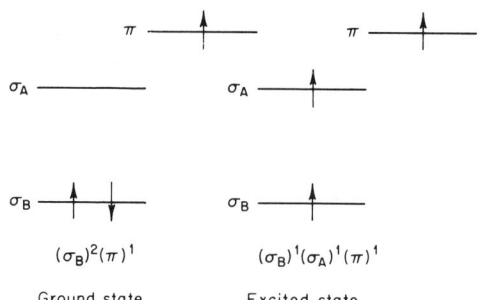

Fig. 4. Ground state and excited state electron configurations of the CH fragment.

perturbed by the Coulombic interaction energy acting between the p_z electron and the electrons in the σ_B bonding orbital, the true ground state is written as

$$\phi = \phi_0 + (\textstyle\int \phi_1 \hat{V}_r \phi_0 \, d\tau / \Delta E)\phi_1 = \phi_0 + \lambda \phi_1 \tag{23}$$

where ΔE is the energy difference between the states ϕ_0 and ϕ_1, $\hat{V}_r$ is the Coulombic repulsion energy operator expressed by Eq. (6), and ϕ_1 is the excited wave function

$$\phi_1 = (1/\sqrt{6})(2\varphi_1 - \varphi_2 - \varphi_3) \tag{24}$$

By using variational calculation the λ parameter of mixing the states is expressed as (Memory, 1968)

$$\lambda = (3/2\sqrt{6}) \cdot [(K_{ph} - K_{ps})/\Delta E] \tag{25}$$

where K_{ph} and K_{ps} are the exchange integrals between orbitals p, h and p, s, respectively,

$$K_{ph} = \int p(1)h(2) \frac{e^2}{r_{12}} h(1)p(2) \, d\tau \tag{26}$$

$$K_{ps} = \int p(1)s(2) \frac{e^2}{r_{12}} s(1)p(2) \, d\tau \tag{27}$$

where r_{12} is the distance between electrons 1 and 2.

The λ parameter is thus a measure of how the unpaired electron in the p orbital is exchanged with the electrons in the hybrid (h) and hydrogen (s) orbitals.

The hyperfine energy is the expectational value of the Fermi interaction Hamiltonian

$$E_F = \langle \hat{H}_F \rangle = \textstyle\int \phi \hat{H}_F \phi^* \, d\tau / \int \phi \phi^* \, d\tau \tag{28}$$

In the ground state ϕ_0 there is no unpaired spin density at the hydrogen, $\langle \hat{H}_F \rangle = 0$. The admixture of the excited state results in an exchange with the unpaired electron resulting in a finite unpaired spin density at the proton. The splitting constant, i.e., the difference between two subsequent hyperfine energy levels, is

$$a_H \approx \sqrt{3} \, \frac{K_{ph} - K_{ps}}{\Delta E} \int \phi_0 \hat{H}_F \phi_1 \, d\tau + \int \phi_1 \hat{H}_F \phi_1 \, d\tau \tag{29}$$

It can be shown that the second integral is negligible and $K_{ph} \gg K_{ps}$. By using Eq. (17) for the Hamiltonian operator, the following expression for the hyperfine splitting constant can be derived:

$$a_H = (8\pi\hbar/3\Delta E)\gamma_e\gamma_n K_{ph} \, |s(0)|^2 \rho_c \tag{30}$$

where $|s(o)|^2$ is the $1s$ electron density at the proton and ρ_c is the unpaired spin density at the carbon atom.

From this it follows that the hyperfine splitting caused by the proton in a $>\dot{\text{C}}$–H fragment a_h is proportional to the spin density at the carbon $2p_z$ orbital ρ_c. The corresponding McConnell equation is

$$a_H = Q_{CH}^H \rho_c \tag{31}$$

where the calculated value for Q_{CH}^C is -23.4 G (Fraenkel, 1962). The negative sign means that the spin state at the carbon sp^2 orbital is opposite to that of the hydrogen $1s$ orbital. This is illustrated in Fig. 5. The situation can be

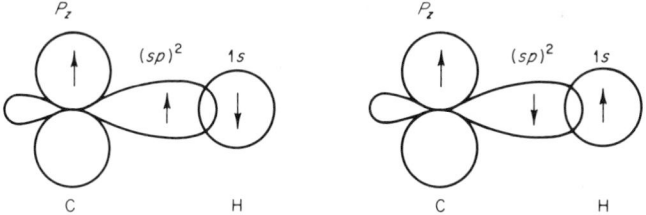

Fig. 5. Spin polarization in the $\dot{\text{C}}$H fragment.

visualized as follows: The unpaired spin density of the carbon $2p_z$ orbital has no direct effect on the proton splitting (nor on the ^{13}C splitting) because the $|\psi|^2$ values at the corresponding nuclei are zero.

As a result of the admixture of the states (configuration interaction), the unpaired electron density is shared between the p_z orbital of the carbon and the bond orbital. The fraction of the unpaired spin density at the bond orbital, which exhibits s character, causes a proton hyperfine splitting which is, at first approximation, proportional to the unpaired spin density at the carbon p_z orbital.

The $Q_{CH}^C = -23.4$ G value refers to an idealized case of an isolated $>\dot{C}H$ fragment. The actual Q_{CH}^H value depends on the neighboring atoms; it can be calculated by considering the actual structure.

The ^{13}C splittings have been considered in detail by Karplus and Fraenkel (1962). The splitting constant is generally expressed as

$$a^C = \left(S^C + \sum_i Q_{CX_i}^C \right) \rho_C + \sum_i Q_{X_iC}^C \rho_{X_i} \tag{32}$$

where S^C is the contribution of the carbon s electrons and X_i is the neighboring atom i to C — it can be carbon, hydrogen, or nitrogen. The Q values for some neighbors are shown in Table I.

When nitrogen nuclei are present the hyperfine interaction with the $I = 1$ spin ^{14}N nucleus can be observed easily. The nitrogen splitting constant in aromatic systems is expressed as

$$a^N = \left(S^N + \sum_i Q_{NX_i}^N \right) \rho_N + \sum_i Q_{X_iN}^{X_i} \rho_{X_i} \tag{33}$$

where ρ_N is the spin density at the nitrogen atom, ρ_{X_i} is that at the neighboring atom X_i, S_N is the contribution of the s electrons of the nitrogen, and $S_N = +11.3$ G. The Q values for some neighbors are shown in Table I.

Equations (26) and (27) have been calculated by different quantum chemical approximations and fitted with the experimental data. The equations are found generally valid for aromatic radicals; the Q values, however, vary between the limits given in Table I, according to the effects of remote neighbors.

VI. Hyperconjugation of the Methyl Group

According to ESR measurements, when a methyl group is connected to an aromatic radical a considerable hyperfine splitting is observed due to the methyl protons, although they are isolated from the system by a C–C σ bond. Similarly, when a methyl group is connected to an aliphatic free radical the splitting due to the methyl protons is significant. This means that the methyl group must have a π character which makes it possible for the hydrogen electrons to be mixed up with the conjugated system. This behavior is termed as hyperconjugation; it can be interpreted on the basis of the MO theory. It appears that the three electrons of the methyl hydrogen atoms cannot be treated separately. From the localized $1s$ atomic orbitals, the following CH_3 group orbitals are constructed

$$\phi_1 = (s_1 + s_2 + s_3)/(3 + 6S)^{1/2}$$
$$\phi_2 = (2s_1 - s_2 - s_3)/(6 - 6S)^{1/2}$$
$$\phi_3 = (s_1 - s_2)/(2 - 2S)^{1/2} \tag{34}$$

where s_1, s_2, and s_3 are the orbitals of the three methyl hydrogen atoms, respectively, and S is the overlap integral defined as

$$S = \int s_i s_k \, d\tau \tag{35}$$

for the overlap between orbitals i and k.

The methyl group orbitals ϕ_1, ϕ_2, ϕ_3 are illustrated in Fig. 6. These orbitals are coupled to the hybrid orbitals of carbon, which is in this case in

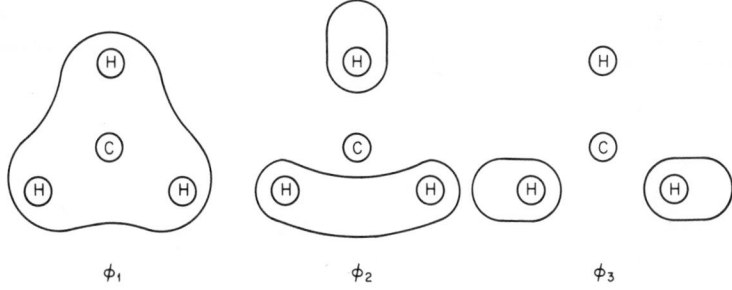

Fig. 6. Illustration of the electronic orbitals of the methyl group.

the sp hybrid state. The two sp hybrid orbitals are coupled to the next carbon atom and to the methyl group orbital ϕ_1 to form σ bonds. The carbon orbitals p_y and p_z are coupled to the p-like methyl group orbitals ϕ_2 and ϕ_3 to form π-type bonds.

This situation is very clearly illustrated in the case of the ethyl radical

$$H-\overset{\displaystyle H}{\underset{\displaystyle H}{\overset{|}{\underset{|}{\dot{C}}}}}-\overset{}{\underset{\displaystyle H}{\overset{}{\underset{|}{C}}}}-H$$

In this case the α-carbon atom is sp^2 hybridized, the three sp hybrid orbitals form bonds with the two hydrogens and with the β carbon, and the unpaired electron is mainly in the p_z orbital. As has been shown above, the β-carbon atom is sp hybridized. This system is conjugated because the p_z orbital of the unpaired electron overlaps with the p-type orbitals of the methyl group. As a result of this the methyl protons contribute appreciably to the proton hyperfine splitting of this radical. The measured splitting values are 22.38 and 26.87 G for the α protons and for the β protons, respectively.

These values have been measured by Fessenden and Schuler(1963) during radiolysis of liquid ethane at $-180\,^\circ$C.

The calculated Q values are 24.35 and 29.25 G for the α and β protons, respectively. The corresponding spin densities according to the McConnell equation are

$$
\begin{array}{ccc}
0.92 & & -0.93 \\
\diagdown & & | \\
& \!\!\text{C} - \text{C} - & \\
\diagup \uparrow & | & -0.93 \\
0.92 \quad 0.919 & -0.93 &
\end{array}
$$

The spin density at the α-carbon atom has been calculated by Chestnut (1958).

This means that in the ethyl radical the unpaired electron is almost evenly delocalized over the whole molecule; i.e., the conventional way of putting the position of the radical at the missing bond at the α-carbon atom is incorrect. In alkyl radicals the hyperfine coupling constants are generally higher for the β protons than for the α protons.

The reason for this is that the steric configuration of the β protons generally results in better overlap with the p_z orbital of the unpaired electron than that of the α protons. According to Bersohn (1956), the hyperfine splitting of the β protons is expressed as

$$a(\beta) = A + B\cos^2\vartheta \tag{36}$$

where A and B are constants, ϑ is the angle between the axis of the p_z orbital of the unpaired electron and the H–C–C plane of the proton. For the alkyl radicals $B \approx 50$ G, $A \approx 2$–4 G. In the case of methyl group due to the rapid rotation $\langle\cos^2\vartheta\rangle_{Av} = 0.5$. This means that the β splitting constants from methyl groups in alkyl radicals should be near to 25 G, which is experimentally observed. The α coupling constants in alkyl radicals are near to 23 G, which is the coupling constant of the methyl radical.

VII. The Structure of Some Higher Alkyl Radicals

The theoretical considerations on alkyl radicals have been experimentally verified by Fessenden and Schuler (1963), who studied the radiolysis of various liquid hydrocarbons by measuring ESR during the course of irradiation. In this case the resolution of the ESR is very good, the linewidths are in the order of 0.5 G. By this technique the structure of many alkyl radicals which are important in radiolysis of macromolecules has been determined.

As an illustrative example let us consider the case when the length of an alkyl radical is increased. This way it is possible to approach the structure of polymeric radicals which can be studied by ESR only at low resolution in the solid (trapped) state. A series of alkyl radicals starting from $\dot{C}H_3$ are collected in Table II. It is seen that by increasing the number of alkyl groups near to the radical the α-proton coupling constants are decreased.

TABLE II

A SERIES OF ALKYL RADICALS[a]

Radical	Proton hyperfine coupling constants (G)		
	α	β	γ
$\dot{C}H_3$	23.04	—	—
$CH_3-\dot{C}H_2$	22.38	26.87	0.5
$CH_3-CH_2-\dot{C}H_2$	22.08	33.2	0.38
$CH_3-CH_2-\dot{C}H-CH_3$	21.8	24.5 CH_3 27.9 CH_2	0.3
$CH_3-CH_2-\dot{C}H-CH_2-CH_3$	21.8	28.8	0.4
$CH_3-(CH_2)_3-\dot{C}H-(CH_2)_3-CH_3$	21.0	24.8	0.3

[a] From Fessenden and Schuler (1963).

It is also seen that the β coupling constants vary quite appreciably, depending on the structure, since these couplings are strongly dependent on the relative orientation of the bonds.

From the ESR data and from MO calculations it can be concluded that in aliphatic organic free radicals the unpaired electron is delocalized over six to eight C–C bonds. As the resolution of the ESR is relatively low, contribution of the γ protons can only be observed directly by the hyperfine structure. By using the electron nuclear double resonance technique, contributions from the δ protons are also observed. A very important general conclusion is that the delocalization is strongly dependent on the steric configuration of the molecule.

VIII. Exchange Interaction

As it is apparent from the spin density concept introduced above, that the unpaired electron carrying the radical property can be imagined as a probability density cloud which extends over six to eight C–C bonds, i.e., over about 10 Å in aliphatic systems and as much as 100–500 Å in aromatic systems. In saturated hydrocarbon polymers correspondingly the unpaired electrons can be regarded as being isolated from each other, their spin densities do not overlap significantly. In conjugated systems, however, the spin densities of the individual radicals may overlap, resulting in a strong interaction (exchange interaction).

The exchange interaction is described by the Hamiltonian operator

$$\hat{H}_{\text{exc}} = - \sum_{i,j} K_{ij} \hat{S}_i \hat{S}_j \qquad (37)$$

where $\hat{S}_i$ and $\hat{S}_j$ are the spin operators of the electrons i and j and K_{ij} is the exchange integral

$$K_{ij} = \int \psi_i(r_i) \psi_j(r_j) \frac{e^2}{r_i - r_j} \, \psi_j(r_i) \psi_i(r_j) \, d\tau \qquad (38)$$

where ψ_i, ψ_j are the wave functions and r_i, r_j are the position vectors of electrons i and j.

The presence of exchange interaction can be detected by analyzing the line shapes of the ESR spectra. In the absence of the exchange interaction the line shapes are determined by inhomogeneities of the local field due to the incomplete averaging of the dipole–dipole interactions and the anisotropy of the G values. These interactions result in Gaussian line shapes. When exchange interaction becomes significant, the line shape changes to Lorentzian. When the exchange interaction among the radicals is stronger than the hyperfine interaction with the nuclei, no hyperfine splitting in the ESR spectra is observed. Indeed, during the course of radiation and thermal degradation of polyethylene, a single ESR line 20 G wide appears (Ohnishi, 1962) as the chains are conjugated. A similar line is observed in degraded polyvinylchloride (Hay, 1970). In highly conjugated polycyclic systems, as in pyrolyzed polyacrylonitrile (Hedvig *et al.*, 1968), the linewidths are reduced to 0.1–0.5 G indicating extremely large delocalization and correspondingly strong exchange interactions.

IX. Problems of Polymeric Radicals

The quantum theory of free radical structures is best developed for aromatic radicals. In these systems the quantum chemical calculations are well correlated with the ESR data obtained by measurements in solution when the resolution is high. The theory is less developed for unsaturated systems. By comparison with the ESR data of simple low molecular weight aliphatic radicals measured in the liquid phase it appears that the main assumptions of the theory are valid. In irradiated polymers the situation is very difficult because the ESR spectra are in most cases measured in the solid phase, the exact structure of which is not known. In such systems the hyperfine splitting is always caused by both the anisotropic and isotropic interactions, and the contributions of these factors are difficult to separate. In polymers the spectrum lines are broadened by the inhomogeneities of the local fields acting on unpaired electrons, and correspondingly, the resolution is poor. Polymers are usually mixtures of macromolecules having different steric configurations; it is to be assumed that the neighborhood of the radicals in these systems is not uniform. Since the hyperfine interaction depends on the bond lengths and bond angles, it is concluded that the measured splitting constants are only average values. All these difficulties show that for a successful study of the structure of polymeric radicals it is necessary to know something about the physical structure of the polymer and about the electronic structure of the low molecular weight segments of six to eight C–C bonds where the unpaired electron is delocalized.

The best way for handling polymeric solids would be to describe their electronic states in terms of crystal orbitals rather than molecular orbitals. Since molecular orbitals are constructed by linear combination of atomic orbitals, atoms arranged in crystal lattices can be constructed in terms of crystal orbitals in which all the electrons in the unit cell of the lattice are taken into account. This method works well for crystalline solids when the crystal structure is known. In these systems the electrons belong to all atoms in the lattice. From this it follows that an unpaired electron in a crystal should be delocalized over the whole unit cell. Similarly, in polymers which are thought to have paracrystalline short-range order even in their amorphous part it would be reasonable to regard the trapped free radicals delocalized over the unit cell or pseudo cell of the system. This concept would be of much help in considering the problems of reactivity of free radicals in polymeric solids.

REFERENCES

Ayscough, P. B. (1967). "Electron Spin Resonance in Chemistry." Methuen, London.

Bersohn, R. (1956). *J. Chem. Phys.* **24**, 1066.

Chestnut, D. B. (1958). *J. Chem. Phys.* **29**, 43.

Das, M. R., and Fraenkel, G. K. (1965). *J. Chem. Phys.* **42**, 1350.

Dewar, M. J. S. (1962). "Hyperconjugation." Ronald Press, New York.

Fessenden, R. W., and Schuler, R. H. (1963). *J. Chem. Phys.* **39**, 2147.

Fraenkel, K. K. (1962). *Pure Appl. Chem.* **4**, 143.

Hay, J. N (1970). *J. Polymer Sci. A-1* **8**, 1201.

Hedvig, P., Kulcsár, S., and Kiss, L. (1968). *Euro. Polymer J.* **4**, 601.

Hedvig, P., and Zentai, G. (1969). "Microwave Study of Chemical Structures and Reactions." Iliffe, London.

Jarrett, H. S. (1956). *J. Chem. Phys.* **26**, 1289.

Karplus, H., and Fraenkel, G. K. (1962). *J. Chem. Phys.* **35**, 1312.

McConnell, H. M. (1956). "Quantum Theory of Magnetic Resonance Parameters." McGraw-Hill, New York.

Memory, F. D. (1968) "Quantum Theory of Magnetic Resonance Parameters." McGraw-Hill, New York.

Ohnishi, S. (1962). *Bull. Chem. Soc. Japan* **35**, 254.

Pilar, F. L. (1968). "Elementary Quantum Chemistry." McGraw-Hill, New York.

Scheffler, K., and Stegmann, H. B. (1970). "Elektronenspin-resonanz." Springer Verlag, Berlin.

5

Molecular Mobilities in Polymers

Péter Hedvig

Research Institute for Plastics, Budapest, Hungary

I. Introduction

The study of molecular mobilities in polymers is important for radiation chemists from two main points of view. One is that the reactions in solids are significantly influenced by structural factors and by molecular mobilities. The other point is that during the course of radiation chemical reactions the physical structure of the polymers may appreciably change, resulting in changes in mechanical, thermal, and electrical properties. These changes evidently influence the course of the reaction. Correspondingly, in the study of radiation-induced reactions, in addition to chemical factors, the physical (structural) factors are to be considered.

The main experimental ways of investigating molecular mobilities in polymers are the thermomechanical methods, including mechanical relaxation, dielectric spectroscopy, broad line nuclear magnetic resonance (NMR) together with the spin-echo method, differential scanning calorimetry (DSC) with the methods based on heat-conductivity measurements, and the dilatometric method. Some information about molecular mobilities of specific systems can be gained by thermoluminescence studies of irradiated samples, electron spin resonance (ESR), and infrared spectroscopy.

The results of these experiments supply information about the mobilities of different parts of the polymer molecules and about rearrangements of the structure of the amorphous and crystalline phases. The measurements are usually performed over a wide temperature and frequency range. The relaxation spectra obtained this way are interpreted in terms of phenomenological and molecular theories.

In this section the basic ideas will be discussed very briefly. Radiation-induced changes in molecular mobilities will be discussed in detail by Lyons and Weir in Chapter 14, Volume II.

II. Multiple Dispersion Regions in Polymers

When a polymer is heated up from very low temperature, its mechanical, thermal, and dielectric properties exhibit significant changes at certain temperature regions referred to as dispersion regions. The most abrupt changes are observed at the glass-transition temperature T_g where the rigid, glassy material becomes viscoelastic. Changes observed below T_g are referred to as glass–glass T_{gg} transitions; those observed above T_g are referred to as liquid–liquid transition T_{ll} (Boyer, 1963).

An alternative way of classifying dispersion regions is that the strongest relaxation is denoted by α. This is, in many cases, the T_g region. The T_{ll}-type transitions are usually denoted by α', the T_{gg} ones by β, γ, δ, in the order of decreasing temperature. This nomenclature is often used in mechanical and dielectric relaxation studies. (See, for example, McCrum *et al.*, 1967.)

It is usually assumed that at temperatures above T_{ll} the polymer molecules as a whole are mobile. In the region between T_g and T_{ll} the molecular motion is somewhat restricted, only submolecules containing about 50–100 C–C bonds are thought to be mobile. Below the glass-transition temperature the motion of the submolecules is frozen; only smaller parts, or side groups, can rotate or vibrate. Usually, several T_{gg}-dispersion regions are observed; these are regarded as the temperatures where different short-segment motions, mainly rotations, become frozen in.

Another very useful approach to interpret dispersion regions is to consider the changes of the thermodynamic variables: entropy, enthalpy, specific heat, and free volume. According to dilatometric measurements, for example, the specific volume of the polymers changes abruptly in the glass-transition region, indicating change in the packing conditions. It has been realized by Illers (1969) that in practical experimental conditions the polymeric glasses are not in thermodynamic equilibrium. There is a structural relaxation process (sometimes referred to as enthalpy relaxation) which represents the time-dependent transition from the nonequilibrium (vitroid) state to the equilibrium vitreous state. The rate of this transformation is very low when the temperature is much below T_g and becomes faster on approaching T_g. The situation is schematically illustrated in Fig. 1. The sample is supposed to have been stored at temperature T_0 for an infinitely long time; thus, thermodynamic equilibrium has been reached. By increasing the temperature the specific

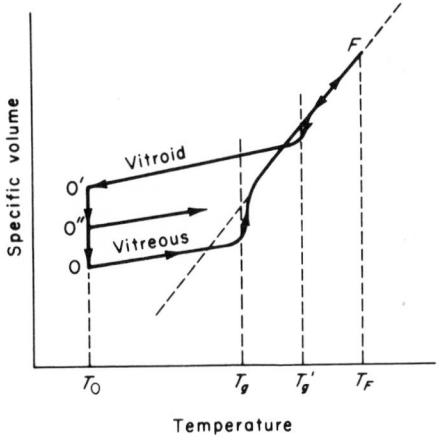

Fig. 1. Change of the specific volume of a polymer by heating and cooling through the glass-transition temperature.

volume (and the free enthalpy) of the sample will exhibit a sharp break at a temperature T_g. The exact position and shape of this change depends on the rate of heating. Reaching the point F the material is transformed to the non-Newtonian liquid state. By cooling, the break is observed at a different temperature, T_g', and such a state is reached which has higher free volume and higher free enthalpy (point $0'$). By storing the sample at a temperature T_0 the free volume is decreased as the system tends to thermodynamic equilibrium. The state 0' is referred to as a vitroid (Eckstein, 1968); it is characterized by higher free enthalpy and free volume than the corresponding equilibrium state 0. In practice, exact equilibrium is never reached; correspondingly, the measurements always start from a nonequilibrium state 0", the corresponding break in the curve will be somewhat shifted from T_g, and its slope will be changed. Indeed, the specific heat $C_p = (\partial H/\partial T)_p$ in the T_g region has been found strongly dependent on the thermal history of the sample. The $C_p(T)$ curves measured by differential scanning calorimetry exhibit large overshoots for samples stored below T_g for a long time, provided that the heating is faster than the cooling.

This effect is shown in Fig. 2 for unplasticized polyvinylchloride, after Illers (1969). It is seen that the longer the sample is stored below T_g (65°C),

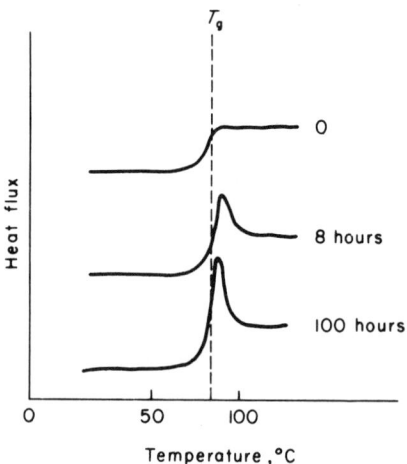

Fig. 2. Effect of storage below T_g on the DSC curves of unplasticized PVC; after Illers (1969). Storage times at 65°C are indicated.

the stronger the overshoot is, indicating that the equilibrium point 0 has been more closely approached. Similar effects have been observed by recording the rate of volume change $(\partial V/\partial T)_0$ as a function of temperature (Kovacs, 1966). An important practical consequence of this behavior is that the rupture strain of polyvinylchloride, stored at 65°C for 100 hours measured at low strain rates decreases by 80%, while the rupture stress increases (Retting, 1969). Another direct effect is that the diffusion rate of gases and liquids is higher in the unannealed samples than that in the annealed ones.

It is to be realized that whenever a reaction in solid polymers is studied, as one does so very often in radiation chemistry, the physical state of the sample should be carefully determined. This can only be done by using the experimental technique and basic theory of relaxation processes.

III. Phenomenological Theory of the Relaxation Processes

The phenomenological interpretation of the mechanical and dielectric relaxation processes is based on macroscopic models; the simplest of them are illustrated in Fig. 3. For considering the response to electrical fields the

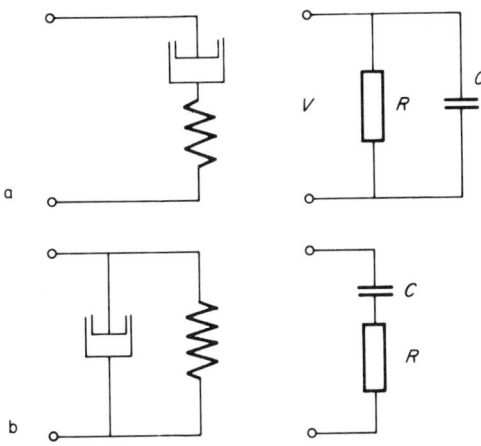

Fig. 3. Phenomenological models for polymers; (a) the Maxwell model, (b) the Voigt model.

polymer is approximated by simple circuits consisting of resistors R and capacitors C. For describing the response to mechanical force fields the system is similarly approximated by combination of springs and pistons. There is a remarkable correspondence between the dielectric and mechanical relaxations. The response of the system to periodical electric fields is described by the frequency (and temperature) dependence of the real (ε') and imaginary (ε'') parts of the complex permittivity. These dependences are expressed by the following simple Debye equations (see, for example, McCrum *et al.*, 1967):

$$\frac{\varepsilon'(\omega, T) - \varepsilon_\infty}{\varepsilon_0 - \varepsilon_\infty} = \frac{1}{1 + \omega^2 \tau^2(T)} \tag{1}$$

$$\frac{\varepsilon''(\omega, T)}{\varepsilon_0 - \varepsilon_\infty} = \frac{\omega \tau(T)}{1 + \omega^2 \tau^2(T)} \tag{2}$$

where ε_0, ε_∞ are the permittivities extrapolated to zero and infinite frequencies, respectively, ω is the angular frequency of the electric field, and $\tau(T)$ is the dielectric relaxation time of the system.

Equations (1) and (2) are crude approximations assuming that the response of the system can be characterized by a single relaxation time τ. In general, there is a distribution of relaxation times $F(\tau)$. In this case the Debye equations are the following:

$$\frac{\varepsilon'(\omega, T) - \varepsilon_\infty}{\varepsilon_0 - \varepsilon_\infty} = \int \frac{F(\tau)}{1 + \omega^2\tau^2} \, d\tau \tag{3}$$

$$\frac{\varepsilon''(\omega, T)}{\varepsilon_0 - \varepsilon_\infty} = \int \frac{\omega\tau F(\tau)}{1 + \omega^2\tau^2} \, d\tau \tag{4}$$

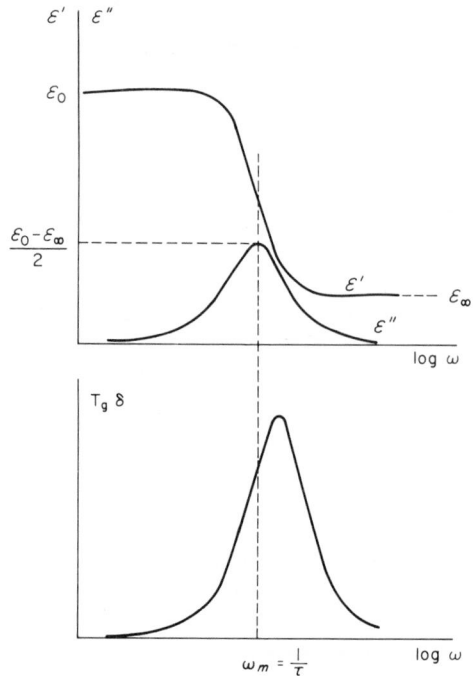

Fig. 4. Theoretical dielectric relaxation curves for the one relaxation time approximation.

In Fig. 4 the shape of the dielectric dispersion $\varepsilon'(\omega)$ and absorption $\varepsilon''(\omega)$ curves are illustrated. It is seen that from the experimental curves the values of ε_0, ε_∞ and τ can be determined. Sometimes the loss tangent $\tan \delta = \varepsilon''/\varepsilon'$

is measured, the corresponding curve is, however, shifted to higher frequencies as shown in Fig. 4.

The distribution of relaxation times cannot be determined exactly; some information about it is gained by analysis of the shape of the $\varepsilon'(\omega)$ and $\varepsilon''(\omega)$ curves. For the study of relaxation time distribution the Cole–Cole representation is very useful (Cole and Cole, 1941). It is shown in Fig. 5. According

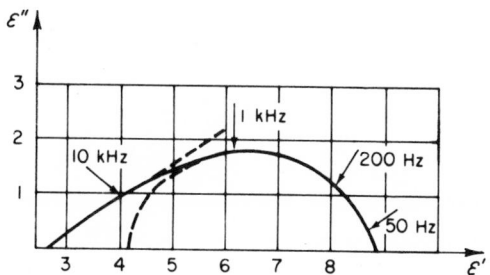

Fig. 5. The complex plane representation of dielectric spectra. (- - -) One relaxation time approximation; (—) according to the modified Debye equation. The frequencies are indicated.

to the Debye equations the dependence of ε'' on ε' is a semicircle corresponding to the single relaxation time approximation when Eqs. (1) and (2) are valid. The skewed curve corresponds to a realistic case [polyvinylacetate]. This realistic, multirelaxation time case is approximated by the following, modified Debye equation (Havriliak and Negami, 1967):

$$[\varepsilon(\omega, T) - \varepsilon_\infty]/(\varepsilon_0 - \varepsilon_\infty) = [1 + (i\omega\tau)^{1-a}]^b \qquad (5)$$

where $\varepsilon(\omega, T) = \varepsilon'(\omega, T) - i\varepsilon''(\omega, T)$ is the complex permittivity, a and b are experimental constants which characterize the distribution of the relaxation times.

The mechanical relaxation behavior of polymers is phenomenologically described by equations analogous to (1) and (2). The frequency dependence of the torsional compliance J, for example, is expressed as

$$\frac{J'(\omega, T) - J_\infty}{J_0 - J_\infty} = \frac{1}{1 + \omega^2\tau^2(T)} \qquad (6)$$

$$\frac{J''(\omega, T) - J_\infty}{J_0 - J_\infty} = \frac{\omega\tau(T)}{1 + \omega^2\tau^2(T)} \qquad (7)$$

where J, J'' are, respectively, the real and imaginary parts of the torsional compliance, ω is the angular frequency of the force field (stress), $\tau(T)$ is the mechanical relaxation time, and J_0, J_∞ are the compliances extrapolated to zero and infinity frequencies, respectively. The relaxation times measured by the mechanical relaxation method are within the experimental error the same as those measured by the dielectric method.

It has been shown recently by Havriliak and Negami (1969) that the mechanical and dielectric relaxation spectra become quite identical if the complex quantities

$$\delta(\omega, T) = [J(\omega, T) - J_\infty]/[J(\omega, T) + \tfrac{2}{3}J_\infty] \tag{8}$$

and

$$\zeta(\omega, T) = [\varepsilon(\omega \ T) - 1]/[\varepsilon(\omega, T) + 2] \tag{9}$$

are used as variables. $\zeta(\omega, T)$ is the complex dielectric polarization, $\delta(\omega, T)$ is the complex mechanical distortability of the material, which is a mechanical analog of the dielectric polarization. For the quantities $\delta(\omega, T)$ and $\zeta(\omega, T)$ the modified Debye equations (5) are equivalent; thus, the parameters a, b and the relaxation time τ are the same.

IV.   The Temperature Dependence of the Relaxation Time

Being based on a thermally agitated rate process, the dielectric and mechanical relaxation time should depend on the temperature as

$$\tau(T) = \tau_0 \exp(-\delta S/k) \exp(\delta H/kT) \tag{10}$$

where δS is the entropy change, δH is the enthalpy change, k is the Boltzmann constant, and T is the temperature. In the T_{gg}-dispersion regions where no order–disorder transitions take place the entropy is assumed to be constant; thus, the relaxation time is expressed as

$$\tau(T) = \tau_0 \exp(E/kT) \tag{11}$$

where E is referred to as the activation energy of the corresponding molecular motion. In the T_{gg}-dispersion regions Eq. (11) is found experimentally verified for all the studied polymers, by mechanical as well as by dielectric relaxation experiments (McCrum et al., 1967; Lewis, 1968).

In the T_g-dispersion regions Eq. (11) is found generally not valid, the $\ln \tau$ versus $1/T$ curves are not straight lines. In the T_g region the following equation holds (Williams et al., 1955):

$$\tau(T) = \tau_0 \exp\{[-A(T - T_g)]/(T - T_g + B)\} \tag{12}$$

where τ_0, A, and B are constants, T is the temperature, and T_g is the glass-transition temperature of the material. Equation (12) is referred to as the WLF (Williams *et al.*, 1955) equation.

The reason for the different temperature dependence of the relaxation times in the glass-transition range is that the activation energy of the process depends on the temperature. Near T_g this dependence is approximated as

$$E(T) = \frac{kAB}{1 - (T_g - B)/T} \tag{13}$$

where A and B are the constants of the WLF equation, T_g is the glass temperature, T is the temperature, k is the Boltzmann constant, and $E(T)$ is the activation energy of the transition. The WLF constants A and B are found very near to $A = 17°C$, $B = 51°C$ for many glass-forming systems while the T_g values change enormously.

Although the Arrhenius equation (11) is not valid in the T_g-dispersion region, it is usual to plot the experimental relaxation times as a function of reciprocal temperature for each dispersion region. As an illustrative example such a plot is shown in Fig. 6 for polyvinylchloride. It is seen that in the T_{gg}-dispersion region (β relaxation) the Arrhenius equation (11) is fulfilled; the

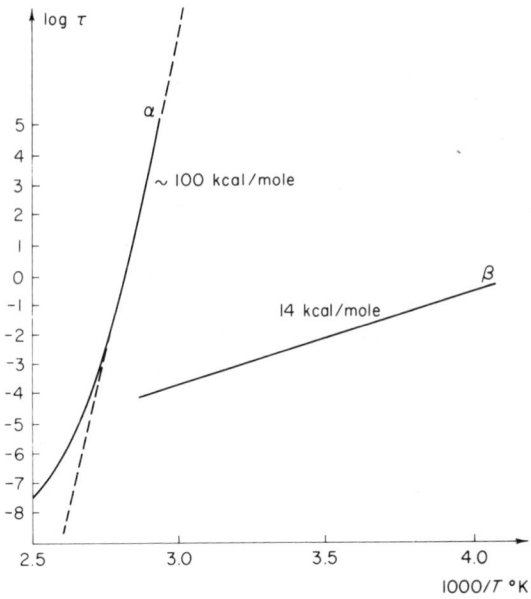

Fig. 6. The temperature dependence of the relaxation times for polyvinylchloride in the Arrhenius representation.

activation energy is 14 kcal/mole. In the T_g-region the $\log \tau$ versus $1/T$ plot is not linear and the apparent activation energy of about 100 kcal/mole has no physical meaning.

Since the relaxation time depends on the temperature, it is possible to measure dielectric and mechanical relaxation spectra by choosing the temperature as an independent variable rather than frequency. This is technically much easier since it is possible to sweep the temperature over a wide range, usually from the temperature of liquid nitrogen ($-196°C$) up to 300°C and record the dielectric or mechanical relaxation spectra continuously. Spectrometers with the corresponding wide frequency sweep are very difficult to construct. In the T_{gg} regions, where the activation energy is constant, this technique works quite well. In the T_g regions, however, as a result of the temperature dependence of the activation energy the temperature plots are difficult to analyze. In basic work, when the distribution of relaxation times is to be analyzed, the frequency is a more suitable independent variable. For practical work when semiquantitative data are sufficient the easier temperature-plots can be used even in the T_g region.

V. Microscopic Theories

The experimental data obtained by different methods are usually processed in terms of the phenomenological theory. As a result of this, one gets the values of the relaxation times corresponding to the dispersion region, and from the temperature dependence one obtains the activation energy or the WLF constants A and B. By analyzing the experimental data, in some cases, information about the shape of the relaxation time distribution function $F(\tau)$ can be obtained (Davidson and Cole, 1951). Very many such data have been collected for homopolymers, copolymers, polymeric blends, and other plastic systems (Lewis, 1968). The experimental results are fairly coherent; values of τ and E obtained by different methods agree rather well. Discrepancies mainly arise from the differences in the sample preparation and in their thermal history.

The experimental data processed in terms of the phenomenological theory are correlated with the molecular mobilities by various theories. This correlation is not yet fully satisfactory since very little is known of the structure of the polymers. The main results of the present microscopic theories is that some typical dispersions can be correlated with certain molecular motions. The situation is relatively easy for the T_{gg}-dispersion regions since in this case the structure of the system does not change. Very difficult to interpret is the glass transition since in this case even in the amorphous phase structural changes are involved. Most problematic are those polymers which have amorphous and crystalline parts. In such polymers no glass transition can be defined in a strict sense for the polymer as a whole.

As has been mentioned in the first part of this section polymeric glass is usually not in thermal equilibrium. For such vitroid systems no temperature can be defined in a strict thermodynamic sense since the thermal energy is not equally distributed among the degrees of freedom of the system. Correspondingly, for vitroids and, thus, for most polymeric glasses the general thermodynamic state equation

$$f(p, v, T) = 0 \qquad (14)$$

is not valid, as the temperature is not defined (Eckstein, 1966). Instead of a general temperature, so-called partial temperatures are introduced (Eckstein, 1968). One of these is the phonon temperature T_{ph}, which is the partial temperature corresponding to the actual lattice vibrations. Other temperature-like parameters $T_c^{(1)}$, $T_c^{(2)}$ ... $T_c^{(4)}$ are introduced which describe different nonequilibrium frozen-in configurations. These parameters are referred to as configurational temperatures; they are observed as breaks in the volume versus temperature curves measured at different cooling rates. Thus, the general thermodynamic equation of state for a vitroid is generally written as

$$f[p, v, T_{ph}, T_c^{(1)} \cdots T_c^{(n)}] = 0 \qquad (15)$$

This way, it is possible to apply thermodynamic considerations to the vitroid state.

VI. The Defect Model

Defects are originally defined for crystalline materials which exhibit long-range ordered structures. It has been realized, however, that some amorphous structures and melts exhibit short range order. These structures are periodic only in a statistical average; the nearest neighbors of each molecule are located near to certain mean values of bond angles and bond lengths. Such structures are referred to as paracrystalline (Hoseman and Bagchi, 1962). In paracrystalline structures defects can be defined in the same sense as in long-range ordered systems. Vitroids are regarded as paracrystalline structures containing a large concentration of defects. Such a system is structurally similar to the melt. The molten state, even in the case of low molecular weight crystalline materials, is regarded as an ordered structure containing so great a defect concentration that the long-range order is destroyed. According to this, the vitroid is regarded as a frozen-in paracrystalline defect structure in which the defect concentration is time and temperature dependent.

The temperature dependence of the relative defect concentration is approximated as (Peibst, 1961)

$$N_d/N = [1 + \exp(E_d/kT)]^{-1} \qquad (16)$$

where N_d is the defect concentration, N is the concentration of the periodic units, and E_d is the activation energy for the defects; the latter is expressed as

$$E_d(T) = H_v + pv_v - T \, \delta S_v \qquad (17)$$

where v_v is the total volume associated with the creation of the defect, p is the pressure, S_v is the entropy, and H_v is the enthalpy.

At the crystalline melting point the activation energy E_d drops suddenly, resulting in a sudden increase in the defect concentration. As a result of this, the long-range order is destroyed by the overlapping defects, and correspondingly, the viscosity decreases and the crystal melts. By cooling the system down the defect concentration decreases continuously. As the defect

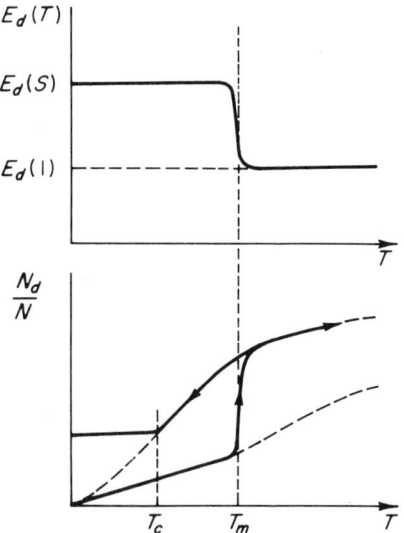

Fig. 7. Schematic illustration of the change of the defect concentration as a function of temperature in a crystalline material. N_d/N = relative defect concentration, T_c = configurational temperature, $T_m \rightarrow$ melting temperature, $E(T)$ = activation energy for the defect formation.

annihilation is time and temperature dependent, at a certain temperature (T_c) where the relaxation for defect annihilation is slowed down, the defect concentration becomes constant (Fig. 7). At this temperature the defect concentration is frozen in. It is seen that by this picture the effects of thermal history mentioned in the introduction can be satisfactorily interpreted. The concept of the vitroid state and defect model developed for inorganic low molecular weight compounds seems to be well applicable to polymeric systems. It is consistent with the free volume theory of Doolittle (1951) and the statistical thermodynamic theories of Gibbs and di Marzio (1958).

The free volume is defined as

$$v_f = (v - v_0)/v \tag{18}$$

where v_0 refers to the closest packed defectless ideal state for which the viscosity is infinite and v is the actual volume. The viscosity is expressed as

$$\eta = \exp\left[a + (b/v_f)\right] \tag{19}$$

where a and b are constants.

The temperature dependence of the free volume is assumed to be

$$v_f(T) = v_f(T_g) + \alpha(T - T_g) \tag{20}$$

where

$$\alpha = (\partial v_f/\partial T)_p$$

is the thermal expansion coefficient.

From Eq. (19) and (20) it follows that

$$\ln \frac{\eta(T)}{\eta(T_g)} = -\frac{[b/v_f(T_g)](T - T_g)}{[v_f(T_g)/\alpha] + T - T_g} \tag{21}$$

This equation has the same form as the WLF equation (12), with the constants $A = b/v_f(T_g)$, $B = v_f(T_g)/\alpha$.

The free volume theory involves the idea that the viscosity is mainly determined by structural factors. Since the free volume is evidently connected with the defect concentration, it appears that the viscosity is also principally determined by the defect concentration and not directly by the temperature. For a recent theoretical analysis, see Bartenev *et al.* (1969).

VII. Mobility Theories

It is evident that the defect model involves molecular motion in a statistical sense and so do the thermodynamic and free volume theories. One should be interested in what particular motions are frozen and liberated at a given dispersion region.

The main possibilities of molecular motion in polymers are (a) rotation of short-chain segments or side groups, (b) vibration of parts of the polymer chains about their equilibrium position, and (c) rearrangement of the structure.

Rotation of short-chain segments and side groups are expected to be frozen at low temperatures, below T_g. Indeed, many of the T_{gg}-type transitions have been identified as being due to this mechanism.

A model involving rotation of short-chain segments has been proposed by Schatzki (1965). The polymer chain segment shown in Fig. 8 forms a crankshaft configuration which can rotate well below the glass temperature. The calculated activation energy for this rotation is 13 kcal/mole. It is assumed that the γ relaxation of many linear homopolymers such as poly-ethylene, polyvinylchloride, and polytetrafluoroethylene is due to crankshaft motion, the observed activation energies are very close to the theoretical value and are independent of the composition of the polymers (Boyer, 1963, 1968).

Another possible source of the T_{gg}-dispersion regions is the rotation of side groups, which are usually hindered only at very low temperatures. In alkyl methacrylate polymers, for example,

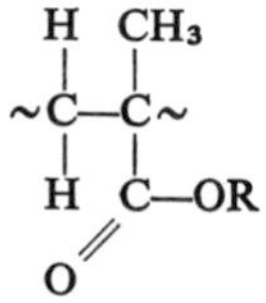

$$\sim\overset{\displaystyle H}{\underset{\displaystyle H}{C}}-\overset{\displaystyle CH_3}{\underset{\displaystyle \underset{O}{C}-OR}{C}}\sim$$

where R is an alkyl group, three T_{gg}-type dispersion regions have been observed and identified as being due to hindered rotation of different side

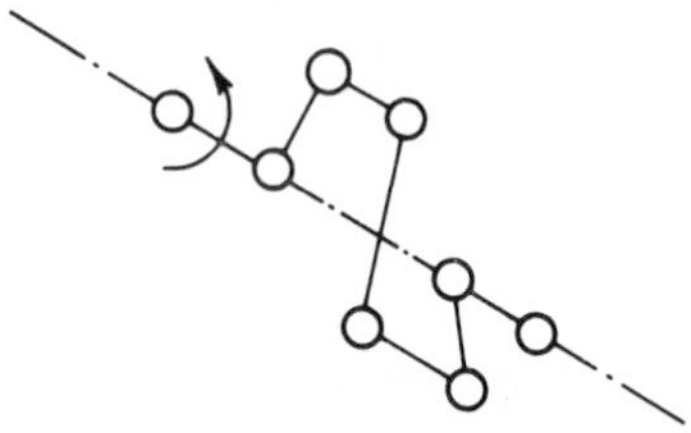

Fig. 8.   Rotation of a crankshaft segment in linear polymers. After Schatzki (1965).

groups. In Fig. 9 the mechanical relaxation spectrum of polymethyl methacry-late ($R = CH_3$) is shown schematically. The strongest peak (α) is due to the glass–rubber transition; it is observed near 100°C at a frequency of 1 Hz. The apparent activation energy of this transition is in the order of 100 kcal/mole.

The β-relaxation peak is attributed to rotation of the –COOCH$_3$ group about the C–C bond which links this group to the main chain. The activation energy for this rotation is about 20 kcal/mole. The γ relaxation is attributed to rotation of the methyl group attached directly to the main chain (α-CH$_3$ group). This results in a weak mechanical loss at low ($-170\,^{\circ}$C) temperature but no dielectric loss. This transition is also observed by nuclear magnetic resonance. From NMR data it has been concluded that the rotation of the methyl group in the ester side chain is frozen at very low temperature around 4°K; it is denoted as δ peak in the figure. For higher alkyl methacrylates this dispersion region is shifted to higher temperatures. Data on alkyl methacrylates are found in McCrum *et al.* (1967), Nielsen (1962), and Lewis (1968).

Hindered rotation of side groups or short segments are effectively interpreted by the barrier theories. The basic principle of these theories is that the interaction of a group with the neighboring molecules is accounted for by potential barriers. The exact potential of the force field is, of course, not

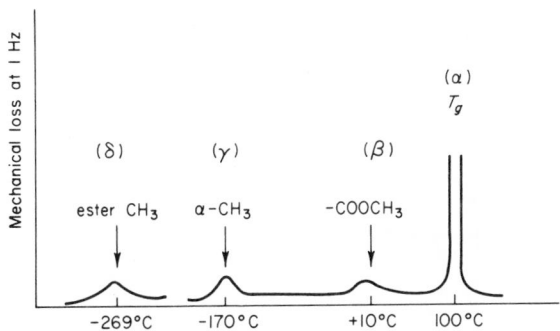

Fig. 9. Schematic illustration of mechanical loss maxima in polymethyl methacrylate at 1 Hz. The rotating groups correlated with the loss maxima are indicated.

known; but it can be approximated quite efficiently. Some typical potential barrier models are illustrated in Fig. 10. The simplest one contains two equilibrium positions 1 and 2 corresponding to rotation by an angle of 180°C. The probabilities for the group to jump between these equilibrium positions are

$$P_{12} = P_0 \exp\left(-V_1/kT\right)$$

$$P_{21} = P_0 \exp\left(V_2/kT\right) \tag{22}$$

where P_0 is a constant and V_1 and V_2 are, respectively, the potential barriers for equilibrium positions 1 and 2. P_{12} and P_{21} are the probabilities for rotation from positions 1 to 2 and 2 to 1, respectively (see Fig. 10a). The corresponding relaxation time is expressed as

$$\tau = 1/2(P_{12} + P_{21}) \qquad (23)$$

This is the single-barrier approximation of Fröhlich (1949) leading to a single relaxation time. In general, there is a distribution of the barrier heights V_1 and V_2, and there may exist several equilibrium positions instead of two. It is also possible that the barrier heights for the different equilibrium positions are different. The model illustrated in Fig. 10b contains four

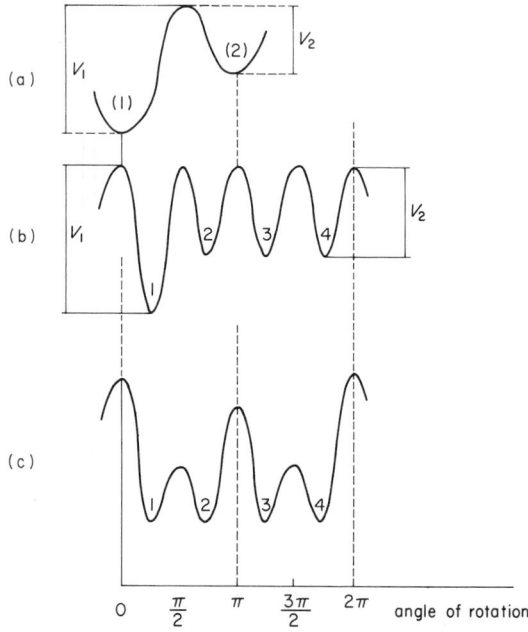

Fig. 10. Potential barrier models; (a) Fröhich (1949), (b) Hoffman (1959), (c) Ishida and Yamafuji (1961).

equilibrium positions with the same barrier heights (Hoffman, 1959). In Fig. 10c such a model is shown where the barrier heights are different (Ishida and Yamafuji, 1961).

In the case of Fig. 10b the transition probabilities are

$$P = P_{12} = P_{14} = P_0 \exp(-V/kT)$$

$$P' = P_{41} = P_{21} = P_{23} = P_{32} = P_{34} = P_{43} = P_0 \exp(-V'/kT) \qquad (24)$$

For the populations of the equilibrium positions the following types of rate equations can be constructed:

$$dN_1/dt = -2PN_1 + P'(N_2 + N_4)$$
$$dN_2/dt = PN_1 - 2P'N_2 + P'N_3$$
$$dN_3/dt = P'N_2 - 2P'N_3 + P'N_4$$
$$dN_4/dt = PN_2 - P'N_3 - 2P'N_4 \tag{25}$$

This model results in four relaxation times.

The model of Fig. 10c can be treated in a similar way. In a particular case hindered rotation of side groups can be at least semiquantitatively interpreted on the basis of the barrier concept and the corresponding relaxation peaks correlated.

The other main type of molecular motion is vibration of chain segments or submolecules about their equilibrium position. The theories of this motion type are referred to as normal mode theories. The basic assumption of these theories is based on Flory's theory of polymer molecules in solution (Flory, 1944). According to this, the polymer molecules can be divided into submolecules of definite chain conformation, and these units are distributed by the Gaussian probability distribution function. Thus, a real polymer chain is approximated by a series of Gaussian subchains. Such a model has been developed by Kargin and Slonimsky (1948) Rouse (1953), and Bueche (1954). The individual Gaussian units are considered to vibrate independently under the action of a friction force produced by the surroundings. At first approximation this friction is considered as a finite viscosity of the medium. Evidently, this model can only be applied strictly to dilute solutions, but some interesting conclusions to solid systems can be drawn. By substituting the Gaussian submolecules with spring and beads it is possible to construct their equations of motion in a viscous medium. The resulting motion can be described in terms of normal modes of vibration. For a quantitative calculation it is to be considered that the velocity distribution in the viscous liquid in which the vibration takes place is changed, and this influences the motion itself (Zimm, 1956).

In Fig. 11 the first three normal vibration modes of a submolecule are illustrated. The relaxation time corresponding to a normal mode n (neglecting the hydrodynamic interaction) is expressed by Rouse (1953) as

$$\tau_n = \langle l^2 \rangle f / 6\pi^2 n^2 kT \tag{26}$$

where $\langle l^2 \rangle$ is the mean square length of the Gaussian submolecule and f the friction coefficient.

In solids the friction coefficient f is structure dependent, as has been discussed in connection with the defect model. The average length of the Gaussian submolecule depends also on the temperature. From quantitative comparison with the experiments it can be shown that the temperature dependence of τ_n is mainly determined by the friction coefficient, i.e., by structural factors. If one could determine the friction acting on the Gaussian submolecules as a result of the interaction with the surroundings, the normal mode vibrations could be quantitatively calculated for solids. This theory has not yet been developed.

Although the friction coefficient cannot yet be determined from molecular parameters, it is possible to evaluate it semiempirically (Ferry, 1956). The

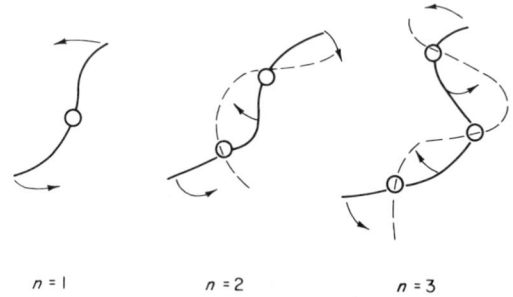

Fig. 11. Normal vibrations of a submolecule.

distribution of relaxation times is expressed approximately for the glass-transition region as

$$F(\ln \tau) = (N\langle l^2\rangle^{1/2}/2)/(kTf/6)^{1/2}\,\tau^{-\mu} \tag{27}$$

where τ is the relaxation time, $\langle l^2\rangle^{1/2}$ is the equilibrium mean square length of the submolecule, f is the friction factor, μ is an exponent which is between $\frac{1}{2}$ and $\frac{2}{3}$, depending on the approximation used, and N is the total number of the submolecules. By measuring the relaxation time distribution $F(\ln \tau)$ correspondingly, it is possible to estimate the value of the friction coefficient f. For this calculation the mean square Gaussian chain length must be known; its value has been taken from measurements in dilute solutions, assuming that they are not significantly changed in the solid (Flory, 1953).

The normal mode concept has been developed to interpret T_{gg}-relaxation effects, too. In some cases the T_{gg}-peaks can be interpreted neither by rotation of side groups nor by the crankshaft mechanism. In these cases the molecular motion is attributed to vibrations of short segments about their equilibrium positions. The corresponding theory is referred to as local mode theory; it has been developed by Yamafuji and Ishida (1962).

Structural rearrangements are very difficult to visualize from the point of view of molecular motion. There are polymers in whose crystalline part the crystal structure changes at certain temperatures. It is known, for example, that in poly(tetrafluoroethylene) the crystal symmetry changes from triclinic to hexagonal at 19°C.

VIII. Conclusions

Radiation chemists working with solid polymers usually do not realize that the physical state of the sample may change appreciably during the course of the radiation chemical reaction. Reactions are very often treated from a purely chemical point of view only, neglecting the effects of the physical structure. From the present very superficial review it is apparent that the study of molecular mobilities in interpreting radiation chemical reactions in solids is absolutely necessary. On the other hand, it is to be realized that the thermal history has an enormous influence on the physical state of a solid polymer. Generally, a polymeric sample below the glass-transition temperature is not in thermodynamic equilibrium; it is in a nonequilibrium vitroid state. This effect is usually not accounted for when polymers are irradiated at low temperatures, and for example, the subsequent reactions of the trapped free radicals are studied. Correspondingly, in order to prepare samples for such studies they should be annealed in such a way that a definite and reproducible physical structure is reached. By studying the temperature dependence on the reaction rate constants in solids, the simultaneous change in the physical structure is to be measured. This can be done by mechanical or dielectric relaxation or by differential scanning calorimetry.

Another important consequence of the nonequilibrium state of solid polymers is that they always contain a high defect concentration which is time and temperature dependent. In these defects gas molecules, electrons, ions, solvent or monomer molecules, or accidental degradation products can be trapped and are released as the structure changes. Very little is known at present about the chemical effects of these trapped species although their presence is experimentally observed.

REFERENCES

Bartenev, G. M., Razumovskaya, I. V., Sanditov, D. S., Lukyanov, I. A. (1969). *J. Polymer Sci. A-1* **7**, 2147.
Boyer, R. F. (1963). *Rubber Chem. Techn.* **36**, 1303.
Boyer, R. F. (1968). *Polymer Eng. Sci.* **8**, 161.

Bueche, F. (1954). *J. Chem. Phys.* **22,** 603.

Cole, R. H., and Cole, K. S. (1941). *J. Chem. Phys.* **9,** 341.

Davidson, D. W., and Cole, R. H. (1951). *J. Chem. Phys.* **19,** 1484.

Doolittle, A. K. (1951). *J. Appl. Phys.* **22,** 1471.

Eckstein, B. (1966). *Glastechn. Ber.* **39,** 455.

Eckstein, B. (1968). *Mater. Res. Bull.* **3,** 199.

Ferry, J. D. (1956). "Die Physik de Hochpolymeren" (H. A. Stuart, ed.). Springer Verlag, Berlin.

Flory, P. J. (1944). *Chem. Rev.* **35,** 51.

Flory, P. J. (1953). "Principles of Polymer Chemistry." Cornell Univ. Press, Ithaca, New York.

Fröhlich, H. (1949). "Theory of Dielectrics." Oxford Univ. Press, London and New York.

Gibbs, J. H., and Di Marzio, E. A. (1958). *J. Chem. Phys.* **28,** 373.

Havriliak, S., and Negami, S. (1967). *Polymer* **8,** 161.

Havriliak, S., and Negami, S. (1969). *Polymer* **10,** 859.

Hoffman, I. D. (1959). *8th Colloq. Ampère* **12,** 36.

Hoseman, R., and Bagchi, S. N. (1962). "Direct Analysis of Diffraction by Matter." North Holland Publ., Amsterdam.

Illers, K. H. (1969). *Makromol. Chem.* **127,** 1.

Ishida, Y., and Yamafuji, K. (1961). *Kolloid Z.* **177,** 97.

Kargin, V. A., Slominski, G. L. (1949). *Dokl. A. N. SSSR.* **62,** 239.

Kovacs, A. J. (1966). *Rheolog. Acta* **5,** 262.

Lewis, O. G. (1968). "Physical Properties of Linear Polymers." Springer Verlag, New York.

McCrum, N. G., Read, B. E., Williams, G. (1967). "Anelastic and Dielectric Effects in Polymeric Solids." Wiley, New York.

Nielsen, L. E. (1962). "Mechanical Properties of Polymers." Reinhold, New York.

Peibst, H. (1961) *Z. Phys. Chem.* **216,** 304.

Retting, W. (1969). *Angew. Makromol. Chem.* **8,** 87.

Rouse, P. E. (1953). *J. Chem. Phys.* **21,** 1272.

Schatzki, T. F. (1965). *Polymer Preprints, Amer. Chem. Soc.* **6,** 646.

Williams, M. L., Landel, R. F., Ferry, J. D. (1955). *J. Amer. Chem. Soc.* **77,** 4701.

Yamafuji, K., and Ishida, Y. (1962). *Kolloid Z.* **183,** 15.

Zimm, B. H. (1965). *J. Chem. Phys.* **24,** 269.

6

Theory of Reactions in the Solid State

H. J. Wintle

Department of Physics, Queens University, Kingston, Ontario, Canada

I. General Theory

The object of this chapter is to set out, with a minimum of detailed mathematics, the main physical ideas, in particular the role of reaction cross sections, behind the formal theory of reactions. A number of specific topics which require individual consideration are given in Section II.

A. LAW OF MASS ACTION

The equilibrium between reactants and products of a chemical reaction is normally expressed in the form of the law of mass action. A general reaction can be written in the two equivalent forms:

$$n_1X_1 + n_2X_2 + \cdots \underset{k_2}{\overset{k_1}{\rightleftharpoons}} n_1'X_1' + n_2'X_2' + \cdots \tag{1}$$

$$\Sigma v_i A_i = 0 \tag{2}$$

where in Eq. (1) the X_i are the reactants and the n_i are the stoichiometric coefficients for those reactants, while the primed symbols are the corresponding quantities for the products. In Eq. (2) some economy is achieved by labeling all the participants A_i in the reaction alike and giving the stoichiometric coefficients v_i positive values for products, negative values for reactants. If we deal with dilute solutions or perfect gases, the equation for equilibrium then has the form

$$\Sigma v_i \ln x_i = \ln K(T, p) \tag{3}$$

where x_i is the mole fraction of the constituent A_i and $K(T,p)$ is the equilibrium constant. The mole fractions x_i are given by

$$x_i = \exp\left[(\mu_i - \mu_i^\circ)/kT\right] \tag{4}$$

where μ_i° is the molecular chemical potential of the pure substance and μ_i is that of the substance in the reacting mixture. The standard free energy for the reaction is

$$\Delta G = \Sigma v_i \mu_i = kT \ln K(T, p) \tag{5}$$

assuming that we are taking the standard states to be identical to the pure states. The derivation of these equations appears in many textbooks and will not be repeated here.

There are two somewhat different points of view which lead to these results. The first is obtained by considering the balance between the forward rate of reaction k_1 and the reverse rate k_2. Elementary kinetic theory leads to the collision frequency Z for a given molecule in a gas of one molecular species

$$Z = 2^{1/2}\pi\, d^2 n v \tag{6}$$

where d is the molecular diameter, n the concentration, and v the mean velocity of molecules in the gas. In a mixture, the collision frequency between a single molecule of species i with molecules of species j will be

$$Z_{ij} = 2^{1/2}\pi\langle d^2 v\rangle n_j \tag{7}$$

where the brackets indicate that a suitable average must be taken. The number of ij collisions per unit volume per unit time is therefore

$$C_{ij} = n_i Z_{ij} = 2^{1/2} \pi \langle d^2 v \rangle x_i x_j n^2 \tag{8}$$

where n is the total concentration of molecules per unit volume. The reaction rate is proportional to the collision rate, and conceptually, it is possible to extend Eq. (8) to reactions of any molecularity. We consequently obtain the rates of reaction

$$R_1 \propto x_1^{n_1} x_2^{n_2} \cdots$$

and

$$R_2 \propto x_1'^{n_1} x_2'^{n_2} \cdots \tag{9}$$

and on equating the forward and reverse rates, these equations yield the law of mass action. It is clear from this calculation that if the cross section σ for the reaction

$$\sigma = \pi \langle d^2 \rangle \tag{10}$$

for any reason changes with concentration, then the rates R_1 and R_2 will not have the straightforward concentration dependence given by Eq. (9) and the simple law of mass action will break down.

The second approach to Eq. (3) is a purely thermodynamic one. In equilibrium the Gibbs free energy is a minimum, and as a result

$$\Sigma \mu_i \, dn_i = 0$$

So long as the changes in the n_i are due only to the reaction (2), then this becomes

$$\Sigma \mu_i \nu_i = 0$$

and by using (4) we obtain (3). In this case the crucial step is embodied in the assumption of ideality. If the mixture is not ideal, then (4) no longer holds and again the simple form of the law of mass action is invalidated.

Now if we turn to the crystalline solid state, we find that these equations can still be employed to describe the equilibrium concentrations of various entities. The calculation of vacancy concentrations in ionic crystals (Mott and Gurney, 1948; see also Chapter 5), for example, employs a statistical calculation of the entropy involved, but the final result is obtained by minimizing the free energy of the system and is thus analogous to the procedure outlined above. The major difference is that the material now contains a fixed number of lattice sites per unit volume, upon which the ions, atoms, or molecules composing the solid may be placed, and a fixed number of interstitial sites, into which foreign atoms or diffusing components may be put in positions of stable equilibrium. Each site therefore provides a potential

well for the particular species it can accommodate, and although in general
an atom located in a particular site may have thermal vibration energy, the
number of such thermally excited states is approximately constant per unit
energy width and they are separated by some vibrational quantum ω. This
is to be contrasted with a kinetic theory gas, in which there is a *continuum* of
thermal levels which increase in number per unit energy width as the kinetic
energy increases as shown in Fig. 1. The consequence is that, compared with
a gas, a far greater proportion of atoms reside in the ground state, and we

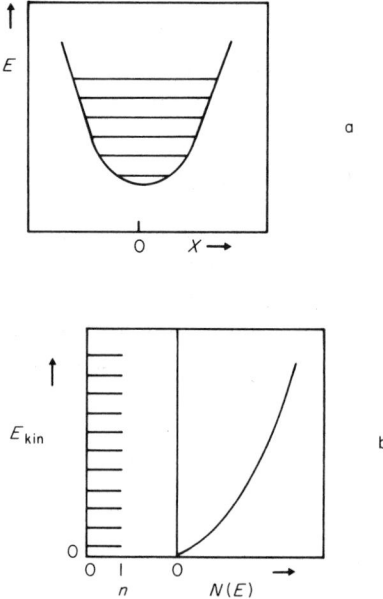

Fig. 1.   (a) Schematic diagram showing a particle (atom, ion, or electron) in a potential
well. The available energy levels are separated by an energy $\hbar\omega$, where ω is the characteristic
angular frequency of the oscillating particle. The spatial extent of the oscillation is shown
by the length of the lines, and the bounding curve is approximately parabolic. The energy
is E and the displacement from the mean position is x. (b) Density of states. The left-hand
portion shows the number of energy levels n (the degeneracy) at each allowed level for a
one-dimensional harmonic oscillator. This corresponds directly to (a). The right-hand
portion shows the number per unit bandwidth $N(E)$ for a kinetic theory gas. It is a continuous distribution which increases monotonically with the kinetic energy E_{kin}.

may ignore, to a first approximation, the thermally excited members of the
population and write

$$n_i/N_i = \exp(-g_i/kT) \qquad (11)$$

where n_i is the number of items in the available locations N_i and g_i is the

Gibbs function taken with respect to the perfect crystal as datum. The law of mass action can then be written in terms of these quantitites. As an example, consider the reaction between a divalent metal ion M and an alkali ion vacancy V in an alkali halide matrix. The reactants can form a stable impurity-vacancy (IV) complex:

$$IV \rightleftharpoons M + V$$

Using an obvious set of subscripts, we can write at once that $x_m = n_m/N_m$, $x_v = n_v/N_v$, but in the alkali halide fcc structure there are $z = 12$ possible orientations of the complex per metal ion site, so that $N_c = 12N_m$, and $x_{iv} = z(n_{iv}/N_{iv})$.

Thus, we find that

$$x_{iv}/x_m x_v \equiv K(T, p)$$
$$= z \exp\left[(-g_{iv} + g_m + g_v)/kT\right]$$
$$= z \exp(g_A/kT)$$

where $g_A = (g_v + g_m - g_{iv})$ is the Gibbs free energy of association with the configurational entropy term $kT \ln z$ removed and expressed explicitly.

The behavior of electrons and holes must be treated carefully. It is well known that they obey Fermi–Dirac statistics and that the occupancy of a particular level of energy E is given by the Fermi distribution function $f(E)$:

$$n_n/N_n = f(E) \equiv \{1 + \exp[(E - E_f)/kT]\}^{-1}$$
$$n_p/N_p = 1 - f(E) \tag{12}$$

where subscripts n and p stand for electrons and holes, respectively. E_F is the Fermi energy which is measured from the same zero as E and is determined by writing down the equations expressing conservation of mass and (in the simplest cases) electroneutrality of the system. In practice, unless a species with occupancy $f(E) \approx \frac{1}{2}$ is under consideration, it is possible to replace the concentration n_n by the constant value N_n [$f(E) \to 1$, $E < E_F$, degenerate case] or by the Boltzmann approximation [nondegenerate, $E > E_F$, $f(E) \to 0$]

$$n_n = N_n \exp[-(E - E_F)/kT] \tag{13}$$

and similarly for hole states. In the case of defect levels which may or may not be occupied by a single electron (and spin is of no importance) a factor of 2 should strictly appear with the exponential, but in practice, it is usually absorbed by modifying the value of E slightly. When dealing with the motion of free electrons, it is necessary to realize there is available to them a continuum of levels, just as for a Maxwellian gas. For the purposes of equilibrium calculations, it is possible to replace the continuum by an "effective density of states" N_c. The necessary restrictions are that the system be nondegenerate and that the conduction band be parabolic, thus giving rise to the ideal behavior $N(E_{kin}) \propto E_{kin}^{\frac{1}{2}}$ where E_{kin} is the kinetic energy of a free electron, which has an effective mass m^*. Under these conditions the free electron concentration n_c becomes

$$n_c = \int_{E_c}^{\infty} f(E)N(E)\,dE$$

$$\simeq N_c \exp[-(E_c - E_F)/kT]$$

where

$$N_c \equiv 2(2\pi mkT/h^2)^{3/2} \simeq 2.5 \times 10^{19}\,\text{cm}^{-3} \tag{14}$$

Thus, the number of electrons can be calculated as if there were a concentration N_c of ficticious levels just at the band edge E_c. The same is true for the holes, for which the effective density of states N_v in the valence band is again a fictitious concentration located at the energy of the valence band edge E_v.

When excited states of various sorts are involved or when irradiation with quanta of various energies is considered, it is clear that normally only a dynamic equilibrium can be maintained, and even this depends for its existence upon a continuing supply of excitation energy. Such cases are therefore better considered as rate processes rather than equilibrium processes.

B. THEORY OF RATE PROCESSES

The rate at which a particular reaction proceeds is characterized by the rate constant. If we restrict our attention to a mixture of i and j molecules, then the rate of reaction is given by

$$R_1 = k_1 n_i n_j$$

Now from Eq. (8) we have

$$R_1 = pC_{ij}$$

where p is the probability that a reaction occurs at each single collision between i and j molecules. Comparing the two, we obtain

$$k_1 = pC_{ij}/n_i n_j \tag{15}$$

It is usually found that the process follows an Arrhenius law so that

$$p = p_0 \exp(-\Delta g_1/kT) \tag{16}$$

where Δg_1 has the nature of a Gibbs function. This indicates that there is a barrier to the reaction taking place and that an activation energy of the order of Δg_1 must be brought by the participating molecules before the reaction can take place. The Boltzmann factor merely reflects the relative frequency with which this necessary energy is found within the ensemble of all colliding i, j pairs. Unless steric factors are important, the quantity $p_0 \simeq 1$ (Dewar, 1965). This situation is conventionally described by a reaction co-ordinate diagram. If, for example, we consider the jumping of an ion in a crystal from one lattice site to an adjacent vacancy, illustrated in Fig. 2,

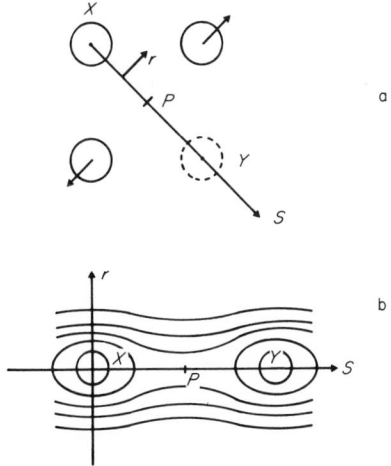

Fig. 2. (a) The atom at X can make a jump into the vacancy at Y, forcing the adjacent atoms apart during the process. For clarity, the atomic radii are not to scale. S is the co-ordinate along the direct path and r the radial displacement. (b) Contours of constant energy for the jumping process. There is a saddle point at P.

then to a first approximation we can take as reaction coordinates the real position coordinates of the moving ion: the length S along the ion-vacancy lattice direction and the cylindrical radius r with respect to this axis. The energy of the system (ion-vacancy pair + local surroundings) is plotted in a contour map form. It is seen that in jumping from X to Y the energy increases and then falls again. Any radial deviation leads to a sharp increase in the energy required so that the reaction path XPY which involves the least energy increase is the favored route. The point P is the saddle point of the energy diagram and is a position of equilibrium. It is a state of stable equilibrium as far as radial displacements are concerned, but unstable for displacements along XY. The fact that it is a state of equilibrium at all allows us to consider it as a state, well defined in the thermodynamic sense; and it is known as the transition state and will have a Gibbs function ΔG with respect to the initial state at X.

A less constricted reaction will not have the high symmetry displayed by this example, but the general features remain. There will be two Gibbs free energies associated with the reaction path, ΔG_1 ($X \to P$) and $\Delta G_2(Y \to P)$. In addition, the reaction coordinates need not be the position coordinates of one of the reactants. We can formally break up the i, j reaction into two steps

$$X_i + X_j \to X^*$$

$$X^* \to \text{product}$$

where X^* is the (excited) transition state, the concentration x_* of which is determined by thermal equilibrium considerations [Eq. (5)]:

$$x_* = x_i x_j \exp(-\Delta G_1/kT) \qquad (17)$$

while the rate at which the product is obtained is given by the unimolecular rate of conversion of the transition state $k_* n_*$, where k_* is the rate constant of this reaction. Thus,

$$k_1 n_i n_j = k_* n_* = k_* x_* N_*$$

where N_* is the possible number of transition states available to the system and $n_* = x_* N_*$ is their actual concentration. Formally, we may identify Δg_1 with ΔG_1 and p_0 with

$$k_* N_*/2^{1/2}\pi n^2 \langle d^2 v \rangle$$

It can be shown that $k_* = kT/h$ (Weiner, 1968; Golden, 1969), and since we expect $p_0 \simeq 1$, then

$$\langle d^2 v \rangle \simeq (N_*/n^2)(kT/h)$$

and also

$$k_1 \simeq \langle d^2 v \rangle \exp(-\Delta G_1/kT) \qquad (18)$$

This last relation, in particular, illustrates the strong dependence of the reaction rate upon d^2 and, thus, upon the cross section σ.

It must be recognized that the excited intermediate will form products or revert to its original constituents only because the excess energy is carried off thermally by further collisions with surrounding molecules. In this development we have avoided the complications introduced by adiabatic reaction paths, and have assumed that there is a unique reaction diagram for a given reaction. A detailed discussion based on the stochastic properties of the lattice (Glyde, 1967) shows that these simplifications do not lead to any great errors in the calculation of rate constants.

II. Application to Solids

A. Diffusion in Solids

When reactants of any form (foreign atoms, atoms of the host, electrons, or more exotic species) are incorporated in a solid matrix and are free to move, then the gas-kinetic theory outlined in the opening paragraphs can be used directly, as it is in the case of solutions. In many cases, the participants will not be free to move but will have to diffuse from site to site, spending relatively long times in the positions of minimum potential energy such as are shown schematically in Fig. 2. In such circumstances, the rate at which a reaction proceeds may be governed by the rate at which the reactants approach one another, that is, by the diffusion itself. Another way of viewing the situation is to note that $\langle d^2 v \rangle$ contains v, which is now the mean velocity of the faster moving reactant. A somewhat better approach is to recognize that the diffusion step illustrated in Fig. 2 can itself be regarded as a unimolecular reaction. It is usual to write this reaction in terms of the jumping probability γ (frequently k is used in the literature), which is the probability per unit time that a single i center will reach the transition state [see Chapter 5, Eq. (22)]. If N is the concentration of unit cells and z the number of equivalent saddle-point configurations per cell, then the number of jumps performed per unit volume per unit time is

$$k_1 n_i z N = \gamma n_i$$

Therefore

$$\gamma = k_1 z N$$
$$= z \langle d^2 v \rangle N \exp(-\Delta G_1/kT) \tag{19}$$

Now, $N = 1/a^3$ where a is the lattice constant (or a suitable average for noncubic lattices) and the cross section for this reaction is clearly of the order of the lattice dimension so that $d^2 \simeq a^2$. The characteristic velocity v is equal to the jump distance, itself approximately equal to a, divided by the time required for atomic rearrangement to be completed. This time will be roughly equal to the reciprocal of the lattice oscillation (phonon) frequency v, again suitably averaged. Thus, we find

$$\gamma \simeq z v \exp(-\Delta G_1/kT)$$
$$\equiv v_{esc} \exp(-\Delta G_1/kT) \tag{20}$$

where v_{esc} is the attempt-to-escape frequency and is defined as the pre-exponential coefficient of γ.

The diffusion coefficient can be related to γ rather simply. Consider two parallel planes of atoms P, P' with coordinates x, $x + a$, and concentrations n_1, $n_1 + \Delta n_1$. Then the atoms jumping from P to P' will be drawn from a thickness of a in the x direction, and the number jumping per unit time will be $\frac{1}{6}\gamma n_1 a$ per unit area. The number in the reverse direction will be $\frac{1}{6}\gamma(n_1 + \Delta n_1)a$, and the difference gives the net flux:

$$\mathscr{F} = \frac{1}{6}\Delta n_1 a$$

But by definition, the diffusion coefficient is given by $\mathscr{F} = -D(dn_1/dx)$. Since $dn_1/dx = \Delta n_1/a$, we find that

$$D = \tfrac{1}{6}\gamma a^2 = \tfrac{1}{6}v_{\mathrm{esc}}a^2 \exp(-\Delta G_1/kT) \tag{21}$$

to within a factor of order unity. There will be additional minor corrections due to correlation effects (Peterson, 1968). The same type of relation will apply to any species (atoms, vacancies, radicals, excitons) for which a strong localization and, therefore, a hopping type of motion occurs. Particular examples have been discussed by Glyde (1967).

B. Reaction Rates during Irradiation

Of particular interest in this book is the behavior of materials during irradiation with either penetrating radiation or with optical quanta. In

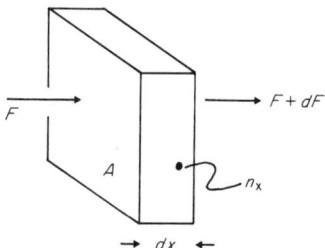

Fig. 3. Absorption of radiation of flux F by a thickness dx of material. Note that dF will be numerically negative. The number of unit cells in the slice will be $NA\,dx$, where $(1/N)$ is the volume of a unit cell. The concentration of photons is n_x.

each case a nonequilibrium situation occurs, though frequently a dynamic equilibrium is set up. The rate of production of excited species is given formally by the same equations as before. In the case of X-rays and of quanta the concept of a cross section σ for the interaction between the irradiating particle and the ground state of the solid is a useful one (Murrell, 1961). Since the strength of the irradiation is usually monitored by the incident flux F (or the incident intensity I) rather than by its local concentration n_x, the rate of reaction is best expressed in terms of them. If σ is the cross section for reaction per unit cell, then the rate of production of excited species in a slice of thickness dx in the direction of irradiation and of frontal area A (Fig. 3) is given by

$$k_1 NA \, n_x \, dx \equiv FAN\sigma \, dx$$

Since in most cases there is no activation energy involved, $k_1 \simeq \sigma v$ and $v = c$, the velocity of light. Thus, $n_x = F/c$, as might have been anticipated on elementary grounds. We can relate the cross section to the absorption coefficient. Treating the simple case in which the process of interest is the only one causing absorption, we have from the previous equation

$$A \, dF = -FAN\sigma \, dx$$

whence

$$F \propto \exp(-N\sigma \, x)$$

and the flux or intensity absorption coefficient α is given by

$$\alpha = N\sigma \tag{22}$$

In the case of electron radiation this theory is far too naive because of the nonuniform nature of the energy loss from the beam. Perhaps the best that can be done is to compute the energy deposition rate as a function of depth (see, for example, Kobetich and Katz, 1968), then make an assumption about the yield of the species of interest as a function of energy density and test the experimental results against this assumption. An example of a purely experimental approach has been given by Salovey and Rosenzweig (1963).

C. DIFFUSION-CONTROLLED REACTIONS

The theory of diffusion-controlled reactions has been developed in a number of papers (Waite, 1957, 1958, 1960; Noyes, 1961). Such reactions are very relevant to radiation chemistry because they provide a detailed description of the reaction kinetics in a system containing a random distribution of reactants in an inert matrix at time $t = 0$. Such a system exists in a material immediately after exposure to a burst of penetrating radiation. The most elementary view to take is that there will be a distribution of reactants throughout the irradiated material, some in close pairs and others further apart. It is most likely that collisions will take place between the close pairs rather quickly. If the probability p of reaction upon collision is close to unity, then these close pairs will be preferentially removed, with an artificially high rate of reaction corresponding to the effectively large local concentration in the neighborhood of these pairs. As the closest pairs are removed, the rate of reaction slows down, and it depends upon how quickly the remaining reactants reach one another; i.e., it depends upon the diffusion coefficient of the more mobile reactant.

In the type of chemical reaction usually considered, p is small and any local disparities in concentration are smoothed out by multiple collisions which take place without reaction. (Technically there are always concentration fluctuations at the atomic level, and in fact it is the pair distribution function which is held infinitely close to the equilibrium value it would have in the absence of reaction.) By contrast, a high value of p prevents any such leveling of the concentrations. The mathematical details have been given in the references above. The chief effect is that the "rate constant" k_1, taken as an average over the whole reacting volume, now appears to be a function of time. In the extreme limit one finds that

$$k_1 = k_{10}[1 + 1/(\pi\tau)^{1/2}]$$

$$k_{10} = 4\pi r_0 D$$

$$\tau = Dt/r_0^2 \tag{23}$$

where D is the diffusion coefficient, r_0 the capture radius, and t the time. Thus, if the diffusion coefficient can be measured separately by tracer methods, then the long-time reaction rate yields a value for r_0 a quantity not otherwise accessible.

When the probability of reaction p falls below unity for reactant separations less than r_0, then rather less abrupt time dependences are found (Waite, 1957, 1960). If long-range forces exist between the reactants, as may well occur if they are ionized or dipolar fragments, then considerable modifications may occur. The elementary approach of introducing an effective capture radius r_{eff} (Waite, 1958) is not even qualitatively accurate since, if the forces are repulsive, the rate of reaction may tend to zero (Wintle, 1963). Similar calculations have been presented for cylindrical symmetry (Bullough and Newman, 1970) which can be adapted to a three-dimensional system.

D. CHARGED-PARTICLE EFFECTS

One aspect of the behavior of charged particles has already been considered: The long-range forces associated with them may lead to "rate constants" for recombination which vary in a quite complicated manner with time. Even after times such that the steady state rate constant has been reached, the formulas given above are only correct when activities are used in place of mole fractions. To a first approximation a solid can be treated as a continuum in which the charged species move, and so for ionic reactions the

Debye–Hückel theory of electrolytes can be taken over unchanged (Lidiard, 1957) since even in a wholly ionic solid, the material forming the undisturbed lattice can be regarded as a supporting medium for the defects (mobile ions and mobile vacancies) which are causing the reactions of interest. In such cases the activity a_i is given by

$$a_i = x_i f_i$$
$$f_i = \exp\{-q^2/[8\pi\varepsilon\varepsilon_0 kT\lambda(1 + r_0/\lambda)]\}$$
$$\lambda^2 = V\varepsilon\varepsilon_0 kT/2q^2 x_i M \tag{24}$$

where f_i is the activity coefficient, λ the Debye screening length, q the electronic charge, M the molecular weight, and V the molar volume of the pure solid. It is clear that the correction terms f_i are of the form $\exp(-E/kT)$, and it turns out that the energy E may be a few tenths of an electron volt. Measured activation energies may therefore differ from theoretically computed saddle-point energies by such an amount.

The Debye–Hückel atmosphere also results in a reduction of the mobility of an ion and, thus also, of its diffusion coefficient. Hence, in the case of diffusion-controlled reactions and in conductivity measurements the ideal diffusion coefficient must be reduced by a factor g. Of various expressions available, Lidiard (1957) favors

$$g = 1 - \frac{q^2\lambda}{12\varepsilon\varepsilon_0\pi kT(1 + \sqrt{2})(1 + r_0/\lambda)(\sqrt{2} + r_0/\lambda)}$$

The rate of a reaction has been shown in the previous section to be given by $k_{10} = 4\pi N_0 D$ after the transient portion of the rate constant has decayed, and this can be formally identified with $\langle d^2 v\rangle \exp(-\Delta G_1/kT)$. These relations no longer apply if there are long-range forces between the reactants, such as the Coulomb interaction between charged species. It has been shown by Antonov-Romanovskii (1968) that these quantities must be replaced by $\sigma_{\mathrm{eff}} v$, where

$$\sigma_{\mathrm{eff}} = \sigma_k M_k$$

$\sigma_k \simeq d^2$ is the "'gas-kinetic" cross section or physical target area, and M_k is an enhancement factor which varies strongly with the carrier mean free path l, the radius r_{eff}, and which σ_k. Since the combinations of these parameters for different materials can lead to vastly different values for M_k, no generalizations can be made, though it should be noted that under certain circumstances an apparently negative activation energy can be found.

E. Problems of Surfaces and Dislocations

It is clear that when two phases coexist, such as crystalline and amorphous regions in polyethylene, different reactions may occur in the two phases because of differences in density, mobility of excited species, and concentration of impurities. It is also obvious that when the surface has been subjected to some special treatment, intentional or otherwise, then the reactions on the surface and in the subsurface layers may well differ from those in the bulk. Since most materials will be more oxidized at the surface then in the interior, one must always be wary of such effects, particularly when thin films or powders are employed. A less obvious source of complication is that the concentration of an impurity or a deliberate additive may change from bulk to surface, and that either depletion or accumulation may occur. This is most marked for ionized species in which the effect is electrostatic in origin (Kliewer and Koehler, 1965), though it will occur for any species in which the binding energy to the surface differs from the binding energy to the bulk (Lifshits and Geguzin, 1965; Burton, 1969). Since dislocations present internal free surface, they too can be surrounded by impurity concentrations which differ from the bulk values, and they may act as sinks for various products of irradiation. The kinetics of diffusion into dislocation cores are similar in mathematical form to the diffusion-controlled reaction mechanism considered in the previous section and have been given in detail by Bullough and Newman (1970). For charged species, the characteristic depth associated with deviations from bulk behavior is a few Debye lengths λ [Eq. (24)] and may amount to several microns in materials with concentrations $\approx 10^{-6} M$.

A more subtle effect in poor conductivity materials is that they may not be in thermal equilibrium at all. Studies of persistent internal polarization and of electrets indicate that the time constants associated with equilibrium of charged species may be several years (Roos, 1969). Since polymer films can be charged by corona effects during unrolling, it is advisable to discharge the material (Zichy 1967) whenever possible and to be careful to run blank tests when irradiated materials are studied by color center, thermoluminescence, and similar trapped-electron methods.

REFERENCES

Antonov-Romanovskii, V. V. (1968). *Phys. Status Solidi* **26**, 173.
Bullough, R., and Newman, R. C. (1970). *Rep. Progr. Phys.* **33**, 101.
Burton, J. J. (1969). *Phys. Rev.* **177**, 1346.
Dewar, M. J. S. (1965). "An Introduction to Modern Chemistry." Athlone Press, London.
Glyde, H. R. (1967). *Rev. Mod. Phys.* **39**, 373.

Golden, S. (1969). "Quantum-Statistical Foundations of Chemical Kinetics." Oxford Univ. Press (Clarendon), London.

Kliewer, K. L., and Koehler, J. S. (1965). *Phys. Rev.* **140**, A1226.

Kobetich, E. J., and Katz, R. (1968). *Phys. Rev.* **170**, 391.

Lidiard, A. B. (1957). "Handbuch der Physik" (S. Flügge, ed.), Vol. 20, pp. 246–349. Springer, Berlin.

Lifshits, I. M., and Geguzin, Ya. E. (1965). *Sov. Phys.-Solid State* **7**, 44.

Mott, N. F., and Gurney, R. W. (1948). "Electronic Processes in Ionic Crystals," 2nd ed. Oxford Univ. Press (Clarendon), London.

Murrell, J. N. (1961). "Symposium on Electrical Conductivity in Organic Solids" (H. Kallmann and M. Silver, eds.), pp. 127–145. Wiley (Interscience), New York.

Noyes, R. M. (1961). "Progress in Reaction Kinetics" (G. Porter, ed.), Vol. 1, Chapter 5, pp. 129–160. Pergamon, Oxford.

Peterson, N. L. (1968). *Solid State Phys.*, **22**, 409.

Roos, J. (1969). *J. Appl. Phys.* **40**, 3135.

Salovey, R., and Rosenzweig, W. (1963). *J. Polymer Sci. A.* **1**, 2145.

Waite, T. R. (1957). *Phys. Rev.* **107**, 463.

Waite, T. R. (1958). *J. Chem. Phys.* **28**, 103.

Waite, T. R. (1960). *J. Chem. Phys.* **32**, 21.

Weiner, J. H. (1968). *Phys. Rev.* **169**, 570.

Wintle, H. J. (1963). *Nature* **198**, 478.

Zichy, E. L. (1967). "Static Electrification," pp. 52–59. Inst. of Phys. and the Phys. Soc., London.

7

Theory of the Electrical Conductivity of Polymers

H. J. Wintle

Department of Physics, Queen's University, Kingston, Ontario, Canada

I. Introduction

The electrical conductivity of organic materials in general has been studied in considerable depth during recent years. A number of books containing much background information are available (Fox *et al.*, 1963; Kallmann and Silver, 1961; Brophy and Buttrey, 1962) while a vast number of references have been given in the compendium by Gutmann and Lyons (1967). Papers on high polymers have been published in the *Journal of Polymer Science C*,

17 (1967) and in the proceedings of the Annual Conference on Electrical Insulation and Dielectric Phenomena (NAS-NRC, Washington), in addition to the usual journal literature.

There are two different aspects of conduction which concern us, dark conductivity and photoconductivity. The latter may be induced optically or by X-, γ-, β-, or cathode-rays. At the time of writing the sign of photo-carriers in a few high polymers has just been ascertained, while the sign of the carrier and the conduction mechanism in the dark has not been satisfactorily settled for any polymer. There is a flowering of interest in amorphous materials, particularly in transition metal oxides, and it is probable that when these relatively simpler systems have become well understood, the validity of the currently developing theories for polymers can be assessed. In the circumstances, the present article reviews the available theories, but it should be borne in mind that, although various authors claim that their experimental data fit a particular theory, it is usually the case that the same data will fit two or three theories equally well, and truly definitive experiments have not yet been made. It is impossible to generalize and each material, in some cases each specimen, has to be taken on its own merits. Double injection, impurity conduction, and polaron theory do not seem at the moment to be very relevant to high polymers and have been omitted.

II. Band Theory

When an isolated atom is electronically excited, an electron increases in energy and occupies a previously empty state, leaving the orbital it came from unoccupied or, as it is alternatively put, occupied by a hole. If N noninteracting atoms are involved, then both the excited electron states and the hole states are N-fold degenerate as long as the isolated atom states are nondegenerate. The fundamental theorem of solid state physics is that when these atoms are coupled together by being placed on a regular lattice, the degeneracy is lifted and the N states form a group of energy levels very closely spaced in energy and with finite upper and lower energies. This situation is shown in Fig. 1A. Normally N is so large that the splitting is very small and this set of states can be regarded as a continously distributed in energy. Such a set of states is called a band (Kittel, 1966, or any similar textbook). There are two bands in particular which are of overriding importance in conductance work. They are the valence band, which is the analogue of the highest level normally occupied by an electron in an isolated atom or molecule, and the conduction band, which is the analog of the first ionization level of the isolated species. Since in idealized (free-electron

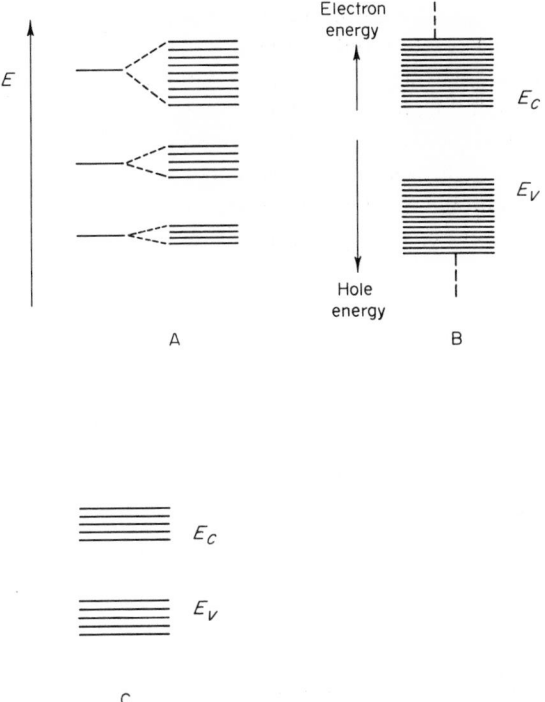

Fig. 1. (A) The atomic levels are split under the influence of adjacent atoms into closely packed bands of energy levels. E is the energy. (B) The conduction band extends from E_c towards higher energies, and contains few electrons. The hole band (valence band) extends from E_v towards lower electron energies and is largely filled (i.e., it contains few holes). Note that the hole kinetic energy increases in the nonconventional downwards direction. In a highly idealized solid, these bands extend indefinitely. (C) The transport bands of an organic solid are finite in width and may be less than kT wide.

approximation) metals the ionized electron can carry any amount of kinetic energy, the conduction band is frequently represented as extending from an energy E_c, the conduction band edge energy, to infinity. It is a result of the continuity of the lattice that a hole, that is a site from which an electron has been stripped, can also migrate, again carrying with it any amount of kinetic energy. Thus, the holes themselves form a band which extends from the valence band edge E_v to infinity, as shown in Fig. 1B. Though in regular semiconductors and in transition metals we have to consider more complex situations in which there is overlapping of a number of free carrier bands, the present state of knowledge of polymers is too insecure to warrant consideration of such refinements. Calculations of band structure have been made for anthracene and its homologs (Katz *et al.*, 1963) and for one-dimensional chains of H_2 (André, 1969) and polyethylene (McCubbin and Manne, 1968), and it seems clear that in molecular solids the free-electron approximation is far from satisfactory. There is a set of conduction and valence bands, the lowest energy members of which are finite rather than infinite in·extent, as shown schematically in Fig. 1C. The calculations yield not only the energies but the density of states $N(E)$, which is the concentration of available levels per unit energy width, shown in Fig. 2A. Similar calculations

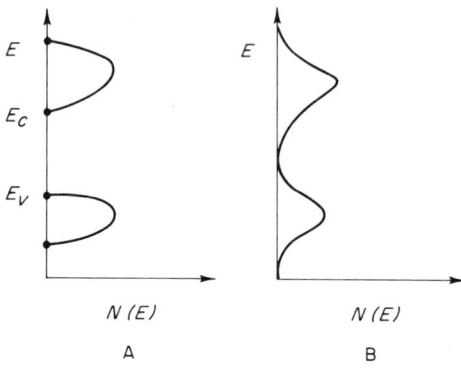

Fig. 2. (A) Density of states for perfect organic solid, such as is shown in Fig. 1C. E is the energy and $N(E)$ the density of states. (B) Density of states for a disordered solid (schematic). The states in the band tails may or may not be localized. Both diagrams should be compared with Fig. 1b of Chapter 6.

have been made for the exciton bands in polyethylene (Partridge, 1968, 1970). Since these organic materials are covalently bonded, the valence band is filled except for a few electrons promoted thermally into the conduction band (see Chapter 6) and the materials are insulators. The carriers behave as if they are free electrons but have a mass different from that of a cathode ray. This effective mass m^* is given in an isotropic material by

$$\frac{1}{m^*} = \frac{1}{\hbar^2} \frac{\partial^2 E}{\partial k^2}$$

where k is the wave vector and E the energy of the particle. Now, in a narrow band material $\partial^2 E/\partial k^2$ will be smaller than for an idealized metal and m^* may consequently be quite large. When the bands become comparable in width to kT, even this simple effective mass concept must be avoided and each state within the band dealt with separately.

There is, at present, considerable debate as to whether an amorphous material can be represented as having a band structure. The general concensus is that, as the random deviations of lattice spacing from the purely crystalline value become larger, so the states close to the crystalline band edge become displaced into the energy gap, leading to a distinct tailing of the band edges as illustrated in Fig. 2B (Wannier, 1949; Moorjani and Feldman, 1964; Hiroike, 1965). According to Mott (1969a), these tail states will be largely localized, though this does not appear to be the case for the other treatments. When the perturbation is due to the incorporation of irregularities, such as impurities or chain branches, then the calculation by Herzenberg and Modinos (1964) suggests again that delocalized band tailing will occur, and a calculation has been made for polymers (Wintle, 1967). A somewhat unusual approach by Williams and Mathews (1969) carries by implication the contrary view. It is apparent that the somewhat picturesque description by Fowler (1956) of contiguous domains in each of which the band energy differs from that of its neighbors is inadequate but that a considerable amount of work is required to disentangle the real situation in polymers. In addition, it is clear that the calculations done to date refer to propagation intramolecularly along the polymer chain, whereas in a real material there are discontinuities at chain ends, at the boundaries of crystalline and amorphous regions, and elsewhere that chain folding occurs. To have continuous dc current flow in a real material there must therefore be intermolecular carrier jumps.

III. Mobility Theory

The relative importance of inter- and intramolecular transfer can be assessed if the resistance provided by each process can be determined. Since series resistances add, the more resistive process dominates in the conductivity of the material. In the case of molecular crystals it has been shown (Katz *et al.*, 1963) that the usual metallic approximation of a relaxation time τ_{rel}, representing the mean free time between successive inelastic collisions of the free carrier with the lattice, provides a satisfactory agreement with the experimental values. Now,

$$\mu = q\tau_{rel}\langle v^2 \rangle / kT \simeq q\tau_{rel}/m^* \tag{1}$$

where μ is the carrier mobility, q the charge, m^* the effective mass, and $\langle v^2 \rangle$ a suitably weighted mean square velocity. Since m^* or $\langle v^2 \rangle$ can be determined from the detailed band structure (Kittel, 1966), the only disposable parameter is τ_{rel}, and it is found that times of the order of 10^{-12} sec (or mean free paths of ≈ 50 Å) or less are appropriate. In the absence of any direct calculations for linear chains, one must suppose that, since the band calculations for molecular crystals and for saturated chains are similar, the mobilities are also likely to be similar and to show a weak temperature dependence inversely proportional to a small power of the absolute temperature due to phonon scattering. Strictly, this model can only apply to materials with mobilities in excess of 1 cm^2 V^{-1}sec^{-1}.

Other work on molecular solids indicates at least two other possible transport mechanisms, the hopping model and the tunneling model. Much care must be taken in transferring such ideas to long-chain polymers. When the transport bands are very narrow, there is a strong interaction between an excess electron and the lattice and the small polaron theory applies. Although such an electron is normally regarded as localized and its transport has been called a hopping process (Glarum, 1963; Glaeser and Berry, 1966), it is in fact more of a tunneling process and the mobility is again likely to be a rather weak function of temperature (Tredgold, 1962). The mobilities in this case are expected to be less than 1 cm^2 V^{-1}sec^{-1}. Although developed for intermolecular transfer, there seems to be no bar to extending this case to intramolecular transfer along a chain.

The transfer of an electron from one molecule to the next may possibly proceed by the same type of thermally assisted tunneling process, but if the molecular separation is at all large, it is likely to be relatively insensitive to thermal fluctuations and the approximate calculation of McCubbin (1963) will apply. Tunneling times of the order of 10^{-14} sec are associated with this process (Kemeny and Rosenberg, 1970). Alternatively, if the potential barrier between molecules is sufficiently opaque, carriers are likely to jump by thermal activation rather than by tunneling. In this case we have (see Chapter 6) that the time t_{esc} taken to complete the transfer will be

$$t_{esc} = v_{esc}^{-1} \exp(\Delta E/kT) \tag{2}$$

If we adopt a simplified view that conduction in a long-chain polymer is due to travel along a chain of average length l, followed by an intermolecular jump of length d, then the overall mobility μ_{mic} (the microscopic mobility) will be given by

$$\mu_{mic} = \frac{d + l}{d/\mu + t_{esc}\mathscr{E}} \tag{3}$$

where d is the chain length, l the intermolecular jump distance, and $\mathscr{E}$ the applied electric field. This mobility is apparently a function of the applied field, but it is likely that, in fact, any intermolecular jumps will be field assisted (Mott and Gurney, 1948), so we must replace Eq. (2) by

$$t_{esc} = v_{esc}^{-1} \exp[(\Delta E - q\Delta V)/kT] \tag{4}$$

where $\Delta V = a\mathscr{E}$, the potential drop across the jump distance a. If this electrostatic barrier is operative, μ_{mic} can become only slightly field dependent. This behavior will occur only if the intramolecular mobility μ is large enough for carriers to accumulate preferentially at one end (downfield for positive carriers) of a molecule. If μ is small, then the number of carriers making jumps in the reverse direction across the intermolecular barrier will become significant and we have

$$t_{esc}^{-1} = v_{esc} \{\exp[-(\Delta E - \Delta V)/kT] - \exp[-(\Delta E + \Delta V)/kT]\}$$

$$\simeq v_{esc} \exp(-\Delta E/kT) \cdot 2\Delta/VkT \tag{5}$$

whence again it follows that μ_{mic} is independent of field but varies strongly with temperature.

In any real polymer system, it is likely that carrier traps occur in great numbers. Shallow traps will capture carriers, hold them for some time, and subsequently thermally release them back into the free carrier band. The *microscopic mobility* is due to the interaction of the carrier with the perfect material and continues to operate. The traps slow down the overall motion, and one obtains an effective *macroscopic mobility* μ_{eff}. If the mean free path in the drift direction between trapping events is λ and the trap is described by an escape frequency and depth v_t, E_t, the time taken for a carrier to cover a distance x is

$$\mu_{mic}^{-1}\mathscr{E}^{-1}x + (x/\lambda)v_t^{-1} \exp(E_t/kT)$$

and the apparent mobility is therefore given by

$$\mu_{eff}^{-1} = \mu_{mic}^{-1} + (\mathscr{E}/v_t\lambda)\exp(E_t/kT) \tag{6}$$

Again there is a marked temperature dependence, and as before, the field dependence may be ameliorated in a real material by a significant reverse jump contribution. In many polymers it would seem that trap distributions occur, rather than single trapping levels (Rose, 1955, 1963), and then a weighted average over the whole distribution must be employed.

There are few reliable measurements of mobilities in high polymers. Recent indications are that, as expected with a trapping mechanism, the macroscopic mobility is thermally activated and very small, in the range 10^{-6} to 10^{-11}cm^2 V^{-1}sec^{-1} or less at room temperature (Davies, 1967; Martin and Hirsch, 1969, 1970; Wintle, 1970).

The conductivity of conjugated systems presents a rather different aspect, since the π-electron system provides molecular orbitals which extend over the complete length of the molecule. Calculations by Tobin (1962) indicate an intramolecular mobility of 1.4 cm^2 V^{-1}sec^{-1} and a conductivity of 0.12

$\Omega^{-1}cm^{-1}$ for polyacetylene. The experimental situation is much less certain. Pohl and Chartoff (1964) indicate resistivities of $\approx 10^{10} \Omega$ cm for such materials, with much lower values for polyacene quinone radical (PAQR) polymers in which dimensional stability is assured by multiple bonding, and use the term "ekaconjugation" to describe these and similar systems. Their case that high spin concentration and high conductivity are closely connected is persuasive but must be regarded with caution since similar arguments advanced for anthracene and for pyrocarbons are not entirely satisfactory. Careful work by Berets and Smith (1968) indicates a resistivity of $10^6 \Omega$ cm for polyacetylene and suggests strongly to the present author that surface conduction is being observed in their pressed powder under all conditions. There is therefore no satisfactory evidence against which the theory of Tobin (1962) can be tested.

IV. Carrier Production and Transport

The resistivity of high polymers is high because both the mobility and the carrier concentration are low. The concentration of carriers produced intrinsically by thermal ionization will be very small since the band gaps appear to be several electron volts (see Chapter 6). It seems more likely that ionization of impurities will be responsible for any standing concentration of carriers (Partridge, 1967), and since purification of high polymers is rarely attempted before electrical measurements are made, it is hardly surprising that the results of absorption current and dark conduction measurements are quite variable even from sheets of nominally the same material. Impurities may also provide carriers by internal field emission (Poole–Frenkel effect) (Jonscher, 1967) and a fairly convincing demonstration of this effect in the presence of gross doping has been given by van Beek and van Pul (1962).

Carriers are also introduced into dielectrics by injection from the electrodes, and this is probably the main source in high polymers. If the temperature is large enough then carriers are excited thermally over the potential barrier at the electrode–polymer interface (Schottky emission), while if the field is large, tunneling into the conduction band occurs (field emission, Fowler–Nordheim regime). Intermediate cases have been discussed by Good and Müller (1956) and Hill (1967). In very thin films, such as those formed by glow discharge, it is possible to achieve direct quantum mechanical tunneling between the electrodes, in a manner quite analogous to one of the inter-molecular mechanisms noted above. There is now very strong evidence (Davies, 1969) that injection will occur under zero applied field, due simply to the contact potential.

If carrier emission occurs freely enough, the current is controlled not by the rate of motion of carriers in the bulk (ohmic regime), nor yet by the rate of supply, but by the inhibiting effect of carriers already introduced, and a space charge limited current (SCLC) is obtained. Under these conditions trapping is quite important, and a welter of information can be obtained from materials with mobilities in the region of 1 cm^2 V^{-1} sec^{-1} (Many et al., 1961), using a combination of optical excitation and transient techniques. Evidence for similar effects in less conductive materials has been given by Wintle (1970) and by Caserta et al. (1969).

The energy required for injection from metal electrodes has been discussed by Kallmann and Pope (1960) and by Kommandeur (1961). Hole injection will occur without assistance from external fields if

$$(\phi - I) + X > 0 \tag{7}$$

where ϕ is the electrode work function, I the ionization energy of the dielectric, and X the external energy supplied optically or thermally. Electron injection occurs if

$$X + (A - \phi) > 0 \tag{8}$$

where A is the electron affinity of the dielectric. The quantities A and I will differ from the corresponding gaseous molecule energies by a polarization term which typically may amount to 1–2 eV (Lyons and Mackie, 1962), though direct application of this correction to high polymer systems (Wintle, 1964) leads to polarization terms which are clearly too large. Mehl and Funk (1967) have suggested that the bandgap and Fermi energy are parameters which are of importance, but their computations would appear to be invalid for *free* carrier injection. In cases for which $X \simeq 0$, we may expect space charge limited conduction to occur, but if $X > 0$, the current will be injection limited.

There are repeated suggestions that ionic conductivity occurs in high polymers, though in many cases the data can be accounted for on both electronic and ionic theories and no clear-cut distinction can be made (e.g., the analysis of polypropylene by Foss and Dannhauser, 1963). In view of the recent charge-transfer work by Davies (1969 and unpublished) it seems highly likely that electronic behavior is dominant in most high polymers. Ionic motion is probable in polyethylene terephthalate (Smith and Scott, 1966). In the case of nylon, a clear contribution from protons has been identified by electrolysis (Seanor, 1968).

A list of the equations corresponding to the simplest limiting cases is given in Table I. Although it is, in theory, feasible to determine the conduction

TABLE I[a]

CURRENT-FIELD CHARACTERISTICS FOR DIELECTRICS

Mechanism	Relation	Reference
Ohmic	$j = \sigma\mathscr{E}$ $\sigma = pq\mu$	

Space charge limited current
  a. Shallow traps
  b. Distributed traps
	$j = 9\mu\varepsilon\varepsilon_0\theta V^2/8d^3$
	$j \propto V^{l+1}/d^{2l+1}$
	$l > 1$
		Many et al. (1961)
		
		Lampert (1964)

Richardson–Schottky thermal emission	$j = AT^2\exp[(-q\phi + \beta_s V^{1/2})/kT]$ $\beta_s = (q^3a/4\pi\varepsilon\varepsilon_0 d)^{1/2}$	Lengyel (1966)
Poole–Frenkel internal thermal emission	$j = \sigma_0\mathscr{E}\exp(\beta_{PF}V^{1/2}/kT)$ $\beta_{PF} = 2\beta_s$	Mark and Hartman (1968) Pulfrey et al. (1970)
Fowler–Nordheim field emission (tunnelling)	$j \propto \mathscr{E}^2\exp(-c/\mathscr{E})$ $c = 4(2m)^{1/2}\phi^{3/2}/3\hbar q$	
Internal field emission (Zener effect, surface state emission, trap emission)	As above with modified value of c	
Ion jumping (hopping)	$j = j_0\exp(-\Delta G/kT)\,\mathrm{sh}(\lambda q\mathscr{E}/2kT)$ $j_0 = 2\nu q/\lambda^2$	Mott and Gurney (1948)
Impurity conduction	Hopping with an activation energy $(E_i - \text{const.}\,N^{1/3})$ Tunnelling also possible	Hill (1967) Mott (1969b)

[a]Abbreviations:

j	current density	ϕ	work function of electrode
$\mathscr{E}$	applied field	d	specimen thickness
V	applied voltage	a	space charge correction factor, bringing the nominal field V/d to the actual field adjacent to the cathode
p	carrier concentration		
q	elementary charge		
μ	carrier mobility	λ	ionic mean jump distance
ε	dielectric constant	ν	attempt to escape frequency
θ	shallow trapping parameter	ΔG	Gibbs free energy of (formation + motion) of ion vacancy
A	Richardson–Dushman constant ($120\ \text{Å cm}^{-2}\ \text{deg}^{-2}$ in theory, frequently less in practice)	E_i	impurity ionization energy
		N	impurity concentration

mechanism by a careful study of the dependence of current upon field, temperature, and thickness, it turns out, in practice, that it is extraordinarily difficult to make unambiguous assignments. At the time of writing it would appear that much of the existing work can be accounted for on the basis that at low fields ($< 10^4$ V cm^{-1}) the current observed is due to localized motion of the space charge introduced by the contact potential difference while at much higher fields an emission-controlled mechanism occurs.

V. Photoelectric Effects

Illumination of a material can in principle lead to bulk carrier production and hence photoconduction, carrier injection at the electrodes if Eq. (7) or Eq. (8) is satisfied, photovoltages if the material is asymmetric or if the light is heavily absorbed, and photoemission. Although these various effects are well established in anthracene and in charge-transfer complexes, information on their occurrence in polymers has been sparse until recently. Observations of ultraviolet photovoltaic effects in polyethylene have been given by Wintle and Charlesby (1962) and Wintle (1965), and for a number of other materials by Herspring and Oster (1968), Kryszewski et al. (1968), Ofran et al. (1969), and Binks et al. (1970). Infrared effects have been reported by Tanaka and Inuishi (1966, 1967), Lakatos and Mort (1968), and Vermeulen and Wintle (1970). The release of carriers by ionizing radiation will be considered in Chapter 8.

In principle, light incident on the bulk with a quantum flux f and absorbed with an extinction coefficient α will generate carriers at a depth x with a rate

$$g = \eta \alpha f \exp(-\alpha x)$$

where η is the quantum efficiency for carrier production. If the carrier concentration (assumed positive) is p and the lifetime τ, in the absence of both diffusion and electric fields one finds that

$$\frac{dp}{dt} = g - p/\tau$$

There are hidden difficulties in this equation since, if the absorbing centres are being ionized directly and not merely transferring their energy to other ionizable centres, then α will vary with both position and time. There may be several absorbing processes of which only one leads to carrier formation, and the carrier production may be second order in the light intensity (a double quantum process). In addition, the lifetime is governed by the occupancy of recombination centres which may, in turn, vary with p and t, so the build-up and decay of carriers may well not be simple exponential processes (Heijne, 1963, 1966). Including conduction and diffusion and dealing only with planar geometry, we find

$$\frac{\partial p}{\partial t} = g - \frac{p}{\tau} + D\frac{\partial^2 p}{\partial x^2} - q\mu\frac{\partial(\mu\mathscr{E})}{\partial x}$$

where D is the diffusion coefficient and $\mathscr{E}$ the electric field. At the same time, Poisson's equation is obeyed

$$\varepsilon\varepsilon_0(\partial\mathscr{E}/\partial x) = p - n$$

where n is the concentration of ionized centers. In semiconductors it is possible to make plausible approximations, such that the field gradients are not too large. In insulators this is not so and the solution of these equations, even for steady illumination conditions, is formidable and cannot be attempted here. Various possibilities have been discussed by Murrell (1959, 1961). The photovoltage developed across a specimen due to the (differential) diffusion of carriers is called a Dember emf.

The boundary conditions are also not clear except in special cases. The electrodes can cover the whole range from being good sinks to being good sources, and there is evidence for both electron (Vermeulen *et al.*, 1971) and hole (Lakatos and Mort 1968) photoinjection into polymers. In anthracene it has been established that carrier formation occurs through the destruction on the electrodes of excitons which are formed in the bulk (Mulder, 1968a), and similar processes may well occur in other organic materials. In addition, certain adsorbates on the surface can act as efficient sources of carriers (Mulder, 1968b).

The majority of high polymers form heterogeneous systems, and one has to deal almost by definition with internal potential barriers as well as barriers at the surface (see Chapter 6). A single barrier will give rise to a photovoltage

$$V = (kT/q)\ln(1 + cf/j_{\text{sat}})$$

where j_{sat} is the reverse saturation current of the junction and c is a constant which includes the minority carrier diffusion length and the quantum efficiency for pair production as parameters (Heijne, 1968). In any real material, barriers are likely to be randomly distributed so that no net emf will occur under uniform illumination, though the lowering of the impedance may well be important (Wintle, 1965). There are likely to be barrier layers of some sort at the metal–insulator contacts, and if the illumination is directed through one of the contacts, the front and rear barriers will in fact be differently illuminated and a net photovoltage will be produced.

A wide variety of effects can occur in organic systems, of which the least likely would seem to be the straightforward increase of conductivity with illumination such as is seen in inorganic materials.

VI. Absorption Currents, Electrets, and Trapping

On applying a dc field to a dielectric, it is found that there is a rapid charging spike which corresponds to the capacitative charging. This capacitance can be deduced by extrapolating the usual audio- and radiofrequency loss measurements to zero frequency. There, is in addition, a long-term slowly decaying current flow which varies with time roughly as t^{-n} where n is an exponent near unity. This long-lasting current is known as the absorption current or the anomalous current. This current flow frequently obscures the real conduction current (Adamec, 1970), and clearly, the 1-min electrification time prescribed in standard tests (ASTM, 1966) often leads to erroneous treatment of the data. On removal of the field, a similar desorption current occurs, and normally the principle of superposition is obeyed. The apparent dielectric constant ε_{app} is given by

$$\varepsilon_{app} = \int i \, dt / V C_{geom}$$

where i is the current, V the applied voltage, and C_{geom} the capacitance of a vacuum capacitor with the same electrodes. Since ε_{app} frequently amounts to several tens, while many of the materials tested are only weakly polar, this effect cannot arise from slow dipolar orientation as has been suggested in the past. It is more likely that it arises from heteropolar space charge separation, possibly due to ionic rather than electronic motion, though the observed linearity with field tends not to support this view. The interfacial polarization (Maxwell–Wagner) effect has also been proposed as the cause of absorption currents, but the heterogeneous sample should then also give Debye relaxation peaks at low frequencies, and the experimental evidence is inconclusive. It may be that space charges set up by the contact potential differences (Davies, 1969) are slowly distorted under the applied field, thus inducing a change in the charge density on the electrodes and a corresponding external current flow.

In the previous paragraphs it was assumed by implication that there was no dependence of specimen behavior on previous history. In fact, there is normally a considerable electrical hysteresis, and when the time constants associated with the decay of the memory become of the order of days the material is referred to as an electret. This type of charge storage has also been called persistent internal polarization in the literature. There are two main effects:

1. A heterocharge, in which an electric polarization P is set up in the body of the dielectric due both to orientation of dipoles and to separation of the free charges (usually taken to be ions in the discussion of electrets though electronic charges could equally well be involved) into space charges, positive near the cathode and negative near the anode
2. A homocharge, which is due to the injection from the electrodes of charges of like sign into the material

The work of Davies (1969) shows clearly that charge exchange always exists at the electrodes so that both hetero- and homocharge formation occur at the same time. These processes, particularly the homocharge formation, are assisted by illumination at appropriate wavelengths (Goodman, 1968) and can give rise to strong internal fields. Electron bombardment can also yield electrets (Sessler and West, 1969), as well as the more dramatic treeing phenomena. Since most polymers show strong trapping, the space charges will be relatively stable and thus form long-lived electrets (Fridkin and Zheludev, 1966). The collapse of these charges can give valuable information on the trap distribution, though a single trapping level may not give any discharge current (Lindmayer, 1965).

It has widely been assumed that the distribution of traps in polymers follows the exponential dependence on energy depth proposed by Rose (1951):

$$N(E) = A \exp(-E/kT_1)$$

where $N(E)$ is the concentration per unit energy width, E the depth of the trap below the transport band, and A, T_1 constants describing the particular sample studied. It is implicitly assumed that random conformations of the long-chain molecules provide a selection of potential wells which serve as traps. Typical values of A and T_1 derived from radiation-induced conductivity measurements are 10^{18} cm^{-3} and $1000°K$, but there is unfortunately no theoretical background to support these figures or to estimate the cross sections. Recent observations by Schmidt and Allen (1970) on liquids tend to cast doubt on the idea of potential wells. At the same time thermoluminescence studies have revealed only a set of four discrete traps in polyethylene but no trap distribution. Of these, one level arises from O_2^- which is bound to the hydrocarbon chain by a charge–polarizibility interaction (Boustead, 1970), while the remainder appear to come from the crystalline region, the amorphous region, and the crystallite surface and correspond to trapping of electrons on methylene groups (Partridge, 1965), with the activation energy of $\simeq 0.4$ eV above the second structural transition corresponding to carrier mobility along the chain. While this energy agrees well with the apparent activation energy of the radiation-induced conductivity, a direct identification of the two energies is not compatible with the intensity dependence of the latter process. Thus, even in this well-investigated material, the nature of the trapping sites and their energy distribution must still be regarded as only partially resolved. There is evidence (Zichy, 1967) that small organic molecules act as efficient charge scavengers, thus supporting the view that charges in polymer films are, on the whole, rather weakly bound.

VII. Related Processes

One of the most important industrial aspects of polymers concerns the formation and decay of static electrification. It is clear from recent work (Davies, 1969) that the production of static charge may well be largely controlled by contact potential differences and that large static charges are dissipated by time-varying space charge limited currents into the bulk (Wintle, 1970). The longevity of static charges is closely related to the electret process (Zichy, 1967).

As with inorganic compounds, organic materials containing luminescent centres can be stimulated electrically, and the electroluminescent output from several polymers has been studied by Hartman and Armstrong (1967). There seem to be no essential differences between these and more conventional phospors.

Much information has been derived about the conduction mechanisms in molecular crystals by using various electolytic contacts to change the carrier injection conditions. Swan (1967a, b) has observed enhanced conductivity in polyethylene exposed either to iodine-rich solutions or to iodine vapor, but there seems to be no associated injection. It appears possible that the iodine occupies the carrier trapping sites, thus reducing the carrier trapping and so increasing the macroscopic mobility [Eq. (6)], or alternatively that iodine ions act as carriers, though this is somewhat surprising if the spectroscopic evidence of an iodine–alkane interaction is accepted. Some remarkable current oscillations occur, and it is clear that a fertile field of research has been opened up by this work.

A number of interesting experiments have been performed on glow discharge films. Naturally these films are not as well characterized as regular polymer materials, so it is hard to give any theory to account for the effects observed, except to note that direct interelectrode tunneling (Hill, 1967) is likely to occur in the thinnest specimens. The observations of dark conductivity (Mann, 1964), photoconductivity (Colburn and Christy, 1969), secondary emission (Buckman and Bashara, 1966), and photoemission (Kronick, 1968) are indicative of the range of work being undertaken.

REFERENCES

Adamec, V. (1970). *Koll.-Z. Z. Polymere* **237**, 219.
André, J. M. (1969). *J. Chem. Phys.* **50**, 1536.
ASTM (1966). *Standard Methods of Test for D–C Resistance or Conductance of Insulating Materials*, D257–66, Sect. 10.2. ASTM, Philadelphia, Pennsylvania.
Berets, D. J. and Smith, D. S. (1968). *Trans. Faraday Soc.* **64**, 823.
Binks, A. E., Campbell, A. G., and Sharples, A. (1970). *J. Polymer Sci. A2* **8**, 529.
Boustead, I. (1970). *Nature* **225**, 846.
Brophy, J. J., and Buttrey, J. W. (eds.) (1962). "Inter-industry Conference on Organic Semiconductors." Macmillan, New York.
Buckman, A. B., and Bashara, N. M. (1966). *Phys. Rev. Lett.* **17**, 577.
Caserta, G., Rispoli, B., and Serra, A. (1969). *Phys. Status Solidi* **35**, 237.
Colburn, R. H., and Christy, R. W. (1969). *J. Appl. Phys.* **40**, 3958.
Davies, D. K. (1967). "Static Electrification," pp. 29–36. Inst. of Phys. and the Phys. Soc., London.
Davies, D. K. (1969). *J. Phys. D.* **2**, 1533.
Foss, R. A., and Dannhauser, W. (1963). *J. Appl. Polymer Sci.* **7**, 1015.
Fowler, J. F. (1956). *Proc. Roy. Soc. London* **A236**, 464.
Fox, D., Labes, M. M., and Weissberger, A. (eds.) (1963). "Physics and Chemistry of the Organic Solid State," Vols. 1, 2, 3. Wiley (Interscience), New York.
Fridkin, V. M., and Zheludev, I. S. (1966). "Photoelectrets and the Electrographic Process." Van Nostrand, Princeton, New Jersey.
Glaeser, R. M., and Berry, R. S. (1966). *J. Chem. Phys.* **44**, 3797.
Glarum, S. H. (1963). *J. Phys. Chem. Solids* **24**, 1577.
Good, R. H., and Müller, E. W. (1956). "Handbuch der Physik" (S. Flügge, ed.), Vol. 21, pp. 176–231. Springer, Berlin.
Goodman, A. M. (1968). *J. Electrochem. Soc.* **115**, 276C.
Gutmann, F., and Lyons, L. E. (1967). "Organic Semiconductors." Wiley, New York.
Hartman, W. A., and Armstrong, H. L. (1967). *J. Appl. Phys.* **38**, 2393.
Heijne, L. (1963). *Philips Tech. Rev.* **25**, 120.
Heijne, L. (1966). *Philips Tech. Rev.* **27**, 47.
Heijne, L. (1968). *Philips Tech. Rev.* **29**, 221.
Herspring, A., and Oster, A. (1968). *Koll.-Z. Z. Polymere* **226**, 103.
Herzenberg, A., and Modinos, A. (1964). *Biopolymers* **2**, 561.
Hill, R. M. (1967). *Thin Solid Films* **1**, 39.
Hiroike, K. (1965). *Phys. Rev.* **138**, A422.
Jonscher, A. K. (1967). *Thin Solid Films* **1**, 213.
Kallmann, H., and Pope, M. (1960). *Nature* **186**, 31.
Kallmann, H., and Silver, M. (eds.) (1961). "Symposium on Electrical Conductivity in Organic Solids." Wiley (Interscience), New York.
Katz, J. I., Rice, S. A., Choi, S., and Jortner, J. (1963). *J. Chem. Phys.* **39**, 1683.

Kemeny, G., and Rosenberg, B. (1970). *J. Chem. Phys.* **52**, 4151.
Kittel, C. (1966). "Introduction to Solid State Physics," 3rd ed. Wiley, New York.
Kommandeur, J. (1961). *J. Phys. Chem. Solids* **22**, 339.
Kronick, P. L. (1968). *J. Appl. Phys.* **39**, 5806.
Kryszewski, M., Szymanski, A., and Włohowicz, A. (1968). *J. Polymer Sci. C* No. 16, 3921.
Lakatos, A. I., and Mort, J. (1968). *Phys. Rev. Lett.* **21**, 1444.
Lampert, M. A. (1964). *Rep. Progr. Phys.* **27**, 329.
Lengyel, G. (1966). *J. Appl. Phys.* **37**, 807.
Lindmayer, J. (1965). *J. Appl. Phys.* **36**, 196.
Lyons, L. E., and Mackie, J. C. (1962). *Proc. Chem. Soc.* February, p. 71.
Mann, H. T. (1964). *J. Appl. Phys.* **35**, 2173.
Many, A., Simhony, M., Weisz, S. A. and Levinson, J. (1961). *J. Phys. Chem. Solids.* **22**, 285.
Mark, P., and Hartman, T. E. (1968). *J. Appl. Phys.* **39**, 2163.
Martin, E. H., and Hirsch, J. (1969). *Solid State Commun.* **7**, 783.
Martin, E. H., and Hirsch, J. (1970). *J. Non-Cryst. Solids* **4**, 133.
McCubbin, W. L. (1963). *Trans. Faraday Soc.* **59**, 769.
McCubbin, W. L., and Manne, R. (1968). *Chem. Phys. Lett.* **2**, 230.
Mehl, W., and Funk, B. (1967). *Phys. Lett.* **25A**, 364.
Moorjani, K., and Feldman, C. (1964). *Rev. Mod. Phys.* **36**, 1042.
Mott, N. F. (1969a). *Contemp. Phys.* **10**, 125.
Mott, N. F. (1969b) *Phil. Mag.* **19**, 835.
Mott, N. F., and Gurney, R. W. (1948). "Electronic Processes in Ionic Crystals," 2nd ed. Oxford Univ. Press, London.
Mulder, B. J. (1968a). *Philips Res. Rep. Suppl.* No. 4.
Mulder, B. J. (1968b). *Philips Res. Rep.* **22**, 553.
Murrell, J. N. (1959). *Discuss. Faraday Soc.* **28**, 36.
Murrell, J. N. (1961). "Symposium on Electrical Conductivity in Organic Solids" (H. Kallmann and M. Silver, eds.), pp. 127–145. Wiley (Interscience), New York.
Ofran, M., Oron, N., and Weinreb, A. (1969). *J. Chem. Phys.* **50**, 3131.
Partridge, R. H. (1965). *J. Polymer Sci. A* **3**, 2817.
Partridge, R. H. (1967). *Polymer Lett.* **5**, 205.
Partridge, R. H. (1968). *J. Chem. Phys.* **49**, 3656.
Partridge, R. H. (1970). *J. Chem. Phys.* **52**, 2485.
Pohl, H. A., and Chartoff, R. P. (1964). *J. Polymer Sci.* **2**, 2787.
Pulfrey, D. L., Shousha, A. H. M., and Young, L. (1970). *J. Appl. Phys.* **41**, 2838.
Rose, A. (1951). *R.C.A. Rev.* **12**, 362.
Rose, A. (1955). *Phys. Rev.* **97**, 1538.
Rose, A. (1963). "Concepts in Photoconductivity and Allied Problems." Wiley (Interscience), New York.
Schmidt, W. F., and Allen, A. O. (1970). *J. Chem. Phys.* **52**, 4788.
Seanor, D. A. (1968). *J. Polymer Sci.* A2 **6**, 463.
Sessler, G. M., and West, J. E. (1969). *Polymer Lett.* **7**, 367.
Smith, F. S., and Scott, C. (1966). *Brit. J. Appl. Phys.* **17**, 1149.
Swan, D. W. (1967a). *J. Appl. Phys.* **38**, 5051.
Swan, D. W. (1967b). *J. Appl. Phys.* **38**, 5058.
Tanaka, T., and Inuishi, Y. (1966). *Jap. J. Appl. Phys.* **5**, 974.
Tanaka, T., and Inuishi, Y. (1967). *Jap. J. Appl. Phys.* **6**, 1371.
Tobin, M. C. (1962). *J. Chem. Phys.* **37**, 1156.
Tredgold, R. H. (1962). *Proc. Phys. Soc. London* **80**, 807.

van Beek, L. K. H., and van Pul, B. I. C. F. (1962). *J. Appl. Polymer Sci.* **6**, 651.
Vermeulen, L. A., Wintle, H. J., and Nicodemo, D. A. (1971). *J. Polymer Sci.* A2 **9**, 543.
Vermeulen, L. A., and Wintle, H. J. (1970). *J. Polymer Sci.* A2 **8**, 2187.
Wannier, G. H. (1949). *Phys. Rev.* **76**, 438.
Williams, F. W., and Matthews, N. F. J. (1969). *Phys. Rev.* **180**, 864.
Wintle, H. J. (1964). *Photochem. Photobiol.* **3**, 249.
Wintle, H. J. (1965). *Photochem. Photobiol.* **4**, 803.
Wintle, H. J. (1967). *Photochem. Photobiol.* **6**, 683.
Wintle, H. J. (1970). *J. Appl. Phys.* **41**, 4004.
Wintle, H. J., and Charlesby, A. (1962). *Photochem. Photobiol.* **1**, 231.
Zichy, E. L. (1967). "Static Electrification." pp. 52–59. Inst. of Phys. and the Phys. Soc., London.

8

Electrical Conductivity of Irradiated Polymers

Péter Hedvig

Research Institute for Plastics, Budapest, Hungary

I. Radiation-Induced Conductivity in Polymers

A. INTRODUCTION

The electrical conductivity of insulators is known to increase upon exposing them to high-energy radiation (see, for example, Charlesby, 1960). This is generally termed as radiation-induced conductivity or X-ray, γ-ray, neutron "photoconductivity." The reason of this effect is obvious: It is due to the mobile charge carriers, ions, electrons, and holes, created by irradiation.

The primary effect of high-energy irradiation is evidently the production of photoelectrons and Compton electrons. The velocity distribution of these electrons is anisotropic; velocities are highest in the direction of the incident gamma photons. An oriented beam of charge carriers is formed this way, representing a directed current which can be observed without the application of an external electrical field. This zero-field radiation-induced conductivity is only a part of the total induced conductivity since the primary photo-electrons and Compton electrons lose their energy in a series of ionization processes resulting in a large number of δ electrons having isotropic velocity distribution in the sample. The charge carriers released from traps by the thermal energy also exhibit random distribution. The contribution of these charge carriers can be measured by applying an external electric field.

The general response to a radiation pulse of a polymeric isulator subjected to a constant dc voltage is illustrated in Fig. 1. The dark conductivity level

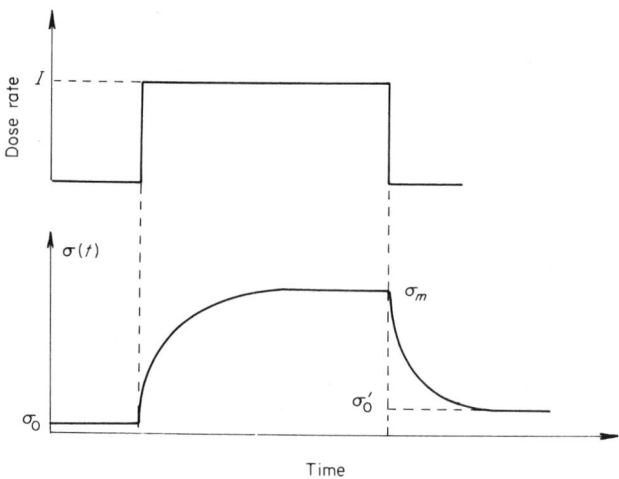

Fig. 1. Response of a polymer to a radiation pulse.

σ_0 increases continuously to a quasi-stationary level σ_m by irradiating the sample and decreases to a level σ_0' when irradiation is stopped. If the total radiation dose is not too high σ_0' equals the dark conductivity σ_0. At high doses, usually above 10 Mrad, the dark conductivity level may change appreciably, indicating irreversible structural changes in the polymer. The shape and height of the induced conductivity response pulse of Fig. 1 is determined by the rate of formation, recombination, and trapping of the charge carriers. The problem can be treated phenomenologically by using the formalism developed for describing photoconductivity in solids [see, for example, Rose (1951), Voul (1961), Hedvig (1964)] without the necessity of using any particular conductivity models. By analyzing the induced conductivity response pulses measured at different radiation dose rates, preirradiation doses, temperatures, and radiation pulse lengths, important information about the trapping of charge carriers is deduced. Correspondingly, this method is interesting not only in such cases when polymers are used in radiation fields, as in nuclear reactors or in space research, but in general when the chemical and physical effects of electrons, holes, trapped ions, and radicals are studied.

In this chapter a very brief review of the experimental technique and main results of radiation-induced conductivities will be given without detailed theoretical discussion.

B. EXPERIMENTAL TECHNIQUE

The general scheme for measuring induced conductivities is shown in Fig. 2. The radiation source can be an X-ray tube, an electron accelerator, an isotope γ-ray source such as ^{60}Co, or a neutron source (a reactor or neutron generator). In order to produce definite radiation pulses a shutter

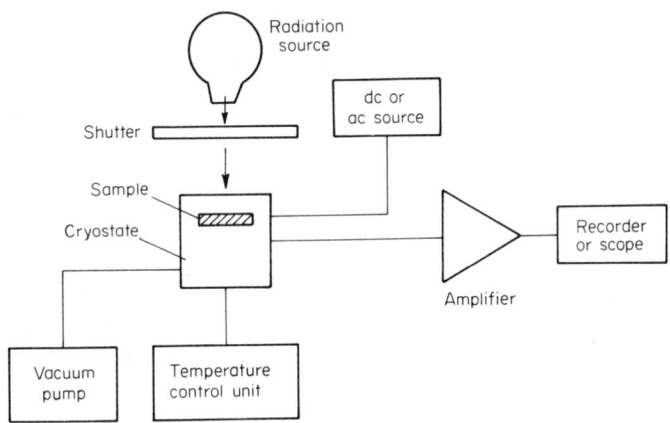

Fig. 2. General scheme for measuring radiation-induced conductivities.

or chopper is used. In the case of X-rays and neutrons (Coppage *et al.*, 1963) such a chopper is easy to construct. ^{60}Co γ-ray pulses are usually produced by using a pneumatic system which lifts the source in position and removes it in a fraction of a second (Harrison, 1963).

The pulsed radiation passing through an adequate window (usually aluminum or beryllium) reaches the sample, which is in high vacuum in order to minimize ionization currents, and is placed on a thermostated copper or aluminium block. A particular sample holder developed in the author's laboratory (Hedvig, 1969) is shown in Fig. 3. This device can be easily thermostated between $-150°$ and $+200°C$ with an accuracy of $\pm0.2°C$. The current flowing through the sample is measured by a recording electrometer. When short X-ray pulses are applied, the response pulses are displayed by a cathode-ray oscilloscope.

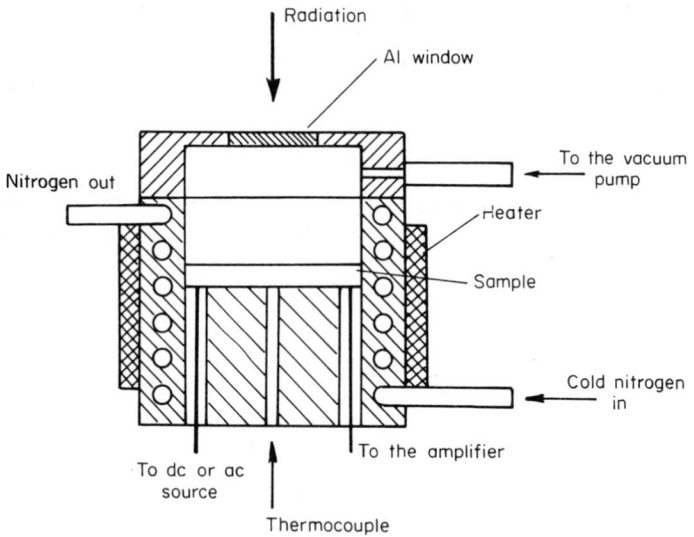

Fig. 3.   Cell for measuring X-ray-induced conductivities in a wide temperature range.

This scheme can be used with any radiation source; just the construction and material of the shutter are to be changed accordingly. In the case of linear accelerators, no shutter is required since radiation is electronically pulsed.

It is essential to guard the voltage leads from radiation carefully and to prevent ionization of air near the surface of the sample. This is why a vacuum chamber is needed. When measurement in vacuum is inconvenient, the sample can be immersed in silicon oil or covered with silicon grease (Hedvig, 1967). For electrodes, vacuum-evaporated aluminum, gold, silver, or indium can be used. The equipment shown in Figs. 2 and 3 is also useful for measuring time-dependent electrical polarization currents during the course of irradiation. It is also possible to measure dielectric spectra and ac conductivities under irradiation this way by replacing the electrometer by a recording dielectric spectrometer (Hedvig, 1969) and the dc source by a low frequency ac generator. Dielectric spectra and ac conductivities in this case can be easily recorded as a function of temperature. The ac technique is very useful because in some cases it is hard to distinguish between the dielectric polarization currents arising from trapped ions and ohmic (electron–hole) currents. By measuring the temperature dependence of the induced ac conductivities, the dielectric and ohmic parts of the conductivity can be separated. The total ac conductivity is expressed as follows:

$$\sigma(\omega, T) = \sigma_0(T) + \frac{\omega}{4\pi}\,\varepsilon''(\omega, T) \tag{1}$$

where ω is the angular frequency and $\sigma_0(T)$ is the ohmic conductivity which is an exponential function of the temperature

$$\sigma_0(T) = \sigma_0^0 \exp\left[-\frac{E_0}{kT}\right] \tag{2}$$

where E_0 is the activation energy of the conduction, k is the Boltzman constant, σ_0 is the ohmic conductivity extrapolated to infinite temperature, and $\varepsilon''(\omega, T)$ is the imaginary part of the dielectric permittivity of the polymer, which exhibits maxima at the dispersion regions (see Chapter 5).

By using the experimental setup of Fig. 1 it is also possible to measure persistent polarization, electret depolarization currents (Murphy *et al.*, 1963), and thermostimulated currents (TSC) after irradiation (Talroze and Frankevich, 1959).

For separating the zero-field-induced conductivity from the total, Meyer *et al.* (1956) developed a three-sample sandwich method shown in Fig. 4. The photoelectrons and Compton electrons formed by the incident gamma photon $h\nu$ move in sample 1 in the direction of the incident beam. This results in a current I_1. In sample 2 the corresponding current is I_2. An external voltage applied to sample (2) produces a current I^+ or I^-, depending on its polarity. The total current recorded by the ammeter is thus for the two polarities of the external field

$$I_{\text{total}}^{(+)} = I_1 - I_2 - I^+ \tag{3}$$

$$I_{\text{total}}^{(-)} = I_1 - I_2 + I^- \tag{4}$$

when the external voltage is zero

$$I_{\text{total}}^0 = I_1 - I_2 \tag{5}$$

By measuring the total currents $I_{\text{total}}^{(+)}$, $I_{\text{total}}^{(-)}$, and I^0 at positive, negative, and zero fields, the individual currents I_1, I_2, I^+, and I^- can be determined from Eqs. (3)–(5). When the three samples are of equal material, the direct

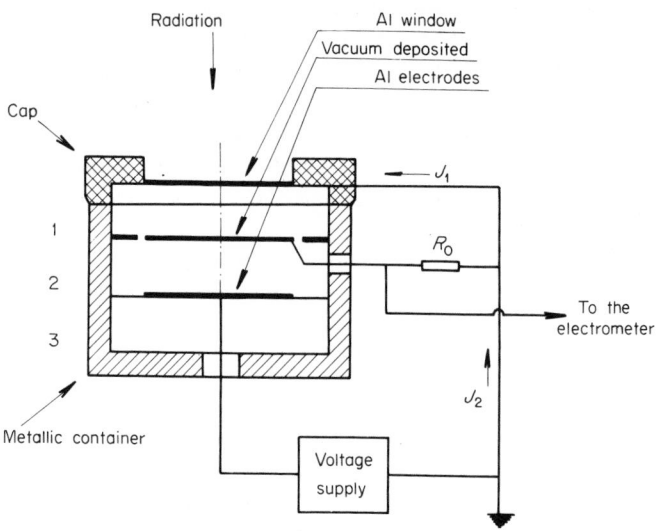

Fig. 4. Sandwich cell for measuring directed (zero field) and undirected induced currents. After Meyer *et al.* (1956). Sample 3 is in contact with the metallic container which is earthed.

currents I_1 and I_2 are compensated; thus, the induced current arising from charge carriers having isotropic velocity distribution can be measured. When the samples are different, the directed components I_1 and I_2 are different and can be determined. Meyer *et al.* (1956), for example, used ceresin wax for samples 1 and 3 and polytetrafluoroethylene for sample 2. This way, a number of polymers can be studied with reference to the same sandwiching material: the wax.

C. REVIEW OF THE EXPERIMENTAL DATA

The first measurement on radiation-induced conductivity of polymers was done by Gemant (1940, 1949) on paraffin irradiated by β-rays. The first systematic study on plastics was made by Fowler and Farmer (1953); their results and the subsequent pioneering work have been reviewed by Charlesby (1960). In this stage of work mainly X-ray- and γ-ray-induced conductivities were studied. Meyer *et al.* (1956) separated the directed Compton current from the undirected currents in polyethylene and in polytetrafluoroethylene. Coppage *et al.* (1963) measured neutron-radiation-induced conductivities in polystyrene and in polyisobutylene. Meyer *et al.* (1956) and Yahagi and Danno (1963) measured the temperature dependence of the ^{60}Co γ-ray-induced conductivities in polytetrafluoroethylene and in polyethylene in a wide temperature range.

Induced conductivities were measured by the author during the solid state polymerization of N-vinylsuccinimide (Hedvig, 1967). Most recently, a series of polymers was studied by Sichkar (1970). A new class of extremely radiation-resistant polymers, the polyimidazopyrrolines, was studied by Reucroft (1970).

The main experimental facts established so far are summarized as follows.

1. The quasi-stationary conductivities σ_m depend on the irradiation dose rate as

$$\sigma_m = \sigma_0 + A \cdot I^\delta \qquad (6)$$

where σ_0 is the dark conductivity, A and δ are constants, and I is the radiation dose rate. The exponent δ is found to vary between 0.5 and 1 for different polymers (Fowler, 1956). At high dose rates a deviation from Eq. (6) is observed (Hedvig, 1964): The induced conductivity $\sigma_m - \sigma_0$ plotted in logarithmic scale against the dose rate exhibits a break at high dose rates. An example of this is shown in Fig. 5.

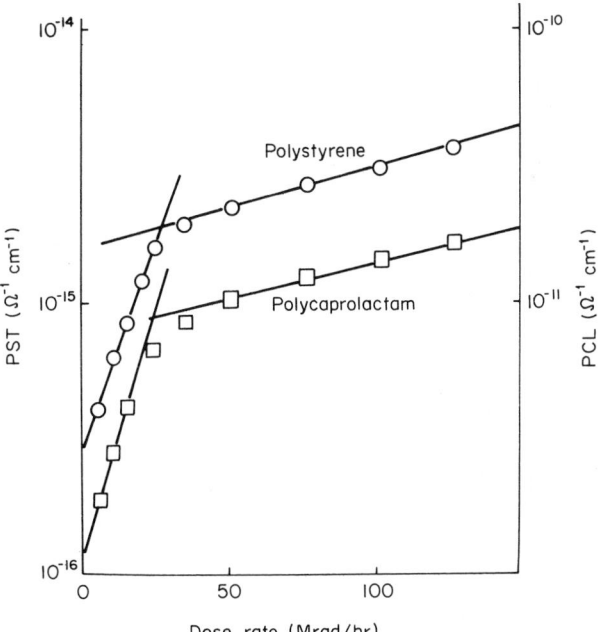

Fig. 5. The dose rate dependence of radiation-induced conductivities in polystyrene and in polycaprolactam. Temperature, 24°C.

2. The directed, zero-field conductivities measured by the sandwich technique of Meyer *et al.* (1956) are higher than the undirected conductivities in the dose rate range of 10–1000 rad/min. In a three-layer polyethylene sandwich cell the radiation-induced currents are found to be strictly proportional to the applied voltage (Meyer *et al.*, 1956). This linearity is observed in single-layer cells, too (Yahagi and Danno, 1963).

3. The conductivity response pulses depend on the preirradiation dose of the sample. It has been shown by the author (Hedvig, 1964) that there is a correlation between the decrease of the quasi-stationary conductivity level σ_m and the concentration of trapped free radicals formed by preirradiation.

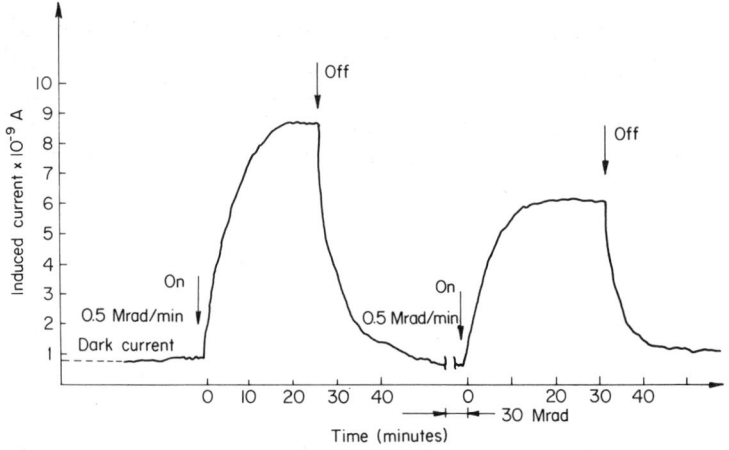

Fig. 6. Effect of preirradiation on the radiation-induced conductivities in polycaprolactam. Temperature, 24°C; dose rate, 0.5 Mrad/min; preirradiation dose, 20 Mrad.

An example for this is shown in Fig. 6 for polycaprolactam. It has also been shown that this change of the σ_m level can be decreased by adding radical acceptors to the system.

The induced conductivity level σ_m is found to increase in time after a high dose preirradiation as the trapped radical concentration is decreased.

4. For the temperature dependence of the quasi-stationary conductivity σ_m the Arrhenius law [Eq. (1)] is obeyed only over a limited temperature range:

$$\sigma_m(T) - \sigma_0 = \sigma_m^0 \cdot \exp(-E_m/kT) \tag{7}$$

where σ_m is a constant, E_m is the activation energy of the induced conductivity, k is the Boltzmann constant, and T is the temperature. Below a certain temperature of about $-50\,°C$ the induced conductivity appears to be independent of temperature (Meyer *et al.*, 1956; Yahagi and Danno, 1963). An example of this is shown in Fig. 7.

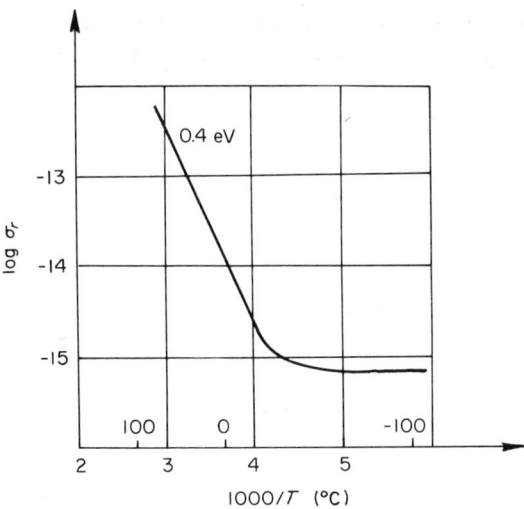

Fig. 7. The temperature dependence of the induced conductivity in polytetrafluoroethylene. After Meyer *et al.* (1956).

The activation energy of the induced conductivity in the high temperature range is found to be on the order of 0.1–0.3 eV for most polymeric insulators, while the corresponding activation energies for the dark conduction are within 1.5–2 eV. In the temperature range between $-196°$ and $-20\,°C$ almost all the induced conductivity is due to charge carriers activated by the irradiation; only a small fraction (about 10%) is estimated as being due to the carriers released from traps by the thermal energy. At higher temperatures and at relatively low radiation dose rates the contribution of the thermally released charge carriers becomes considerable.

5. The decay of the quasi-stationary conductivity σ_m as a function of time depends on the conductivity level at which irradiation has been stopped (i.e., the exposure time) and on the polymer exposed. The X- or γ-radiation-induced conductivity decay is generally expressed by Harrison (1963) as

$$\frac{\sigma(t)}{\sigma(0)} = \sum_{i=1}^{n} k_i \exp\left(-\frac{t}{\tau_i}\right) \tag{8}$$

where $\sigma(t)$ is the conductivity at time t, $\sigma(0)$ is that at $t = 0$, and k_i is the weight factor for the decay constant τ_i.

As an illustrative example the γ-radiation-induced conductivity decay curves of polytetrafluoroethylene are shown in Fig. 8 for different $\sigma(0)$ values, i.e., for different exposure times [after Harrison (1963)]. The time constants and the corresponding weight factors calculated according to Eq. (3) are also shown for the case of 170-sec exposure time, when the quasi-stationary level σ_m is approximately reached, $\sigma_m \approx \sigma(0)$.

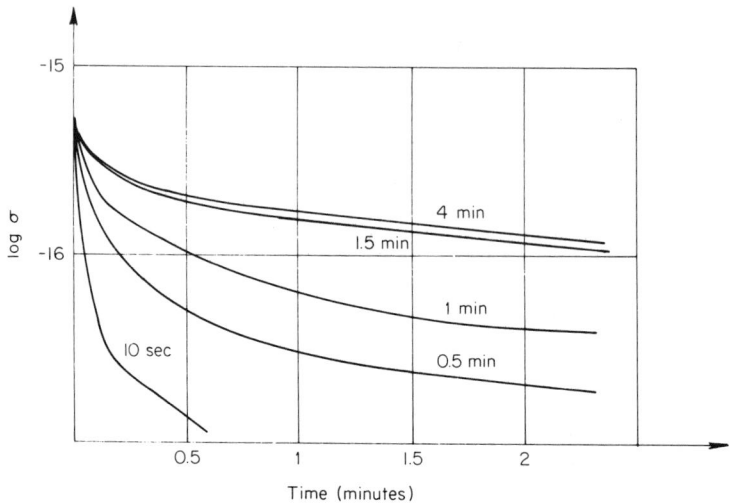

Fig. 8. The decay of γ-radiation-induced conductivity in polytetrafluoroethylene at 60°C. ^{60}Co γ-ray dose rate was 3.3 rad sec^{-1}. After Harrison (1963).

6. It has been shown by Adamec (1964, 1965) that the time-dependent polarization curves exhibit significant changes by irradiation. In these experiments the current flowing through the sample is recorded as a function of time under continuous irradiation at constant temperature. It has been found that the rate of the decay of the polarization current becomes appreciably lower by increasing the irradiation dose rate. This means that not only the ohmic conductivity but the polarization as well is affected by the irradiation.

The art of studying radiation-induced conductivities is not very well developed. Only the basic experimental facts have been established so far and only the most common polymers have been investigated. Since the technique is not very complex, it may be hoped that more experimental data will be collected in the near future which should make it possible to study the nature of charge carrier traps in polymers. Preliminary experiments indicate that the trapped and reactivated electrons and ions may have an important role in the radiation chemistry of polymers. The method of induced conductivity may be developed to become a basic tool for investigating these effects. The polymers investigated so far with this technique are collected in Table I.

TABLE I

POLYMERS STUDIED FOR RADIATION-INDUCED CONDUCTIVITY

Polymer	Reference
Paraffin	Gemant (1940)
	Gemant (1949)
	Frankevich (1967)
Polyethylene	Mayburg and Lawrence (1952)
	Coleman (1953)
	Fowler and Farmer (1955)
	Fowler (1956)
	Wintle (1960)
	Harrison (1963)
	Vannikov (1963)
	Sichkar (1970)
Polyethylene terephthalate	Coppage et al. (1963)
	Adamec (1964)
Polypropylene	Adamec (1964)
	Sichkar (1970)
Polytetrafluoroethylene	Meyer et al. (1956)
	Fowler and Farmer (1955)
	Fowler (1956)
	Yahagi and Danno (1963)
	Harrison (1963)
	Adamec (1963)
	Sichkar (1970)
Polychlorotrifluoroethylene	Sichkar (1970)
Polystyrene	Armistead et al. (1949)
	Coleman and Bohm (1953)
	Coleman (1953)
	Fowler (1956)
	Fowler and Farmer (1955)
	Coppage (1962)
	Harrison (1963)
	Coppage et al. (1963)
	Adamec (1964)
	Sichkar (1970)
Polycarbonate	Coppage et al. (1963)
	Adamec (1964)

TABLE I (*Continued*)

Polymer	Reference
Kel-F	Harrison (1963)
Polyisobutylene	Coppage (1962)
	Coppage *et al.* (1963)
Cellulose acetate	Coppage *et al.* (1963)
Polyvinylchloride	Munick (1956)
	Herwig and Jenckel (1959)
	Harrison (1963)
	Sichkar (1970)
Polyamide	Harrison (1963)
	Hedvig (1964)
	Weisberg *et al.* (1969)
Polymethylmethacrylate	Fowler and Farmer (1955)
	Fowler (1956)
	Murphy *et al.* (1963)
	Sichkar (1970)
Poly-*N*-vinylsuccinimide	Hedvig (1967)
H-film	Coppage *et al.* (1963)
Polyimidazopyrrolone	Reucroft (1970)

II. Permanent Changes of the Conductivity after Irradiation

The physical and chemical structure of a polymeric system evidently will be changed when it is irradiated even with a relatively small dose. The structural changes are reflected in the changes of the electrical conductivity and polarization of the material. The study of such phenomena is interesting from a practical point of view since some polymeric electrical insulators must work under the action of high-energy radiation. The problem is interesting from the point of view of basic research, too, since radiolysis of solid polymers is evidently connected with formation, recombination, and trapping of charged particles.

A. Thermostimulated Currents and Electret Depolarization

When a polymer is irradiated, first of all the physical traps of the structure will be filled with electrons, holes, and ions formed by the irradiation. Some of these charge carriers are trapped at low temperatures and are released by increasing the temperature. The release of the charge carriers can be directly measured by recording the current flowing through the sample without applying any external voltage. This is referred to as thermostimulated current. This technique has been developed by Talroze and Frankevich (1959) and applied to polymeric systems by Kryszewsky *et al.*

(1965, 1967) and by Sichkar *et al.* (1967). A typical thermostimulated current recording for polyethylene is shown in Fig. 9. The release of the charge carriers from traps is indicated by current peaks at certain temperatures which roughly correspond to the structural transition regions of the polymer (cf. Chapter 5).

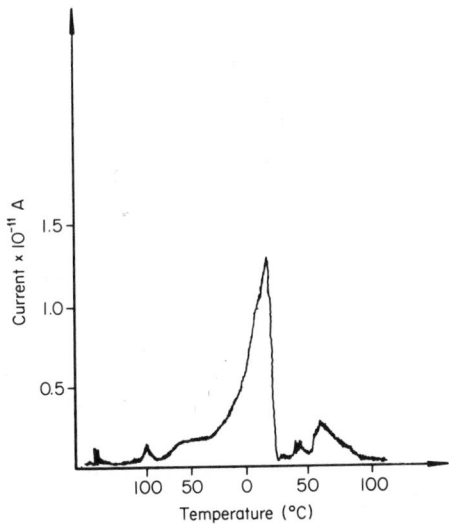

Fig. 9. Thermostimulated current in polyethylene. Irradiation, $-196°C$, 10 Mrad; rate of heating, $8°C/min$.

In some cases charge carriers may remain deeply trapped in the polymer even at high temperatures. This results in a persistent internal polarization of the sample, which can be easily measured by a conventional electrometer. Materials exhibiting persistent internal polarization are referred to as radioelectrets. By further increasing the temperature, the deeply trapped carriers can also be released, resulting in depolarization currents. This process is illustrated in Fig. 10 for a polytetrafluoroethylene radioelectret, formed by irradiating the sample at $-196°C$ under an electrical field of 100 kV/cm by a dose of 10 Mrad. By increasing the temperature, the depolarization current is recorded at a constant field. From the initial exponential rise of the depolarization current the activation energy of the process can be calculated approximately by the following equation:

$$\frac{I(T)}{I(T_0)} = \exp\left(\frac{E_d}{R} \cdot \frac{T - T_0}{TT_0}\right) \tag{9}$$

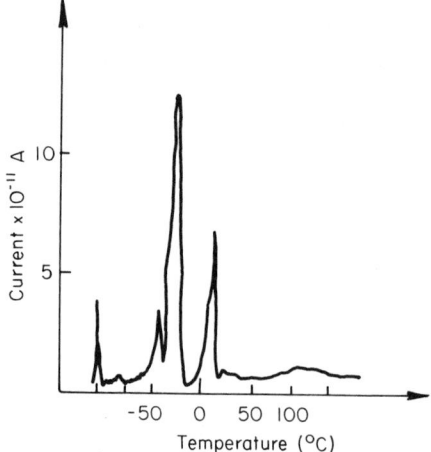

Fig. 10.   Electret depolarization curve of polytetrafluoroethylene irradiated at $-196°$C.

where $I(T)$ and $I(T_0)$ are the depolarization currents at temperatures T and T_0, respectively, R is the gas constant, and E_d is the activation energy for the depolarization, the value of which in this case is 0.9 eV (Murphy *et al.*, 1963).

The electret depolarization process can be more thoroughly studied by recording the depolarization current as a function of time at different temperatures. The time constants of these decay curves are expressed as

$$\tau = \tau_0 \exp(E_d/kT) \tag{10}$$

where the value of τ_0 is in the order of 10^{-10} sec (Fowler, 1956) and the calculated activation energy is again 0.9 eV for the polytetrafluoroethylene radioelectret.

From the data of thermostimulated current and electret depolarization measurements it is concluded that after irradiation quite a high concentration of charge carriers remain trapped in the polymer resulting in change of the polarization as well as in ohmic conductivities (e.g., through the hopping process). The chemical effects of these species have not been studied so far. The mechanism of the persistent internal polarization has been discussed in general by Freeman *et al.* (1961).

B.   CONDUCTIVITY DURING RADIATION DEGRADATION AND CROSS-LINKING

Appreciable permanent changes in the conductivity are expected as a result of changes in the physical and chemical structure of the polymers after irradiation. Most prominent changes are found as a result of such

radiation-initiated degradation processes which lead to formation of conjugated chains. The ac conductivity of unplasticized polyvinylchloride
(PVC), for example, as a function of degradation time can be linearized in
a semilogarithmic coordinate system according to the monomolecular
kinetic scheme. The ac conductivity level increases as the HCl-elimination
degradation of PVC proceeds. The rate constants and activation energies
measured this way agree with those measured by determination of the
evolved HCl. The activation energy of the thermal degradation of a $K = 70$
unplasticized PVC from the conductivity curves is found to be 32 kcal/mole
(Hedvig and Kisbényi, 1969). By irradiation at room temperature with
10 Mrad the conductivity level increases and the activation energy decreases
as thermal degradation is accelerated. As an example, the temperature
dependence of the total ac conductivity of unplasticized PVC is shown
before and after irradiation in Fig. 11. It is seen that the dielectric loss peak
is practically unchanged while the ohmic conductivity is enormously increased
after irradiation with 60 keV X-rays to a total dose of 10 Mrad. A similar
effect has been observed by recording the temperature dependence of the

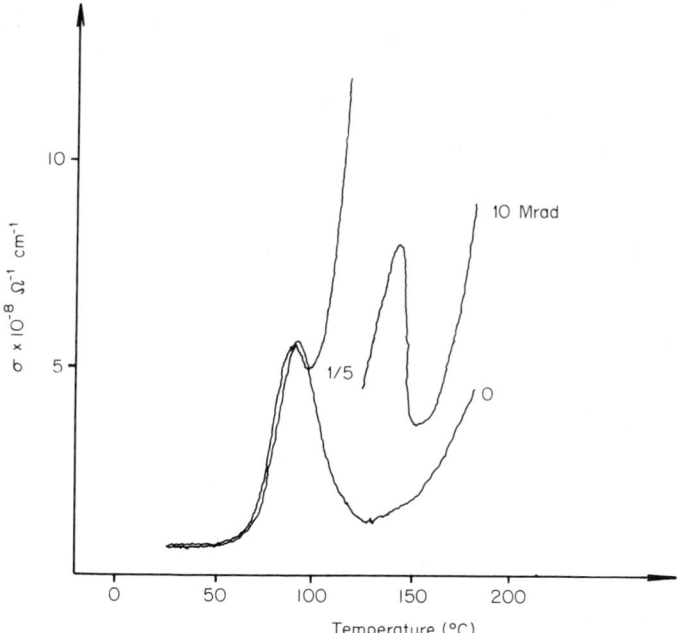

Fig. 11. The effect of irradiation on the total ac conductivity of unplasticized (K-65)
polyvinylchloride. Frequency, 1 kHz; rate of heating, 5°C/min; radiation dose, 10 Mrad at
30°C. Insert curve has values of the ordinate decreased by a factor of 5.

dc conductivities of plasticized PVC after irradiation (Kisbényi and Hedvig, 1969).

Another drastic permanent effect of radiation on the conductivities is observed during the course of thermal or radiation cross-linking of polyester resins.

It is known that polyester resins can be effectively cured by X- or γ-radiation at low temperatures. The cross-linking process results in change of the physical and chemical structure of the polymer leading to a sharp decrease in the conductivity.

This effect is primarily due to the change of the physical structure of the system. By measuring the total ac or dc conductivity of the resin as a function of the irradiation dose, a rapid decrease is observed which is correlated with the cross-linking reaction (Judd, 1965). When the total ac conductivity is recorded as a function of the temperature at different stages of the reaction, the dielectric peak corresponding to the glass-transition temperature of the resin is shifted to higher temperatures and the total conductivity level decreases. This indicates that during the course of cross-linking the mobilities as well as the concentration of the polar groups are changed. The ohmic part of the conductivity [$\sigma_0(T)$ in Eq. (1)] is also reduced by the cross-linking. The reaction can be followed this way practically to the latest stage of the cure when the resin is quite hard and the usual chemical thermal and thermomechanical methods are insensitive.

The radiation-induced permanent physical and chemical changes of the electrical conductivities of polymers have been very little studied. As our general knowledge on the mechanism of conductivity in polymers is slowly but continuously improved (cf. Chapters 7–10) it is hoped that a lot of radiation-induced structural changes will be followed and interpreted by using the combined dielectric spectroscopic and electrical conductivity technique.

REFERENCES

Adamec, V. (1963). *Dielectrics* 1, 159.
Adamec, V. (1964). *Int. J. Appl. Isotopes* 15, 477.
Adamec, V. (1965). *Proc. IEE.* 112, 405.
Armistead, F. C., Pennock, F. C., Mead, L. V. (1949). *Phys. Rev.* 76, 860.
Charlesby, A. (1960). "Atomic Radiation and Polymers." Pergamon Press, Oxford.
Coleman, J. H. (1953). *Nucleonics* 11, 42.
Coleman, J. H. and Bohm, D. (1953). *J. Appl. Phys.* 24, 497.
Coppage, F. N. (1962). AIEE, CP 62–1238.
Coppage, F. N., Snyder, A. W., Peterson, F. C. (1963). Neotron Effectiveness in Producing Photoconductivity in Dielectric Materials. Sandia Corp. reprint No. SCR–670.

Day, M. J., and Stein, G. (1951). *Nature (London)* **168**, 644.
Fallah, E. (1963). *Ind. Plast. Mod.* **15**, 37.
Farmer, F. T. (1942). *Nature (London)* **150**, 521.
Farmer, F. T. (1945). *Brit. J. Radiol.* **18**, 148.
Farmer, F. T. (1946). *Brit. J. Radiol.* **19**, 27.
Feng, P. Y. and Kennedy, J. W. (1955). *J. Amer. Chem. Soc.* **77**, 847.
Fowler, J. F. (1956). *Proc. Roy. Soc. (London)* **A236**, 464.
Fowler, J. F., and Day, F. T. (1955). *Nucleonics* **13**, (12) 52.
Fowler, J. F., and Farmer, F. T. (1945). *Brit. J. Radiol.* **18**, 148.
Fowler, J. F., and Farmer, F. T. (1946). *Brit. J. Radiol.* **19**, 27.
Fowler, J. F., and Farmer, F. T. (1953). *Nature* **171**, 1020.
Fowler, J. F., and Farmer, F. T. (1954a). *Nature* **173**, 317.
Fowler, J. F., and Farmer, F. T. (1954b). *Nature* **174**, 136.
Fowler, J. F., and Farmer, F. T. (1954c). *Nature* **174**, 800.
Fowler, J. F., and Farmer, F. T. (1955a). *Nature* **175**, 516.
Fowler, J. F., and Farmer, F. T. (1955b). *Nature* **175**, 590.
Fowler, J. F., and Farmer, F. T. (1955c). *Nature* **175**, 648.
Frankevich, E. L. (1962). *Izv. Akad. Nauk.* p.000, 1699.
Frankevich, E. L. (1967). *Khim. Vys. Energ.* **1**, 567.
Frankevich, E. L., and Yakovlov, B. S. (1965). *Zh. Fiz. Khim.* **37**, 1106.
Frankevich, E. L., and Talroze, V. L. (1962). *Tr. Vses. Sovesch. Radiocts. Khim. Izd. Akad. Nauk. SSSR. Moscow p.* 2651.
Freeman, J. R., Kallmann, H. P. and Silver, M. (1961). *Rev. Mod. Phys.* **33**, 553.
Gemant, A. (1940). *Phys. Rev.* **58**, 904.
Gemant, A. (1949). *J. Appl. Phys.* **20**, 887.
Gutmann, F., and Lyons, L. E. (1967). "Organic Semiconductors." Wiley, New York.
Harrison, S. E. (1963). Sandia Corp. Reprint, SCR–671.
Harrison, S. E., Coppage, F. N., and Snyder, A. W. (1963). IEEE, CPA 63–5156.
Hedvig, P. (1964). *J. Polymer Sci.* **A-22**, 4097.
Hedvig, P. (1967). *Proc. Tihany Symp. Radiat. Chem.*, *2nd* (J. Dobó, and P. Hedvig, eds.). Akadémiai Kiadó, Budapest.
Hedvig, P. (1969). "Electrical Conductivity and Polarization in Plastics" (in Hungarian). Akadémiai Kiadó, Budapest.
Hedvig, P., and Kisbényi, M. (1969). *Angew. Makromol. Chem.* **7**, 198.
Herwig, H. V., and Jenckel, E. Z. (1959). *Elektrochem.* **63**, 360.
Judd, N. C. W. (1965). *J. Appl. Polymer Sci.* **9**, 1743.
Kallmann, H., and Kramer, B. (1952). *Phys. Rev.* **87**, 91.
Kichinosuke, J., and Akibumi, D. (1963). *J. Appl. Phys.* **34**, 604.
Kisbényi, M., and Hedvig, P. (1969). *Europ. Polymer J.* Suppl. 291.
Kryszewski, M., and Szymanski, A. (1965). *Plaste Kautschuk* **12**, 642.
Kryszewski, M., Patora, J., and Szymanski, A. (1967). *Polymery Tworzywa Wielkoczasteczkove* **12**, 459 (in Polish).
Linder, E. G., and Rappaport, P. (1953). *Phys. Rev.* **91**, 202.
Liversage, W. E. (1952). *Brit. J. Radiol.* **25**, 434.
Mayburg, S., and Lawrence, W. L. (1952). *J. Appl. Phys.* **23**, 1006.
Meyer, R. A., Bouquet, F. L., and Alger, R. S. (1956). *J. Appl. Phys.* **27**, 1012.
Munick, R. J. (1956). *J. Appl. Phys.* **27**, 1114.
Murphy, P. V., Ribeiro, C., Hilanez, F., and de Moraes, R. G. (1963). *J. Chem. Phys.* **38**, 2400.

Nikolskii, V. G. (1963). *Fiz. Tverd. Tela* **5**, 2248.

Ramsey, N. (1953). *Nature (London)* **172**, 214.

Rappaport, P., and Linder, E. G. (1953a). *J. Appl. Phys.* **24**, 1110.

Rappaport, P., and Linder, E. G. (1953b). *Phys. Rev.* **91**, 202.

Reucroft, P. J. (1970). *J. Appl. Polymer Sci.* **14**, 1361.

Rose, A. (1951). *RCA Review* **12**, 363.

Rozman, I. M., and Tsimmer, K. G. (1956). *Zh. Techn. Fiz. USSR* **26**, 1681.

Seanor, D. A. (1968). *J. Polymer Sci.* A-2 **6**, 643.

Sichkar, V. P. (1970). Radiation Induced Electrical Conductivity in Polymers. Thesis in Russian. Moscow.

Sichkar, V. P., Weisberg, S. E., and Karpov, V. L. (1967). *Khim. Vys. Energ.* **1**, 561.

Sichkar, V. P., Weisberg, S. E., and Karpov, V. L. (1969). *Khim. Vys. Energ.* **3**, 438.

Sisman, O., and Bopp, C. D. (1951). *Oak Ridge Rept.* 928.

Sisman, O., and Bopp, C. D. (1953). *Oak Ridge Rept.* 1373.

Stark, H. H., and Garton, C. G. (1955). *Nature (London)* **176**, 1225.

Talroze, V. L. (1959). *Szv. Akad. Nauk SSSR* p. 000, 369.

Talroze, V. L., and Frankevich, E. L. (1959). *Dokl. Akad. Nauk SSSR* **129**, 859.

Vannikov, A. V. (1963). *Dokl. Akad. Nauk SSSR* **139**, 1339.

Voul, B. H. (1961). *Dokl. Akad. Nauk SSSR* **139**, 1339.

Warner, A. J., Muller, F. A., and Nordlin, H. G. (1954). *J. Appl. Phys.* **25**, 131.

Weisberg, S. E., Sichkar, V. P., Stepanov, V. F., and Karpov, V. L. (1969). *Vys. Soed. A* **11**, 2577.

Weisberg, S. E., Sichkar, V. P., and Karpov, V. L. (1969). *Khim. Vys. Energ.* **3**, 454.

Wieder, H. H., and Kaufman, S. (1953). *J. Appl. Phys.* **24**, 644.

Wintle, H. J. (1960). *Int. J. Appl. Radiat. Isotopes* **8**, 132.

Yahagi, K., and Danno, A. (1960). *J. Appl. Phys.* **31**, 734.

Yahagi, K., and Danno, A. (1963). *J. Appl. Phys.* **34**, 804.

9

ESR and Optical Studies of Trapped Electrons in Glasses and Polymers

Ronald M. Keyser,[a] Kozo Tsuji,[b] and Ffrancon Williams[c]

[a] *Union Carbide Corporation, Nuclear Division, Oak Ridge, Tennessee;*
[b] *Central Research Laboratory, Sumitomo Chemical Company, Ltd., Osaka, Japan;*
[c] *Department of Chemistry, University of Tennessee, Knoxville, Tennessee*

I. General Considerations

Macromolecular systems are usually characterized by a high viscosity, especially in the solid state at low temperatures. Consequently, there is a large probability that intermediate species generated during the irradiation of polymers remain trapped in the host material, their reaction rates being essentially limited by the slow diffusion processes that apply in such circumstances. Therefore, the study of trapped species represents an important area of investigation in the radiation chemistry of macromolecules. In this contribution we focus attention on recent work which provides direct spectroscopic evidence for electron trapping in glasses and polymers.

After a brief survey of trapped electrons in organic glasses (Section II), this section gives a comprehensive account of the characterization and properties of trapped electrons in γ-irradiated polymers, with particular reference to polyethylene. Much of the work to be presented here is based on the dissertation by Keyser (1970) which provides the only definitive study in the field. In retrospect, we can attribute the failure to detect trapped electrons in earlier work on polymers both to the ease with which these labile species are photobleached by visible and infrared light and to the need for careful ESR studies at microwave powers considerably lower than those normally used in the investigation of free radicals.

Electrons may be released into condensed materials by a number of different techniques, including γ-irradiation, photoionization, the use of alkali metal donors, and electric field injection. We shall restrict attention to the irradiation methods, and these are particularly suitable for generating electrons throughout a rigid matrix. Whenever excitation by high-energy electrons or photons results in ionization, an electron is released with sufficient kinetic energy to explore the molecular environment surrounding the positive ion. The process of electron thermalization has been discussed in Chapter 2, and here we are concerned specifically with the stabilized electrons that become separated from the positive ions in the condensed phase. Along with other authors (Eiben, 1970), we find it convenient to denote both solvated electrons in liquids and trapped electrons in solids as stabilized electrons. After charge separation has been achieved, the subsequent reactions of stabilized electrons appear to be governed largely by molecular or segmental diffusion. The practical importance of molecular motion is illustrated by the considerable variation in the lifetimes of stabilized electrons according to the viscosity of the medium. In glasses and polymers, these lifetimes are sufficiently prolonged to permit spectroscopic studies at leisure, whereas in liquids it is necessary to use flash photolysis or pulse radiolysis techniques to study these species.

Stabilized electrons are generally produced in condensed systems where the individual molecules do not possess low-lying orbitals capable of accepting electrons. Therefore, stabilized electrons must be clearly differentiated from simple radical anions formed by electron attachment to single molecules, and there are several lines of evidence (Dye *et al.*, 1970) which suggest that the excess electron is associated with several solvent molecules. As mentioned in Chapter 2, it is conceivable that electron stabilization can be achieved through the occupation of a supramolecular orbital built from the antibonding orbitals of several individual molecules. The basis for this suggestion comes from recent ESR studies on the radical anion of sulfuryl chloride (Kerr and Williams, 1971) and on the dimer radical anion of acetonitrile (Takeda and Williams, 1970; Sprague *et al.*, 1971). For both of these species, the ESR results are most satisfactorily interpreted by structures which imply that the unpaired electron resides in a supramolecular bonding orbital derived from the antibonding orbitals of two separate molecules. An extension of this concept to larger molecular aggregates could well provide new insight as to the nature of stabilized electrons.

In contrast to the above remarks, current theories of stabilized electrons (Copeland *et al.*, 1970) do not regard the excess electron as belonging directly to the molecular orbitals of the constituent molecules. Rather, the electron is assigned to a cavity or void and this entity polarizes the surrounding medium, which is regarded as a structureless continuum characterized by the static and high-frequency dielectric constants. This cavity or continuum model has been developed in considerable mathematical detail, and calculations using adjustable parameters can reproduce some of the experimental findings with fair precision. Thus, using a limiting cavity radius approaching zero, reasonable agreement between theory and experiment was obtained (Jortner, 1964) for the electronic transition energy of the hydrated electron corresponding to the maximum in the visible absorption band. However, such agreement obviously does not verify the physical basis of the cavity model. A related discussion on the nature of the trapped electron in polymers is given in Section III, F from which it is evident that at least some of the observations can be explained qualitatively in terms of a cavity model.

II. Trapped Electrons in Glasses

In summarizing recent work on trapped electrons in organic glasses of low molecular weight compounds, we shall emphasize saturated hydrocarbon systems because of their chemical resemblance to polyethylene and polypropylene. Our discussion will be chiefly concerned with studies by ESR and optical spectroscopy. Recent reviews which deal with various aspects of electron trapping in solids are those of Albrecht (1970), Eiben (1970), Ekstrom (1970), Hamill (1968), Kevan (1969), Ershov and Pikaev (1969), Whelan (1969), and Willard (1968).

Although the techniques of ESR and optical spectroscopy now provide the most reliable methods for the identification of trapped electrons, it should be pointed out that neither the optical nor the ESR spectrum really establishes the nature of the species from first principles. In particular, the lack of hyperfine structure in the ESR spectrum means that no information is directly available about the molecular environment of the trapped electron. Despite these inherent limitations, we shall see that in practice there is little or no ambiguity about the assignment of spectra to this "primitive" species on the bases of chemistry and phenomenology although serious mistakes were made in some investigations.

Definitive studies on trapped electrons in γ-irradiated aqueous and organic glasses date from 1962. Eiben (1970) has summarized much of the significant work on aqueous and alcoholic glasses. Methyltetrahydrofuran (MTHF) was the first organic glass to be investigated systematically (Ronayne et al., 1962). A broad structureless absorption band with λ_{max} at ≈ 1200 nm was observed after γ-irradiation at $77°$K, and it was found that the formation of this band could be largely suppressed by the presence of small concentrations (≈ 0.2 mole%) of biphenyl in the MTHF glass. Other additives such as naphthalene and tetracyanoethylene (TCNE) were shown to be about equally effective in depressing the yield of the broad optical absorption, and in each case the distinctive spectrum of the corresponding radical anion was observed instead. Moreover, the addition of carbon tetrachloride as a competitive electron scavenger to the MTHF–biphenyl system reduced the yield of the biphenyl radical anion, thereby confirming the process of electron attachment to biphenyl. From these experiments it was inferred that the broad absorption band induced in undoped MTHF is associated with electrons trapped in the glass. It is now well established that radiation-induced trapped electrons are produced in several aprotic glasses, as evidenced by a broad intense absorption with a maximum in the near-infrared region. This band is photobleached most readily at wavelengths which are significantly lower than the position of the absorption maximum. In one of the few quantitative studies (Dyne and Miller, 1965), the initial quantum efficiency for bleaching trapped electrons in MTHF was found to be less than 0.01 above 900 nm, but the efficiency increases sharply below this wavelength, attaining a value of 0.1 at 500 nm.

The absorption bands of trapped electrons in several saturated hydrocarbon glasses are shown in Fig. 1. These spectra were obtained (Lin et al., 1968) by difference from recordings before and after photobleaching with near-infrared light ($\lambda > 1000$ nm). Qualitatively, it is found that trapped electrons in saturated hydrocarbon glasses are much more sensitive to visible light than those in MTHF, and exposure even to subdued room light can result in the irreversible loss of these centers. As in the case of MTHF, no optical absorption was produced in the near infrared when 0.15 mole%

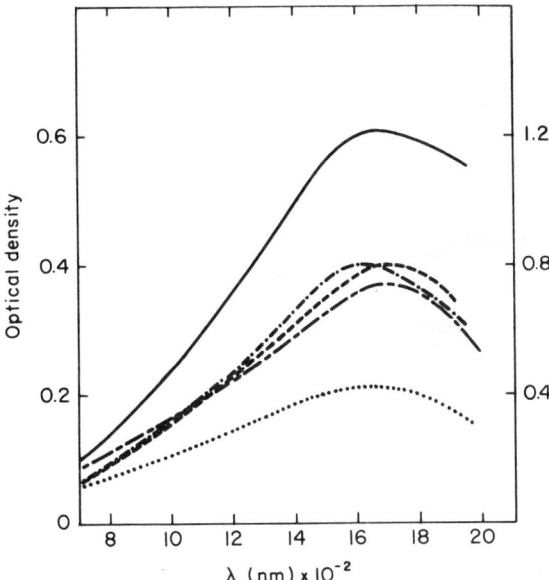

Fig. 1. Optical absorption spectra of γ-irradiated hydrocarbon glasses at 77°K.(—) 3-Methylhexane, λ_{max} 1650 nm, path 1 cm, dose 1.4 × 10^18 eV g^-1; (—·—) 3-methylpentane, λ_{max} 1600 nm, path 1 cm, dose 2.8 × 10^18 eV g^-1; (---) 3-methylheptane, λ_{max} 1700 nm, path 0.2 cm, dose 7.0 × 10^18 eV g^-1; (—·—) 4-methylheptane, λ_{max} 1700 nm, path 0.2 cm, dose 7.0 × 10^18 eV g^-1; (......) methylcyclohexane, λ_{max} 1650 nm, path 0.2 cm, dose 1.4 × 10^19 eV g^-1. The right-hand ordinate scale refers to the spectra of 3-methylpentane and methylcyclohexane. [Reproduced from *J. Amer. Chem. Soc.* **90**, 2766 (1968). Copyright (1968) by the American Chemical Society. Reprinted by permission of the copyright owner.]

biphenyl was present in the hydrocarbon glass, and again, the characteristic absorption band of the biphenyl radical anion (λ_{max} at 408 nm) was observed.

Trapped electrons are also detectable by ESR spectroscopy provided the measurements are carried out at sufficiently low microwave power. In glasses of MTHF, triethylamine, and saturated hydrocarbons, the ESR spectrum consists of a single line having a g factor of 2.0022 and a width (ΔH_{ms}) of about 4 G measured between the peaks of the first-derivative signal. This g factor is close to the free spin value (2.0023), but since many organic radicals have somewhat similar g factors, this parameter alone cannot be made the basis for identification. As illustrated in Fig. 2 by the ESR spectrum of γ-irradiated 3-methylpentane, the 4 G linewidth of the singlet spectrum from the trapped electron is much narrower than the linewidth associated with the anisotropic (powder) spectra of most organic radicals in the glassy

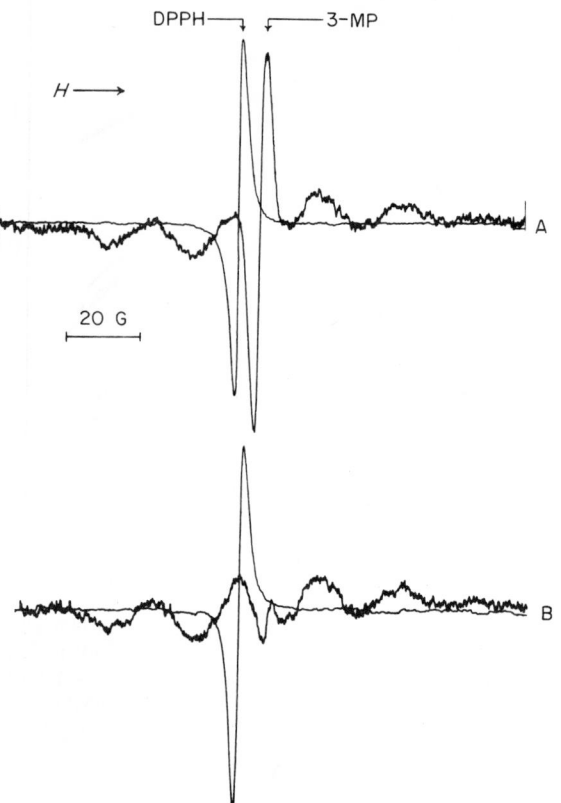

Fig. 2. ESR first-derivative spectra of γ-irradiated 3-methylpentane before (upper) and after (lower) photobleaching with near-infrared light. The spectra were recorded at low microwave power (0.01 mW) in a dual cavity at 77°K with DPPH as the reference signal. The sample was γ-irradiated at 71°K to minimize thermal decay. Reproduced from G. K.-W. Lok, M.S. thesis, Univ. of Tennessee, 1969.

state. Fortunately, this means that the narrow singlet is rather easily discriminated from the broad underlying spectra of other paramagnetic species, even when the latter are present in much higher concentration. As a result, yield measurements and kinetic studies on trapped electrons can be carried out by ESR with the same order of accuracy as that obtained by optical detection.

Figure 2 also illustrates how the complete removal of the ESR singlet is achieved by photobleaching with near-infrared light, exactly paralleling the observations made by optical spectroscopy (Hamill, 1968; Lin *et al.*, 1968) on γ-irradiated 3-methylpentane. In Table I we summarize the optical and

TABLE I

SPECTROSCOPIC PARAMETERS AND 100-EV YIELDS OF TRAPPED ELECTRONS IN γ-IRRADIATED
ORGANIC GLASSES AT 77°K

Glass	Optical λ_{max} (nm)	(eV)	Optical $G(e_t^-)$	ESR $G(e_t^-)$	ESR g	ESR $\triangle H_{ms}$	Ref.
Methyltetrahydrofuran	1250	1.00	2.6	2.6	2.0022	4.2 $\pm$ 0.2	[a,b,c]
3-Methylpentane	1600	0.78	0.65	0.69	2.0022	4.21 $\pm$ 0.05	[c,d]
3-Methylhexane	1650	0.75	0.89	0.87	2.0022	4.07 $\pm$ 0.05	[c,d]
Methylcyclohexane	1650	0.75	0.30	0.38	2.0022	4.38 $\pm$ 0.05	[c,d]
3-Methylheptane	1700	0.73	0.58	0.68			[c]
4-Methylheptane	1700	0.73	0.55	0.44			[c]

[a] M. R. Ronayne, J. P. Guarino, and W. H. Hamill, *J. Amer. Chem. Soc.* **84**, 4230 (1962).
[b] D. R. Smith and J. J. Pieroni, *Can. J. Chem.* **43**, 876 (1965).
[c] J. Lin, K. Tsuji, and F. Williams, *J. Amer. Chem. Soc.* **90**, 2766 (1968).
[d] G. K.-W. Lok, M.S. thesis, Univ. of Tennessee (1969).

ESR spectroscopic parameters of trapped electrons in hydrocarbon glasses; also included is a compilation of $G(e_t^-)$ values evaluated independently from ESR and optical measurements on a number of γ-irradiated organic glasses (Lin *et al.*, 1968; Lok, 1969). For most of these glasses, there is quantitative agreement between the ESR and optical determinations of $G(e_t^-)$. Other evidence which strongly confirms the assignment of both the near-infrared absorption and the ESR singlet spectrum to the same trapped electron species comes from the parallel effects of added electron scavengers. A particularly convincing demonstration was provided by an ESR experiment with 1 mole% methyl iodide in a 3-methylpentane glass (Lin, 1968). After γ-irradiation of this glass, the singlet spectrum attributable to the trapped electron was not observed under appropriate conditions for its detection, whereas the easily identifiable quartet spectrum of the methyl radical was prominently displayed, thus providing evidence for dissociative electron capture by methyl iodide.

The dependence of trapped electron concentration in glassy 3-methylhexane at 77°K on the γ-irradiation dose is shown in Fig. 3. It is remarkable that the concentration increases to a maximum of 7.5 $\times$ 10^{-4} M at a dose of $\approx$ 1 $\times$ 10^{20} eV g^{-1} and then decreases steadily to a value below 10^{-4} M at a dose of 5 $\times$ 10^{20} eV g^{-1} (Lin *et al.*, 1968; Tsuji and Williams, 1968). There is very little thermal decay of trapped electrons in 3-methylhexane at 77°K, so the electrons must disappear by some radiation-induced mechanism. A similar effect has been observed for a number of other glasses (Ekstrom *et al.*,

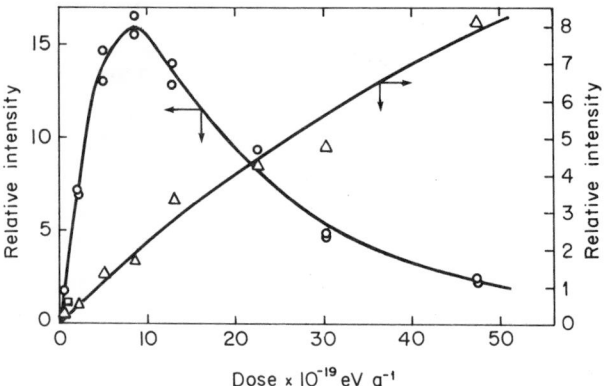

Fig. 3. Dose dependence of trapped electron and free-radical concentrations in γ-irradiated 3-methylhexane at 77°K. The ordinate scales are arbitrary and unrelated. Trapped electron, $\bigcirc$; free radical, $\triangle$; trapped electron in a sample which was preirradiated to a dose of 3×10^{20} eV g^{-1} and bleached before irradiation, $\square$. [Reproduced from *J. Phys. Chem.* **72**, 3884 (1968). Copyright (1968) by the American Chemical Society. Reprinted by permission of the copyright owner.]

1970) and in polymers (Keyser, 1970). By contrast, the free radical concentration in Fig. 3 increases progressively with dose over the entire range. Reasons are advanced in Section III, C for believing that trapped electrons disappear from the system by reaction with the free radicals which build up in the sample during irradiation.

Photoionization of suitable molecules in rigid glasses provides an independent method of generating trapped electrons. Original work by Lewis and Lipkin (1942) indicated the potentialities of this method in organic glasses, but the emphasis in the early work was on the details of the photochemical transformations undergone by the solute molecule. More recently, attention has been directed at the phenomena of photoconductivity and recombination luminescence which are observed in these rigid organic solutions (Albrecht, 1970), with particular reference to the role of mobile and trapped electrons. The optical and ESR techniques which have been used for the identification of trapped electrons in γ-irradiated glasses are just as relevant in this instance, and therefore, it is interesting to compare the results of photoionization and γ-irradiation studies in model systems. Again we restrict the discussion mainly to saturated hydrocarbon glasses.

A solution of ≈ 0.02 mole% of *N,N,N',N'*-tetramethyl-*p*-phenylenediamine (TMPD) in a rigid hydrocarbon glass constitutes a very suitable system for UV photoionization experiments and has been widely investigated by different techniques. Gallivan and Hamill (1966) showed that after photoionization of TMPD in a 3-methylpentane glass at 77°K, the broad absorption spectrum in the near-infrared region is virtually identical to that of the trapped electron in a γ-irradiated 3-methylpentane glass. ESR studies (Tsuji and Williams, 1968, 1969a) provide further evidence for electron trapping in these photoionized glasses. The narrow singlet in the ESR spectrum of photoionized TMPD in 3-methylhexane at 77°K has the same *g* factor and linewidth parameters as those of the trapped electron singlet observed in the ESR spectrum of γ-irradiated 3-methylhexane (Lin *et al.* 1968). Figure 4 shows that in each case the singlet is entirely photobleached

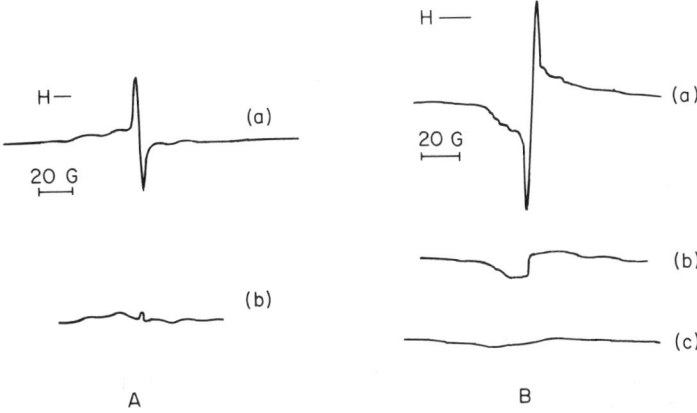

Fig. 4. (A) ESR first-derivative spectra of γ-irradiated 3-methylhexane (dose, 2.9×10^{18} eV g^{-1}) at 77°K before (a) and after (b) bleaching with infrared light. The spectra were recorded at the same sensitivity and a power of 0.01 mW with modulation frequency of 100 kHz. (B) Effect of photobleaching on ESR signal from photoionized TMPD in 3-methylhexane at 77°K: (a) immediately after irradiation; (b) after photobleaching with $\lambda > 1000$ nm; (c) after photobleaching with $\lambda > 640$ nm; microwave power, 0.01 mW. The same gain settings were used throughout. [Reproduced from *J. Amer. Chem. Soc.* **90**, 2766 (1968) and *J. Phys. Chem.* **72**, 3884 (1968). Copyright (1968) by the American Chemical Society. Reprinted by permission of the copyright owner.]

by near-infrared light. Moreover, the radiofrequency (rf) power saturation characteristics of these ESR signals in photoionized and γ-irradiated glasses were found to be very similar (Tsuji and Williams, 1969a), as illustrated in Fig. 5 by the power saturation curves for 3-methylpentane glasses. Still further evidence for the essential identity of the trapped electron species derived by these two methods comes from the scavenging action of carbon dioxide. In both γ-irradiated and photoionized glasses containing CO_2, the ESR spectrum is largely that of CO_2^{-} (Johnson and Albrecht 1966a), which can be easily differentiated from the authentic spectrum of the trapped

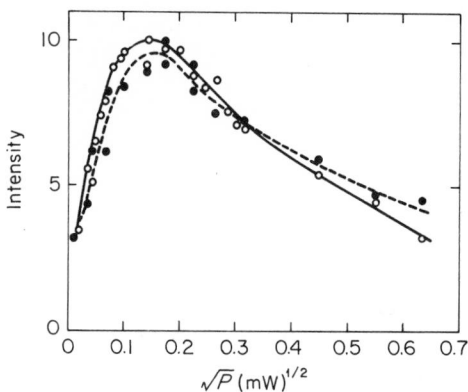

Fig. 5. Microwave power saturation curves for ESR signal from trapped electrons. (●) γ-Irradiated (dose, 1.2×10^{19} eV g^{-1}) 3-methylpentane at 71°K; (○) 3-methylpentane-TMPD during photoionization at 77°K; modulation frequency, 400 Hz. Reproduced from *Trans. Faraday Soc.* **65**, 1718 (1969).

electron by virtue of its markedly different photobleaching and power saturation characteristics (Tsuji and Williams, 1967; Lin *et al.* 1967a, 1968).

Once it had been established that trapped electrons could be detected by ESR spectroscopy in photoionized organic glasses (Lin *et al.*, 1967a), the method was used to monitor the trapped-electron population during as well as after illumination (Lin *et al.*, 1967b). This was accomplished by *in situ* irradiation of the sample in the spectrometer cavity, the spectrum being scanned repeatedly after the exciting light had been turned on. Typical

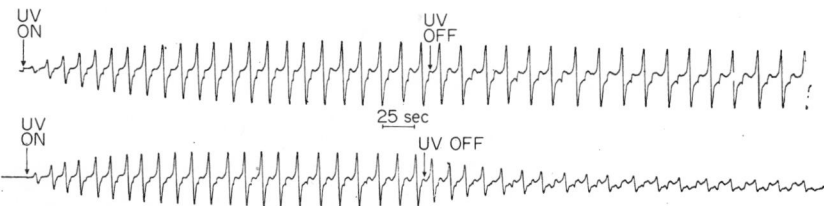

Fig. 6. Photograph of recorder traces showing repeated scans during the growth and decay of the ESR signal from the UV irradiation at 77°K of 3-methylpentane (lower trace) and triethylamine (upper trace) glasses each containing $\approx$ 0.02 mole% TMPD. Both traces were recorded under identical irradiation conditions with the amplifier gain set at 125 for 3-methylpentane and at 200 for triethylamine. Reproduced from *J. Chem. Phys.* **46**, 4982 (1967).

examples of recorder traces obtained in this manner are shown in Fig. 6 for 3-methylpentane and triethylamine glasses. Initially, there is a rapid growth of the signal amplitude to a steady state. It should be noted here that this growth is not just a function of the rise time for the mercury arc lamp (B-H6) used in these experiments, and the increase of signal intensity to the steady state was much slower in the case of MTHF glass (Tsuji and Williams, 1968). The steady state concentration of trapped electrons attained in these experiments was estimated to be $\approx 5 \times 10^{-5}$ M, which is more than a factor of 10 lower than the peak concentration reached in Fig. 3 for thermally stable electrons in a γ-irradiated 3-methylhexane glass at 77°K. After the light was switched off, the ESR signal decayed at a rate which was generally a function of both the viscosity and polarity of the glass, and these observations paralleled the rates for thermal decay of trapped electrons in γ-irradiated glasses (Lin et al., 1967a). From Fig. 6 it can be seen that the decay from the steady state concentration is much more rapid in 3-methylpentane than in triethylamine at 77°K. Even small concentrations (5 mole%) of amines or other polar molecules incorporated into 3-methylpentane had a most significant effect in retarding the decay rate of trapped electrons in both photoionized and γ-irradiated glasses (Bonin et al., 1968). A particularly striking demonstration of this effect was also observed in isopentane, a very soft glass at 77°K with a viscosity close to 10^6 P as compared to the corresponding value of $\approx 10^{12}$ P for 3-methylpentane (Lombardi et al., 1964). Trapped electrons were not detected during TMPD photoionization in isopentane at 77°K, but the presence of 7 mole% triethylamine in the matrix was sufficient to allow direct observation of the trapped electron signal in the steady state, although the decay of the signal after UV irradiation was extremely rapid (Bonin et al., 1968). This experiment shows clearly that the thermal mobility of the trapped electron is a function of matrix polarity, and this effect operates independently of that due to viscosity (see below).

It is also interesting to compare the ESR studies of photoionization with some of the photoconductivity experiments reported by Johnson and Albrecht (1966b) in 3-methylpentane glasses. For this purpose, the ESR results are represented by the signal profiles in Fig. 7. It can be seen that the growth of the ESR signal to the steady state concentration in the hydrocarbon matrices is largely achieved within ≈ 1 min under constant illumination of the TMPD solutions at 320 nm; this parallels the rise to the steady state value in the photoconductivity signal from 3-methylpentane (Albrecht, 1970). Although the ESR technique measures the concentration of *trapped* electrons in the matrix whereas photoconductivity is directly determined by the

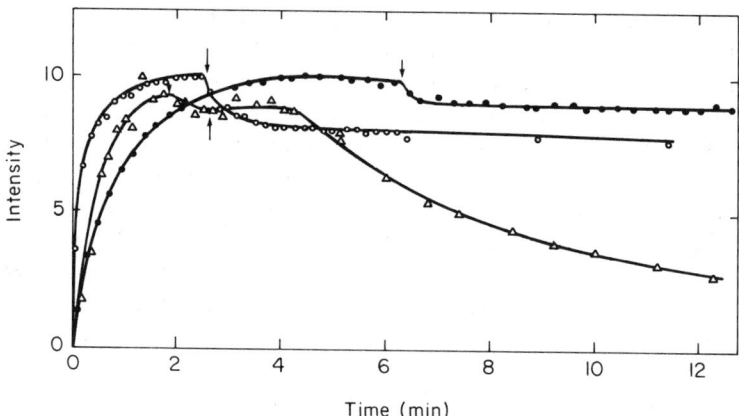

Fig. 7. Typical growth and decay curves of the ESR signal due to trapped electrons by photoionization of TMPD. (●) Triethylamine at 77°K; (○) 3-methylhexane at 77°K; (△) 3-methylpentane irradiated at 70°K, decay at 77°K; (↓) shows the point at which the light was cut off; (↑) shows the point at which helium gas was turned off. for 3-methylpentane the inteival between arrows represents the period during which temperature, cavity Q, and resonant frequency was changing; [Reproduced from *J. Phys. Chem.* **72,** 3884 (1968). Copyright (1968) by the American Chemical Society. Reprinted by permission of the copyright owner.]

number density of *mobile* electrons, the remarkable parallelism between the two observations is reconciled by the idea of a photodynamic ionization equilibrium mediated by mobile electrons (Johnson and Albrecht, 1966b; Tsuji and Williams, 1968). On the assumption that the very short-lived mobile electrons are derived by optical excitation of trapped electrons, the ratio of these two populations should remain constant during the period of illumination, as required by the observations.

The above interpretation is corroborated by a number of additional experimental facts. First, the attainment of a steady state concentration of trapped electrons is a general finding for all matrices irrespective of the subsequent rate of thermal decay in the dark. For example, Fig. 7 shows that, except for a small decrease which occurs immediately after the light is turned off, the thermal decay of trapped electrons is much slower in 3-methylhexane than in 3-methylpentane at 77°K. Therefore, the rapid establishment of the steady state in each case means that the recombination of electrons during illumination is controlled by optical rather than thermal activation. Evidence that mobile electrons are responsible for the photoconductivity comes from experiments (Johnson and Albrecht 1966b), showing that while the conductivity decreases almost to zero immediately after the 320-nm light is turned off, a transient photoconductivity spike is induced on subsequent excitation of the trapped electrons by visible or near-infrared light. This is consistent with the ESR results showing that trapped electrons remain in the matrix after photoionization although they undergo a relatively slow process of thermal decay as determined by the polarity and viscosity of the matrix. Finally, the most significant correlation between the ESR and photo-conductivity results on 3-methylpentane at 77°K is provided by the fact that the time characteristic for the decay of the ESR signal from the trapped electron in the dark (Lin *et al.*, 1967b) is almost exactly the same as that obtained for the decrease in the infrared sensitivity of the matrix as measured from the stimulated photocurrent spikes (Albrecht, 1970). This is gratifying since it provides a firm link between the spectroscopic and electrical measurements.

It has been mentioned several times in passing that the thermal decay of the trapped electron is a function of the matrix (glass) viscosity. A fuller discussion of this topic has been deferred until now because considerable care is needed in the detailed interpretation of results obtained by different experimental methods. The main problem in working with glasses is one of a sample standardization (Tsuji and Williams, 1969b). By its very nature, the glassy state is not easily characterized, and the precise condition of a glass may be difficult to reproduce from sample to sample. Indeed, it is well known in the polymer field (Bueche, 1962) that problems arise in the exact definition of the glass-transition temperature and that different values are obtained by the same experimental method when the sample is cooled at different rates.

Thermal recombination of positive ions and trapped electrons in ir-radiated solids may be regulated by two distinct mechanisms. First, there is the possibility that a trapped electron is raised to the conduction band and moves freely for a certain time before undergoing recombination or retrapping. A model of this type with a suitable distribution of trap depths has been invoked to explain the radiation-induced conductivity of hydrocarbon polymers (Fowler, 1956). Second, the diffusive motion of the trapped electron could be the rate-determining process. It was pointed out by Williams and Hayashi (1966) that the diffusion model of recombination predicts the right order of magnitude for the recombination of ions in liquids and glasses over a range of $\approx 10^{11}$–10^{12} in viscosity, and more recent work has definitely shown that trapped electron recombination in glasses is largely determined by the viscosity of the medium.

A direct illustration of viscosity-controlled recombination is provided by the two ESR traces shown in Fig. 8. In this case, the decay curves of photo-

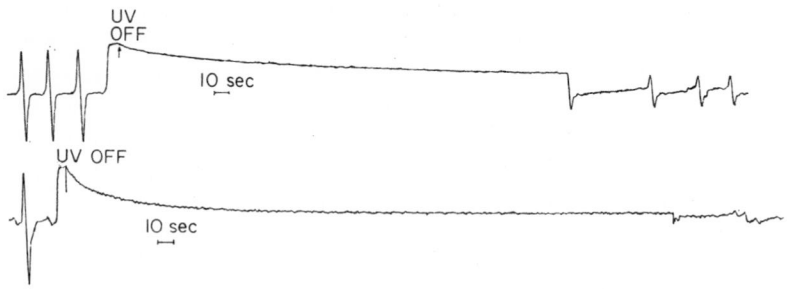

Fig. 8. ESR spectra showing thermal decay for trapped electrons following photo-ionization of 0.02 mole% TMPD in hydrocarbon glass at 77°K. Upper trace refers to 3-methylpentane and the lower to 40 mole% isopentane in 3-methylpentane; gain sensitivity was 160 and 400, respectively. [Reproduced from *J. Amer. Chem. Soc.* **90**, 2766 (1968). Copyright (1968) by the American Chemical Society. Reprinted by permission of the copyright owner.]

ionized samples were recorded by holding the magnetic field constant at the value corresponding to the peak amplitude of the first-derivative signal due to the trapped electron. At 77°K, the time for half decay ($t_\frac{1}{2}$) in the 40 mole% isopentane mixture (viscosity of $\approx 5 \times 10^9$ P) is about 10 sec as compared to a few minutes in 3-methylpentane, which has a viscosity approaching 10^{12} P. Additional measurements of this type indicated that $t_\frac{1}{2}$ depended on the square root of viscosity (Lin *et al.*, 1968), and a similar phenomenological relation was obtained by Leone and Hamill (1968) from studies of recombination luminescence at much lower viscosities.

Since it is well known that there is a considerable change of viscosity in the immediate region of the glass-transition temperature (Williams *et al.*, 1955) a viscosity-controlled decay rate for glasses should be very sensitive to small temperature changes. This is certainly the case as judged by the fact that trapped electrons in 3-methylhexane decay less than 10% in 60 min at 77°K, whereas $t_\frac{1}{2}$ is only 10 min at 84°K (Lin *et al.*, 1968). The correlation between viscosity and thermal decay rate also applies to different hydrocarbon glasses. Thus, the results in Fig. 9 show that decay at 77°K in 3-methylpentane is faster than in 3-methylhexane, as expected, because the latter compound has a somewhat higher glass-transition temperature. It is worth noting that the glass-transition temperature of 3-methylpentane occurs at 77°K (Carpenter *et al.*, 1967), so this glass is not ideal for some purposes at liquid nitrogen

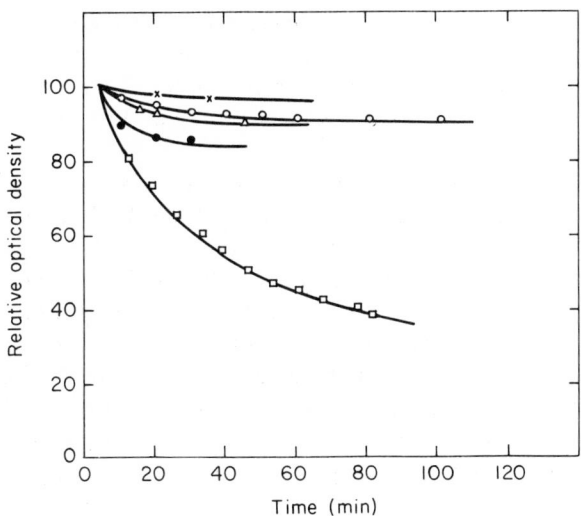

Fig. 9. Thermal decay curves for trapped electrons in various hydrocarbon glasses at 77°K after γ-irradiation at 77°K, as followed by optical spectroscopy. All curves have been normalized to the initial value of the measured optical density at λ_{max}. (□) 3-Methylpentane, dose 2.8×10^{18} eV g^{-1}; (○) 3-methylhexane, dose 1.4×10^{18} eV g^{-1}; (△) 3-methylheptane, dose 7.0×10^{18} eV g^{-1}; (●) 4-methylheptane, dose 7.0×10^{18} eV g^{-1}; (×) methylcyclopentane, dose 1.4×10^{19} eV g^{-1}. Dose rate $= 1.4 \times 10^{18}$ eV g^{-1} min^{-1}. [Reproduced from *J. Amer. Chem. Soc.* **90**, 2766 (1968). Copyright (1968) by the American Chemical Society. Reprinted by permission of the copyright owner.]

temperature, and often, it is useful to employ 3-methylhexane as a harder glass.

The preceding discussion has been concerned with a comparison of decay measurements made by the same technique, either optical or ESR on different materials. Of course, it is most desirable to show that the same decay rate applies to a given material irrespective of the experimental technique. Actually, this proved to be very difficult to establish in the case of 3-methylpentane at 77°K (Lin *et al.*, 1968). The source of the problem was finally traced to the different methods of sample preparation for optical and ESR studies. Optical measurements are conventionally done with 1-cm^2 square cells, whereas ESR samples are generally prepared in 3-mm bore tubes. In each case the sample is prepared by quenching to 77°K. However, the cooling rates in the two instances are unlikely to be identical because the times required for contraction from the liquid to the glassy state are observed to differ. When decay studies were carried out by the same technique (ESR) on samples of 3-methylpentane prepared by different, but controlled, cooling procedures, the results showed that longer decay times were obtained with the more slowly cooled samples (Lok, 1969). From this it would appear that the structure or hardness of the glass depends on the cooling rate. Therefore, the discrepancy between the original optical and ESR results becomes understandable.

It remained to show that agreement in the decay times could be achieved by carrying out optical and ESR experiments on standard samples. By using a large cylindrical microwave cavity, it was possible to do ESR studies on a sample of 3-methylpentane in a standard optical cell (Tsuji and Williams, 1969b; Lok, 1969). Figure 10 shows how the decay curves in this case differ from the results obtained in the narrow tubes. Further careful work (Lin *et al.*, unpublished) showed that the decay curves obtained by optical and ESR experiments on identically prepared 1-cm² 3-methylpentane samples at 77°K were in perfect coincidence. Hence, these experiments furnish final proof that both detection techniques respond to the same population of trapped electrons.

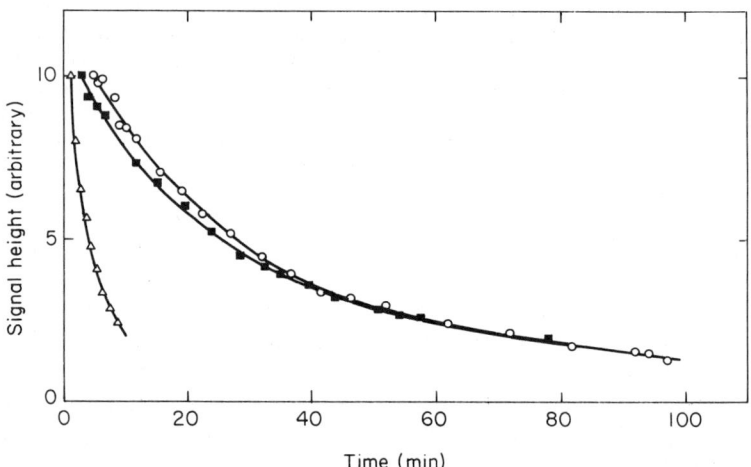

Fig. 10. Decay curves of the ESR signal due to trapped electrons in γ-irradiated 3-methyl-pentane (dose, 2.6 × 10¹⁸ eV g⁻¹) at 77°K. (○, ■) Sample was contained in 10-mm square optical cell and irradiated at 77°K. Time is measured from the end of γ-irradiation. The absolute values of the initial signal heights in these two runs were the same when normalized to a time of 5 min after irradiation. (△) Sample was contained in ≈ 4-mm-i.d. cylindrical ESR tube and irradiated at 71°K. Time is measured from stoppage of helium bubbling through liquid nitrogen. Reproduced with permission from *Int. J. Radiat. Phys. Chem.* **1**, 383 (1969), Pergamon Press.

III. Trapped Electrons in Polymers

A. INTRODUCTION

As discussed elsewhere in this volume, the existence of trapped electrons has been widely invoked in order to explain such facets of radiation effects on polymers as conductivity and thermoluminescence. However, radiation-induced conductivity and thermoluminescence are phenomena which depend on the liberation of electrons from traps. Hence, by their very nature, techniques of this type can only provide an indirect means of investigating the occurrence of electron trapping in polymers. In Section II the uses of ESR and optical spectroscopy in the study of trapped electrons in low molecular weight alkane glasses have been discussed. In view of the success of these methods in providing direct physical evidence for trapped electrons, it might be supposed that similar methods could prove fruitful in identifying and characterizing the behavior of trapped electrons in polymers. This is indeed the case and in the remainder of this section, we consider the evidence for electron trapping in polymers as derived from ESR and optical techniques.

B. ESR SPECTRA OF TRAPPED ELECTRONS IN POLYMERS

1. *Polyethylene*

From the chemical similarity between a polyethylene molecule and one of a low molecular weight alkane, it seems reasonable to expect that if trapped electrons were to be present in polyethylene, their ESR spectra should reflect the narrow linewidth, ease of microwave power saturation, and photobleaching behavior characteristic of trapped electrons in alkane glasses. This has been demonstrated. ESR spectra assigned to trapped electrons in polyethylene have recently been reported by Keyser *et al.* (1968a, b) Keyser and Williams (1969), and Keyser (1970). Typical ESR spectra of bulk-crystallized, antioxidant-free Marlex* and Alathon† polyethylenes obtained by these investigators after γ-irradiation in the dark at 77°K are shown in Figs. 11 and 12. Marlex 6050 and Marlex 50 are representative of linear, high-density polyethylenes, whereas the Alathon 1414 material is a branched, low-density polyethylene. The most prominent feature in the spectra taken before photobleaching is the existence of an intense narrow singlet super-imposed on a broad background spectrum. The g factor and linewidth between derivative maxima, ΔH_{ms}, for the singlet are the same in all these materials, being 2.0022 ± 0.0002 and 3.4 ± 0.2 G, respectively.

The singlet spectra attributed to trapped electrons in these γ-irradiated polyethylenes at 77°K are selectively eliminated by exposure of the samples

---

* Trademark of Phillips Petroleum Co.
† Trademark of E. I. de Pont de Nemours Co.

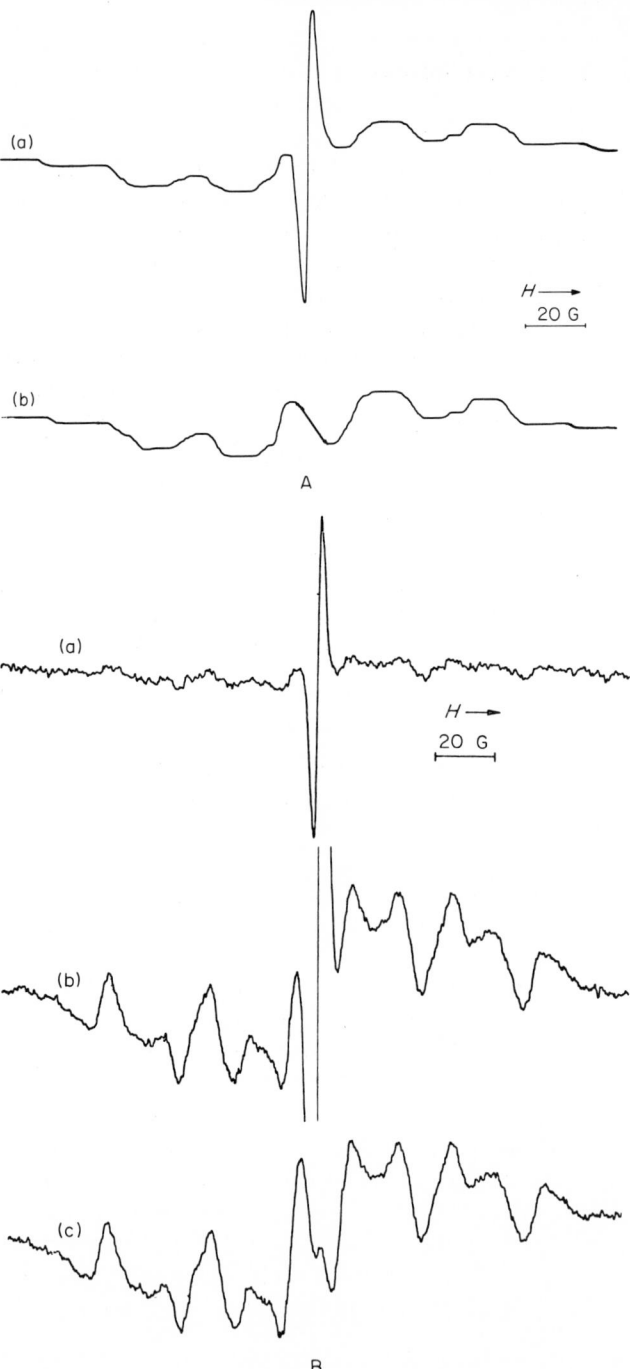

Fig. 11A, B

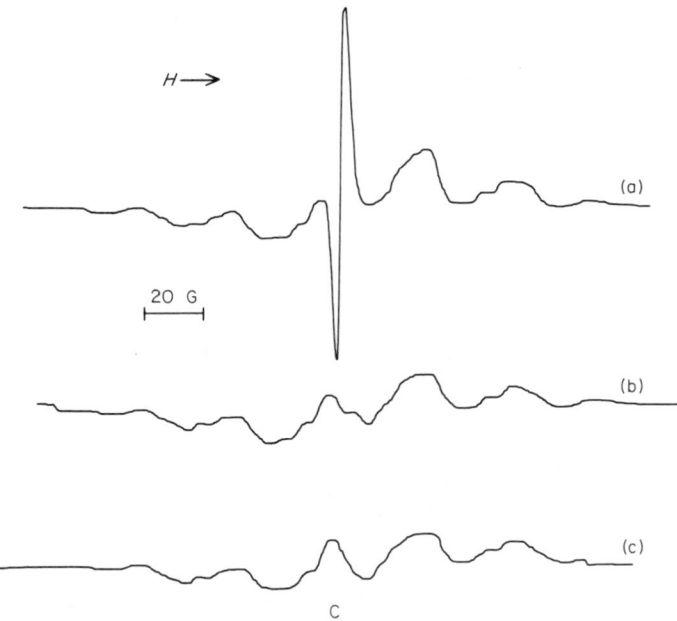

Fig. 11 (A) ESR first-derivative spectra of γ-irradiated antioxidant-free Marlex 6050 polyethylene at 77°K; dose, 1.1×10^{19} eV g^{-1}. Spectrum (a) was recorded before and spectrum (b) after bleaching with near-infrared light ($\lambda > 1000$ nm) at the same gain settings, microwave power, 0.002 mW. (B) ESR first-derivative spectra of γ-irradiated stretched antioxidant-free Marlex 6050 polyethylene at 77°K; the axis of stretch is perpendicular to the applied magnetic field; dose, 1.1×10^{19} eV g^{-1}. Spectra (a) and (b) were recorded in the dark and spectrum (c) was recorded after bleaching with near-infrared light ($\lambda > 1000$ nm). All three spectra were recorded at the same gain setting but the modulation amplitude was increased from ≈ 1.7 G in (a) to ≈ 10 G in (b) and (c). (C) ESR first-derivative spectra of γ-irradiated commercial Marlex 50 polyethylene at 77°K; dose, 0.75×10^{19} eV g^{-1}; sample diameter, 4.8 mm. Spectrum (a) was recorded before and spectrum (b) after bleaching with infrared light ($\lambda > 1000$ nm). Spectrum (c) was recorded after additional photobleaching with red light ($\lambda > 640$ nm); microwave power, 0.002 mW. Spectrometer gain settings were 500, 400, and 500 for (a), (b), and (c), respectively. [Reproduced from *Macromolecules* **1**, 289 (1968). Copyright (1968) by the American Chemical Society. Reprinted by permission of the copyright owner.]

to near-infrared light of wavelength greater than 1000 nm. These photobleaching effects are clearly evidenced by the changes in the ESR spectra shown in Figs. 11 and 12. The spectra shown in Figs. 11A and 11B were obtained using unstretched and stretched samples, respectively, of bulk-crystallized Marlex 6050 polyethylene free from antioxidant. In the stretched sample, there is a preferential orientation such that the polymer chain axis

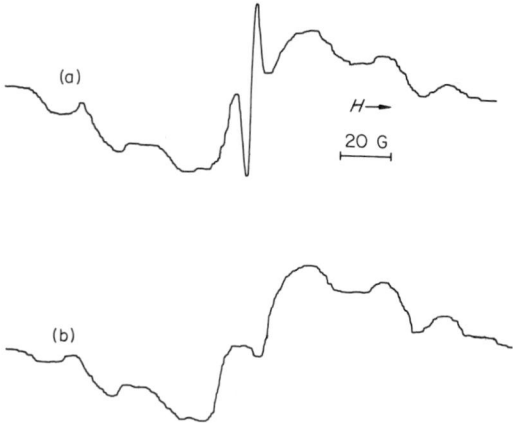

Fig. 12. ESR first-derivative spectra of γ-irradiated Alathon 1414 polyethylene at 77°K; dose, $1.1 > 10^{19}$ eV g^{-1}. Spectrum (a) was recorded before and spectrum (b) after bleaching with near-infrared light ($\lambda > 1000$ nm) at the same gain settings; microwave power, 0.002 mW.

lies in the direction of the stretch. It can be seen that the background spectrum which is insensitive to photobleaching in Fig. 11B differs from the corresponding spectrum of the polycrystalline sample in Fig. 11A, and is much better resolved. This difference is attributable to the effect of magnetic anisotropy on the spectrum of the oriented alkyl free radical $-CH_2\dot{C}HCH_2-$ in polyethylene. (For a recent review of free radical spectra in irradiated polyolefins, see Rånby and Carstensen, 1967.) In contrast, the linewidths of the singlet spectra are almost identical in these two samples, and uniaxial orientation produced no effect of magnetic anisotropy on the spectrum of the trapped electron. Although this experiment provides no positive evidence that the trapped electron is associated with the oriented crystalline regions of the polyethylene, no definite conclusion concerning the location of the trap, i.e., crystalline or amorphous, can be drawn from this result. In view of the narrow linewidth of the ESR singlet in polycrystalline samples, the magnitude of the hyperfine coupling to the individual hydrogen nuclei must be extremely small and consequently any anisotropy may be undetectable in oriented specimens.

Figure 11C shows the results obtained for Marlex 50, a commercial polyethylene which had been pretreated with boiling acetone in an attempt to remove most of the antioxidant originally present in the sample. Again it can be seen that the singlet spectrum produced by γ-irradiation at 77°K was readily removed by photobleaching with near-infrared light ($\lambda > 1000$ nm), but in this case there was some further photobleaching with red light

($\lambda > 640$ nm) as revealed by a careful comparison of the center regions of spectra (b) and (c). For a γ-irradiated sample of untreated Marlex 50, the intensity of the singlet spectrum relative to the background spectrum was lower than that in spectrum (a), and there was additional hyperfine structure which could be removed by exposure to visible light. It is likely that those features which are only photobleachable by visible light come from species that are produced from the antioxidant during γ-irradiation, possibly as a result of electron capture to form radical anions.

The effect of microwave power on the intensity of the singlet in Marlex 6050 polyethylene at a modulation frequency of 200 Hz is shown in Fig. 13.

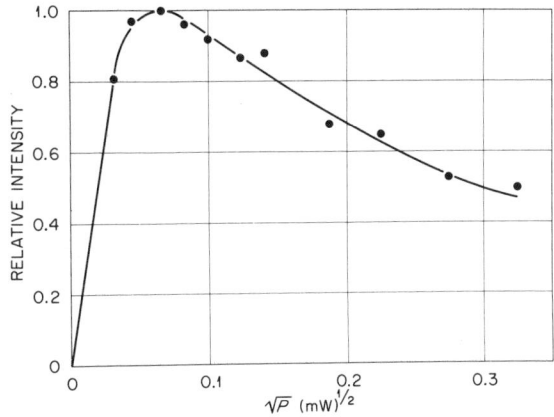

Fig. 13. Microwave power saturation curve for trapped electrons in γ-irradiated Marlex 6050 polyethylene; modulation frequency, 200 Hz.

The pronounced sensitivity of the singlet toward microwave power saturation is readily apparent. The signal height rises to a maximum at 0.004 mW (milliwatts) and thereafter decreases with further increase in microwave power. The saturation behavior for the low-density material, Alathon 1414, is qualitatively similar. Analysis of the saturation data of Fig. 13 according to the method of Castner (1959) yields values for the spin–lattice relaxation time T_1 and the spin–spin relaxation time T_2 of 2×10^{-2} and 5×10^{-8} sec, respectively (Keyser and Williams, 1969). These values should only be regarded as lower limits for T_1 and T_2, however, since a modulation frequency of 200 Hz is too great to satisfy strictly the slow passage conditions upon which Castner's treatment is based.

Before comparing the experimental characteristics of the ESR singlet in polyethylene with those of the similar singlet in alkane glasses, it seems worthwhile to explore first the possibility that the polyethylene singlet results from either an impurity effect or from a free radical species. Impurity effects can be discounted since all efforts to remove impurities likely to be present such as dissolved oxygen or antioxidants lead to enhanced yields of the species responsible for the singlet. A singlet has been reported in the ESR spectra of polyethylene subjected to very large radiation doses and has been assigned to a polyenyl free radical of the type $-CH_2(CH=CH)_n\dot{C}HCH_2-$ (Lawton *et al.*, 1960). The ESR singlet from this species can only be observed at doses above 2×10^{22} eV g^{-1}. Its linewidth is $\Delta H_{ms} = 26$ G and it is thermally stable even at 150°C. This behavior is clearly not in accord with that of the singlet observed in Figs. 11 and 12. The linewidth of the singlet in both Marlex 6050 and Alathon 1414 is 3.4 G, it may be observed at doses even below 10^{18} eV g^{-1}, and it undergoes appreciable thermal decay at temperatures only slightly above -196°C (see Section III, E). Thus, a polyenyl free radical cannot be responsible for the singlet observed in these polyethylenes.

In fact, a number of features of the narrow ESR singlet are not in accord with the characteristics of ESR spectra from free radicals in polymers. The singlet linewidth is much narrower than is typical of polymeric free radicals. The singlet is quite susceptible to microwave power saturation, whereas polymeric free radicals are much more resistant to rf saturation. For example, the singlet in Marlex 6050 displays a maximum in its rf saturation curve at 0.004 mW (Fig. 13), but the underlying alkyl free radical is not appreciably saturated even at the relatively high power level of 0.100 mW (Keyser, 1970). Finally, the singlet is readily removed by exposing the sample to near-infrared light ($\lambda > 1000$ nm). Typical polymeric free radicals are not affected by infrared light. All these differences indicate that some species other than a free radical is responsible for the narrow singlet observed in polyethylene.

When one compares the ESR singlet in the polyethylenes with that due to physically trapped electrons in alkane glasses, a number of close similarities are immediately evident. In both instances the singlet is quite narrow; ΔH_{ms} equals 4.0 ± 0.2 G in a variety of alkane glasses (Lin *et al.*, 1968) and 3.4 ± 0.2 G in polyethylene. The singlet is readily bleached by near-infrared light ($\lambda > 1000$ nm) in both alkanes and polyethylene. Moreover, the singlet in both types of materials is very susceptible to microwave power saturation. The maximum in the saturation curve for the singlet in a typical alkane glass such as 3-methylhexane displays a maximum in the saturation curve at 0.025 mW (Tsuji and Williams, 1969), whereas the corresponding maximum in polyethylene occurs at the somewhat lower value of 0.004 mW.

Thus, despite small differences in linewidth and saturation effects, the main features and behavior of the ESR singlet in polyethylene are entirely similar to the trapped-electron singlet in alkane glasses. In view of this overall correspondence, there can be little doubt that the polyethylene singlet results from electrons physically trapped in the polymer matrix (Keyser *et al.*, 1968a, b). The term "physically trapped electron" is employed in this context to denote an electron trapped in the polymer matrix in such a manner that it can be considered as a distinct entity by itself, i.e., the trapped electron is not localized in a specific molecular orbital belonging to a single molecule or segment thereof. Thus, the narrow linewidth, ease of microwave power saturation, and photobleaching effect all imply that the electron is only weakly coupled to its environment and supports the concept of an electron bound to no single molecule. It is tempting to compare the physically trapped electrons in hydrocarbon-type materials to the familiar solvated electrons in liquid ammonia or to the *F* center in alkali halide crystals although it is best not to push analogies of this type too far. Liquid ammonia is a more polar system than hydrocarbons and the ionic nature of alkali halide crystals stands in contrast to the covalent bonding in hydrocarbon-like materials. Such differences can be expected to exert influences on the mode of electron trapping and also on the precise nature of the electron trap.

It is important to note that the characteristics of the trapped electron singlet in polyethylene allow a clear distinction to be drawn between physically trapped electrons and what might be conveniently termed "chemically trapped electrons." For present purposes, a chemically trapped electron can be regarded as being bound to some particular molecule as a result of an electron capture process such that it resides entirely within the characteristic molecular orbitals of that particular molecule. A chemically trapped electron would therefore be expected to undergo significant hyperfine interaction with protons (or other magnetic nuclei) of its host molecule. This interaction should be revealed either as hyperfine splitting in the ESR spectrum or, if such is unresolved, as an ESR line considerably broader than the uncommonly narrow line characteristic of physically trapped electrons. Moreover, a chemically trapped electron, since it is an intimate part of a molecular system, should be strongly coupled to its environment and, therefore, should not be so markedly sensitive to power saturation effects as are trapped electrons. Finally, while it is known that chemically trapped electrons (those formed as a result of electron capture by aromatic solutes, for example) can be photobleached, the effect normally requires photon energies corresponding to the visible or UV region rather than the near-infrared region characteristic of physically trapped electrons.

The essential experimental features of the trapped electron ESR singlet in polyethylene have recently been confirmed by Campbell (1970). In Campbell's work, however, the trapped electron singlet is not so prominently displayed as in Figs. 11 and 12, due to the fact that his experiments were conducted at a microwave power level of 0.1 mW. In this region, the singlet undergoes quite appreciable power saturation (Keyser, 1970). Alfimov *et al.* (1964) have also attempted to observe ESR spectra of trapped electrons in polyethylene. These investigators also reported that the ESR spectrum of polyethylene after γ-irradiation at 77°K consisted of a singlet superimposed on a broad free radical spectrum. Furthermore, the singlet could be removed by photo-bleaching the sample. The singlet observed by these workers, however, is readily distinguished on the basis of linewidth from that observed in Figs. 11 and 12. As noted above, the linewidth of the trapped electron singlet in both Marlex 6050 and Alathon 1414 polyethylenes is 3.4 G. In the work of the Russian authors, the singlet is much broader, $\Delta H_{ms} \approx 20$ G, as judged by the fact that the resolution of the center two lines of the underlying free radical spectrum is totally obscured by the presence of the singlet.

Alfimov *et al.* (1964) suggest that the singlet they observed could result from an electron "captured in a trap" but the nature of the trap, i.e., physical or perhaps chemical as might result from electron capture by an impurity, was not specified. A broad line in the center of the spectrum could well be the result of unresolved hyperfine structure from a paramagnetic species produced by electron capture. Indeed, the Russian workers also report that radiolysis of both polyethylene containing benzene and aromatic compounds also yielded a broad ESR singlet. Such a result is in accord with the idea of chemical trapping of the electron in their work rather than physical trapping.

2. *Polypropylene*

The presence of physically trapped electrons in isotactic polypropylene has also been demonstrated (Keyser and Williams, 1969; Keyser, 1970). Spectrum (a) of Fig. 14 shows the ESR spectrum from bulk-crystallized isotactic polypropylene (Phillips Petroleum Co.) obtained after γ-irradiation in the dark at 77°K. The material was free of antioxidants, and data from the manufacturer indicate it to be 95% isotactic and 70% crystalline as determined by X-ray diffraction. The singlet in spectrum (a) is characterized by $g = 2.0021 \pm 0.0002$ and $\Delta H_{ms} = 2.9 \pm 0.2$ G. It readily undergoes microwave power saturation in the region of 0.010 mW, and spectrum (b) indicates that it can be photobleached with light of $\lambda > 1000$ nm. These features, again, are just those observed for trapped electrons in alkane glasses, and the assignment of the singlet in isotactic polypropylene to physically trapped electrons follows accordingly. Identical spectra were obtained from an antioxidant-free isotactic polypropylene sample manufactured by Hercules, Inc. This result offers further proof that the singlet spectrum of trapped electrons does not arise from an impurity effect since it seems highly unlikely

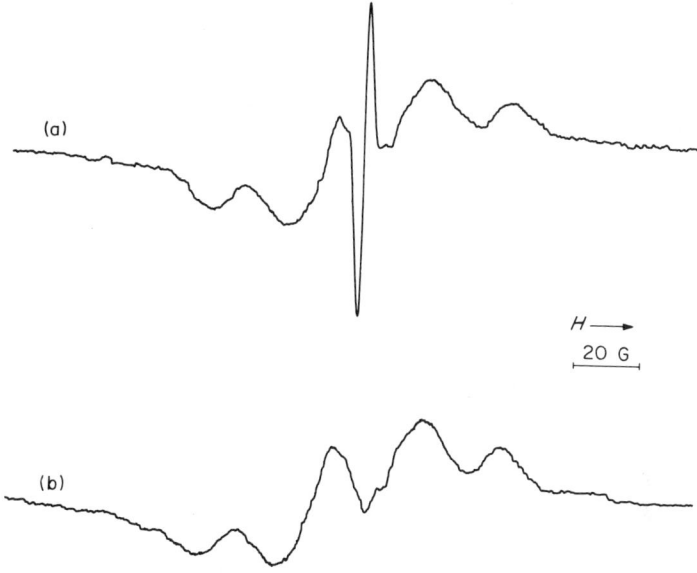

Fig. 14. ESR first-derivative spectra of γ-irradiated isotactic polypropylene at 77°K; dose, 1.1×10^{19} eV g^{-1}. Spectrum (a) was recorded before and spectrum (b) after bleaching with near-infrared light ($\lambda > 1000$ nm) at the same gain settings; microwave power, 0.002 mW.

that the nature and concentration of an impurity would be the same in samples received from different sources.

The assignment of the underlying free radical spectrum in isotactic polypropylene is uncertain. Several different radical species have been proposed in the literature, but so far, it would appear that no general agreement on a specific assignment has been reached. The problem has recently been reviewed (Rånby and Carstensen, 1967), and it seems likely that the spectrum may result from more than one radical structure.

ESR spectra obtained before and after photobleaching for atactic polypropylene (Hercules, Inc.) are depicted in Fig. 15. The small "blip" or singlet in the center of spectrum (a) can be photobleached as shown by spectrum (b) and is assigned to trapped electrons on this basis. It is evident that trapped electron yields in atactic polypropylene are considerably less than in the isotactic material.

3. Poly(4-methylpentene-1)

An ESR singlet characteristic of trapped electrons in antioxidant-free isotactic poly(4-methylpentene-1) is shown in Fig. 16. The material, obtained from Imperial Chemical Industries, Ltd., is ≈40% crystalline. The trapped

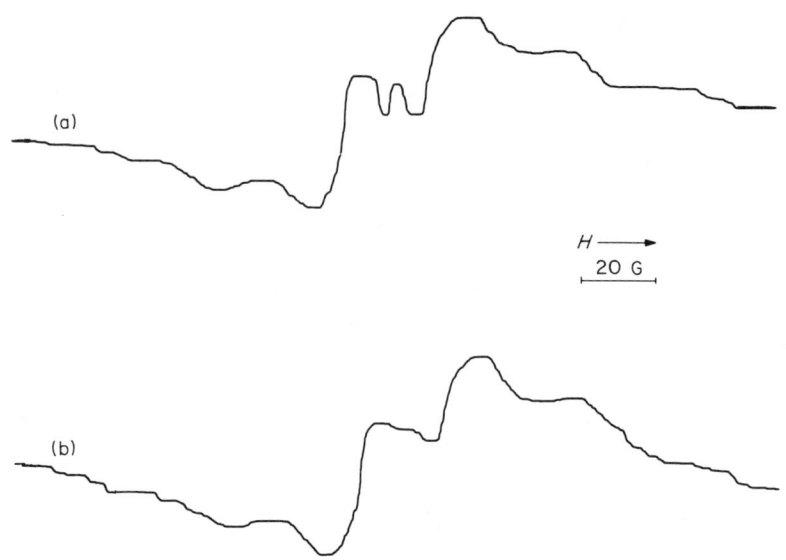

Fig. 15. ESR first-derivative spectra of γ-irradiated atactic polypropylene at 77°K; dose, 1.1×10^{19} eV g^{-1}. Spectrum (a) was recorded before and spectrum (b) after bleaching with near-infrared light ($\lambda > 1000$ nm) at the same gain settings; microwave power, 0.002 mW.

electron singlet has $\Delta H_{ms} = 2.6 \pm 0.2$ G and $g = 2.0021 \pm 0.0002$ (Keyser and Williams, 1969; Keyser, 1970). Whelan and Pinkerton (1970) have recently obtained the ESR spectra of free radicals in isotactic poly(4-methylpentene-1) after γ-irradiation at 77°K. Their published spectra, however, reveal no indication of a trapped electron singlet. Presumably, this could have been the result of working in the presence of light and at high microwave power because the investigators were primarily interested in free radicals. At 77°K, Whelan and Pinkerton find the ESR spectrum consists of an octet with 24 ± 1 G hyperfine splitting. The underlying free radical spectrum in Fig. 16, most clearly revealed in spectrum (b) after photobleaching, consists of six lines with a hyperfine splitting of about 23 G. This spectrum is evidently in agreement with that of Whelan and Pinkerton if it is assumed that the intensity of the outermost lines are below the sensitivity of the spectrometer. Whelan and Pinkerton assign the octet spectrum to the free radical formed as a result of loss of a hydrogen atom at the tertiary carbon atom in the side chain.

4. *Polyisobutylene*

Using the same techniques which have successfully demonstrated the existence of trapped electrons in the polymers mentioned above, Keyser

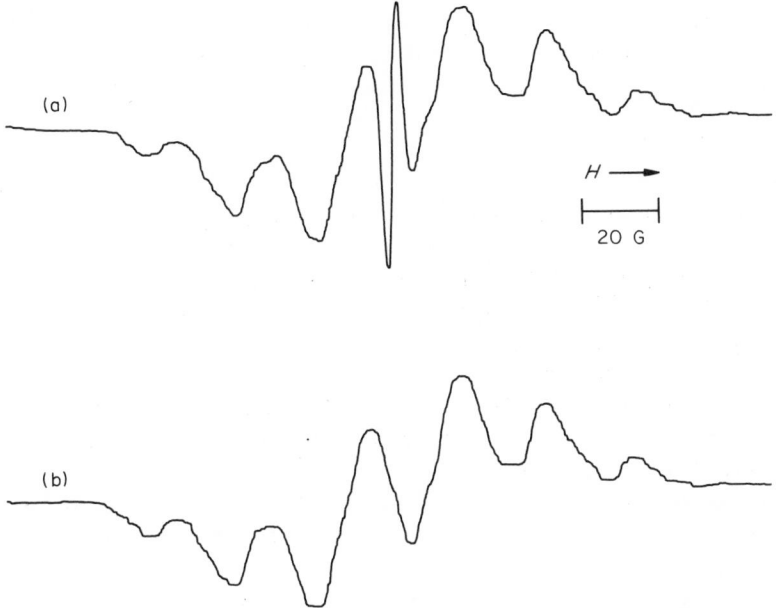

Fig. 16. ESR first-derivative spectra of γ-irradiated isotactic poly(4-methylpentene-1); dose, 1.1×10^{19} eV g^{-1}. Spectrum (a) was recorded before and spectrum (b) after bleaching with near-infrared light ($\lambda > 1000$ nm) at the same gain settings; microwave power, 0.002 mW.

and Williams (1969) and Keyser (1970) have been unable to detect any evidence for trapped electrons in antioxidant-free polyisobutylene. The ESR spectrum after γ-irradiation in the dark consisted of a poorly resolved free radical spectrum which gave no indication of the presence of a narrow singlet and which underwent no change on photobleaching with light of $\lambda > 1000$ nm. Investigation of polyisobutylene prepared by radiation-induced polymerization of highly purified isobutylene also gave no indication of electron trapping (Lin *et al.*, 1968, see footnote 27 of this reference), a fact which rules out the possibility that the absence of trapped electrons results from electron scavenging by spurious impurities. It thus appears that the polyisobutylene matrix is incapable of trapping electrons.

5. Polystyrene

Alfimov *et al.* (1964) and Tochin *et al.* (1966) have reported that after irradiation of polystyrene at 77°K with fast electrons, the ESR spectrum consists of a singlet superimposed on the free radical spectrum. The latter authors observed that the singlet (ΔH_{ms} estimated to be 15–20 G from their published spectrum) could be photobleached by visible light and assigned it to a "stabilized radical ion." These facts suggest a chemically trapped electron, but it appears that the precise nature of the species has yet to be established.

6. Polymethacrylic Acid

To the authors' knowledge, the only other report in the literature regarding ESR identification of trapped electrons in polymeric systems is by Ormerod and Charlesby (1962). These authors observed a singlet in the ESR spectrum of polymethacrylic acid following irradiation at $77°K$. From scavenger studies with TCNE, a well-known electron scavenger, the ESR spectrum of the TCNE⁻ radical anion was obtained. They concluded that an electron was the precursor of the species responsible for the singlet and suggested the singlet was the result of electron capture by monomeric impurities which are difficult to remove (Ormerod and Charlesby, 1964). However, more recently, it has been established (Whelan, 1969) from many ESR studies on irradiated carboxylic acids at $77°K$ that radical anions are produced by electron capture, so it seems highly probable that a similar process occurs in polymethacrylic acid at $77°K$.

C. The Dependence of Trapped Electron Concentration on Irradiation Dose

1. Polyethylene

The concentration of trapped electrons in polyethylene as a function of irradiation dose has been investigated by Keyser (1970), and a plot of trapped electrons per gram of Marlex 6050 linear polyethylene is shown in Fig. 17 for irradiation doses up to 6×10^{19} eV g^{-1}. The concentration rises to a maximum at 3×10^{19} eV g^{-1} and thereafter decreases with further increase in dose. At doses below 0.5×10^{19} eV g^{-1}, the trapped electron concentration increases linearly with dose yielding a limiting value of $G(e_t^-) = 0.46$ trapped electrons/100 ev at low doses. The corresponding value of $G(e_t^-)$ low-density Alathon 1414 polyethylene has been found to be 0.12 trapped electrons/100 eV.

In these experiments, trapped-electron concentrations in the polyethylenes were obtained using the trapped-electron singlet in MTHF, for which $G(e_t^-) = 2.6$ (Ronayne et al., 1962), as a standard. The trapped-electron singlets in both polyethylene and MTHF have Gaussian line shapes so that the product of singlet amplitude and the square of the linewidth can be taken as proportional to trapped electron concentration when comparing the polyethylene singlet to that in MTHF. As a check on the values of $G(e_t^-)$ in polyethylene, the spectra in Fig. 11A were double integrated. Subtraction of the area under

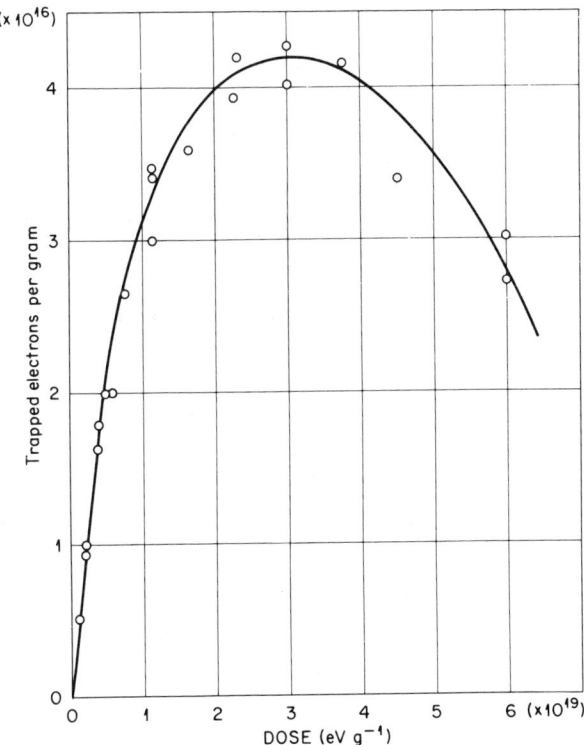

Fig. 17. Concentration of trapped electrons as a function of dose in Marlex 6050 polyethylene γ-irradiated at 77°K.

the integral curve of spectrum (b) from that of (a) gives an area proportional to the trapped electron concentration in Marlex 6050. By comparison of this area with that of the alkyl free radical [spectrum (b)], the measured $G(e_t^-)$ enables the G value for free radical production, $G(R)$, to be calculated. $G(R) = 2.9$ found in this way is in very good agreement with the value of 3.0 reported by Lawton *et al.* (1960). Such close agreement lends strong support to the accuracy of $G(e_t^-)$ in polyethylene.

It is well known that linear polyethylenes such as Marlex 6050 display a higher crystallinity than do branched polyethylenes of the Alathon 1414 type. The fact that Marlex 6050 exhibits a quite significantly higher $G(e_t^-)$ than does Alathon 1414 has led Keyser *et al.* (1968b) and Keyser and Williams (1969) to propose that the trapping site is associated with crystalline regions in the polymer. This conclusion is strengthened by the observation of a correlation of trapped-electron yields with crystallinity for polyolefins in general. This point is discussed further in Section III, F.

The fact that the trapped-electron concentration in Marlex 6050 decreases at doses above 3×10^{19} eV g^{-1} is somewhat puzzling at first sight. A similar decrease has been observed in studies of electron trapping in alkane glasses by Lin *et al.* (1968), Tsuji and Williams (1968), and Shirom and Willard (1968). The latter investigators have proposed that the decrease may be accounted for if the matrix contains only a limited number of trapping sites and that some product of the irradiation reacts with trapped electrons in a manner so as to render the trapping site incapable of further electron trapping.

Nikol'skii *et al.* (1963) have suggested from photoluminescence studies on irradiated polyethylene that free radicals may influence electron trapping. One way in which this might occur and still satisfy the conditions set forth by Shirom and Willard (1968) is represented by the reaction in Eq. (1):

$$R \cdot + e_t^- \rightarrow R^- \tag{1}$$

If it is assumed that the product of this reaction, a carbanion R^-, occupies the trapping site, then further electron trapping at the site would be precluded. Since R^- is diamagnetic, the trapped electron concentration as observed by ESR would be expected to decrease when the free radical concentration has reached a level where reaction can occur to a significant extent. It should be noted that Dole and Bodily (1967) have pointed out that reactions of the type represented by Eq. (1) can be expected to be exothermic by about 1 eV or more, as judged by analogy with the electron affinity of the methyl free radical, and should therefore be thermodynamically feasible.

The possibility that free radicals may affect trapped-electron yields in polyethylene has been examined by Keyser (1970). The essential features of this investigation are as follows. A sample of antioxidant-free Marlex 6050 polyethylene was irradiated at 77°K to a dose of 3.0×10^{19} eV g^{-1}. This produced a maximum concentration of 4.2×10^{16} trapped electrons/g (see Fig. 17). The sample was then photobleached with near-infrared light, eliminating the trapped-electron ESR singlet. No change in the underlying free radical signal was observed to occur as a result of the photobleaching treatment. The sample was then given a second 3×10^{19} eV g^{-1} irradiation dose at 77°K. This second irradiation doubled the intensity of the free radical signal but produced a new trapped-electron singlet whose intensity was only one-half of that which resulted from the initial irradiation, *viz.*, 2.1×10^{61} electrons/g. This result suggests that trapped-electron yields are influenced by the build-up of free radicals in the system, the trapped-electron concentration decreasing with increasing free radical concentration.

The possibility that some other radiation product whose yield is proportional to dose might be responsible for the effect can be ruled out on the basis of thermal annealing experiments. A typical experiment of this type was carried out by irradiating a Marlex 6050 sample at 77°K to a dose of 3 × 10¹⁹ eV g^{-1}, photobleaching the sample to remove trapped electrons, and then allowing the sample to warm to room temperature where it was maintained in a stream of dry nitrogen gas for 7 hr. During this period, 90% of the free radicals decayed out as revealed by the fact that the intensity of the free radical signal subsequent to annealing was only one-tenth of the original intensity prior to annealing. A second irradiation dose at 77°K of 3 × 10¹⁹ eV g^{-1} following this treatment produced a trapped-electron concentration of 3.8 × 10¹⁶ electrons/g, nearly equal to the maximum concentration in Fig. 17. Thus, annealing the sample to the point where virtually all the free radicals have been destroyed restores the electron-trapping ability of the matrix. A second annealing experiment was carried out in a similar fashion except that the annealing period was terminated when approximately half the free radicals had been destroyed. The intensity of the trapped electron signal after a second 3 × 10¹⁹ eV g^{-1} irradiation dose at 77°K corresponded to a trapped electron concentration intermediate between the values of 2.1 × 10¹⁶ and 3.8 × 10¹⁶ electrons/g found in the two preceding experiments.

Thus, the yield of trapped electrons in experiments of this type appears to be strongly correlated with free radical concentration in accord with Eq. (1). Experimental support for the existence of carbanions in irradiated hydrocarbons has recently been obtained by Ekstrom et al. (1970). These investigators have noted that the optical absorption spectrum of 3-methylpentane exhibits, in addition to the trapped-electron absorption at 1600 nm, enhanced absorption in the visible-UV region below 500 nm following γ-irradiation at 77°K. Photobleaching with 1400-nm light removed the trapped electron band and increased the visible-UV absorption. Subsequent photobleaching with 375-nm light decreased the absorption in the visible-UV region and regenerated a trapped electron band at 1600 nm. The visible-UV absorption remaining after this treatment is reminiscent of the absorption spectra of alkane free radicals found by Sauer and Mani (1968). Ekstrom et al. (1970) summarize their observations in terms of the following reaction scheme:

$$e_t^- \xrightarrow[1400 \text{ nm}]{hv} R^- \xrightarrow[375 \text{ nm}]{hv} e_t^- \qquad (2)$$
$$\text{(IR absorbing)} \quad \text{(Vis.-UV absorbing)} \quad \text{(IR absorbing)}$$

Thus, there is abundant experimental evidence for assigning the decrease in trapped-electron concentration above 3 × 10¹⁹ eV g^{-1} in polyethylene to a reaction of free radicals with trapped electrons, the product being a carbanion. It should be pointed out, however, that Eq. (1) can only explain the results if the reaction involves *trapped* electrons. Reaction of free radicals with mobile electrons prior to trapping would account for a plateau in the concentration versus dose curve for polyethylene but not a *decrease* in concentration. For reaction of free radicals with electrons already trapped, it is necessary to postulate either that free radicals have sufficient mobility in polyethylene so as to enable them to encounter trapped electrons, or else that some mechanism operates whereby the trapped electron can tunnel to a nearby free radical. The former situation seems unlikely since the absence of thermal decay of free radicals in polyethylene at 77°K indicates that their mobility must be very low.

2. *Isotactic Poly(4-methylpentene-1) and Polypropylene*

The dependence of trapped electron concentration on irradiation dose in both isotactic poly(4-methylpentene-1) and isotactic polypropylene is qualitatively similar to that in polyethylene. With increasing dose, the trapped-electron concentration in both materials first increases, then reaches a maximum, and finally decreases (Keyser, 1970). The initial $G(e_t{}^-)$ for isotactic polypropylene is 0.17 and for poly(4-methylpentene-1) is 0.08. The maximum concentration of trapped electrons in isotactic polypropylene is 1.3×10^{16} electrons g^{-1} at a dose of 3×10^{19} eV g^{-1}. Respective values for poly(4-methylpentene-1) are 0.5×10^{16} electrons g^{-1} at a dose of 2×10^{19} eV g^{-1}. In view of the fact that the behavior of the trapped-electron concentration with dose in these materials is similar to that observed in polyethylene, it seems reasonable to suppose that the previous comments regarding the effect of free radicals on trapped electron concentration would also be applicable in this instance.

D. OPTICAL SPECTRA OF TRAPPED ELECTRONS

1. *Poly(4-methylpentene-1)*

Isotactic poly(4-methylpentene-1) is optically clear despite being partially crystalline and therefore provides an ideal system for observing trapped electrons in a polymeric matrix by means of optical spectroscopy. Following γ-irradiation at 77°K, enhanced optical absorption can be observed in this material beginning at 700 nm and increasing monotonically out to 2050 nm which was the long wavelength limit of the apparatus employed (Keyser and Williams, 1969; Keyser, 1970). This additional absorption can be photobleached with light of $\lambda > 1000$ nm and is therefore assigned to trapped electrons. Evidently, the band maximum lies beyond 2050 nm. It is of interest to note that λ_{max} for the trapped electron in alkane glasses lies in the region of 1600–1700 nm (Lin *et al.*, 1968). Thus, the band maximum for poly-

Acquisitions Research

INFORMATION SERVICES & ALAN TETELMAN LIBRARY
149 Commonwealth Drive, Menlo Park, CA 94025 • (415) 326-9400

To: Dan Roberts

Date: 7-25-97

From: **Chris Spitzel**

Job number: MT00001.0001/AOTO

- [✓] Materials enclosed
- [ ] Balance will follow
- [✓] Please return to Library by: 8-18-97
- [✓] Borrowed from: UCB
- [ ] Other

If you have
any questions,
please call   **Chris x 7163**

(4-methylpentene-1) is evidently red-shifted relative to the absorption in the alkane glasses.

2. Polyethylene

Partridge (1970) has recently investigated the optical absorption spectrum of low-density polyethylene after γ-irradiation at 77°K. Increased absorption was observed in the region from 240 to 850 nm. Part of this absorption could be photobleached by exposure of the sample to light from an ordinary 100-W light bulb. The difference spectrum, obtained by subtracting the spectrum resulting after photobleaching from that prior to photobleaching, consists of a broad band with a maximum at about 280 nm and a long wavelength tail extending out to the limit of observation, i.e., 850 nm. Partridge assigns the "bulk [of this spectrum] to electrons weakly trapped (probably by polarization forces) in the polymer itself since this type of wide featureless absorption is well known from studies on trapped electrons in organic glasses . . . ." The photobleaching effect is certainly suggestive of some type of ionic recombination process, but it should be pointed out that the presence of known electron scavengers such as oxygen and biphenyl appeared to have no effect on the formation of this absorption band. It is, of course, possible that the absence of effects from electron scavengers results from the fact that the species responsible for the optical band lies only in crystalline regions of the polymer, whereas the scavengers occupy only amorphous regions.

However, even if one accepts that the absorption band results from some type of trapped electron species, several lines of evidence point to the fact that it cannot be a physically trapped electron. First, the maximum of this absorption band at 280 nm is considerably displaced from values near 1600 nm that have been observed in the spectra of physically trapped electrons in alkane glasses (Lin *et al.*, 1968). Second, the optical absorption measurements on poly(4-methylpentene-1) just discussed imply that optical absorption from physically trapped electrons in a polymeric matrix also lies in the near infrared. Finally, it has already been noted in Section III, C that the concentration of physically trapped electrons in γ-irradiated polyethylene as measured by ESR passes through a maximum at a dose of 0.5 Mrad. Such behavior does not correspond to the results of Partridge since, in this instance, the optical density at 360 nm is still increasing at a dose of 6 Mrad. It is of interest to note, however, that the optical density at 670 nm appears to pass through a maximum at ~1.5 Mrad. This result is more in keeping with the ESR results if it is supposed that this portion of the optical spectrum includes a contribution from the high-energy tail of an absorption band centered in the near infrared.

Since the visible-UV band clearly does not correspond to physically trapped electrons, two inferences drawn by Partridge based on the blue-shift of this band relative to that of trapped electrons in alkane glasses would appear to require modification. Thus, Partridge asserts that the blue-shift results from the linear nature of polyethylene compared to the branched structure of alkane glasses. Yet, the ESR singlet spectrum of the trapped electron in polyethylene generally resembles that observed in alkane glasses with respect to linewidth and the susceptibility to both infrared bleaching and microwave power saturation. It therefore appears that the nature of the trapped electron is not drastically altered in going from branched alkanes to linear polymers, and consequently, one does not expect an extraordinarily large blue-shift in the position of the band maximum of the optical spectrum. Secondly, on the basis of a void model for the electron trap, Partridge infers that the blue-shift is indicative of electron trapping in smaller voids in polyethylene than in alkane glasses. However, as will be described in Section III, F, a number of lines of evidence indicate that just the opposite is the case (Keyser and Williams, 1969).

Finally, it is of interest to note that there is evidence to indicate that carbanions absorb in the neighborhood of Partridge's band maximum as already discussed in Section III, C. A number of inconsistencies and contradictions in the assignment would be resolved if the bulk of the visible-UV absorption were to be assigned to carbanions with the long-wavelength tail being part of another band located in the near infrared and resulting from physically trapped electrons.

E. THERMAL DECAY OF TRAPPED ELECTRONS

The thermal decay of the ESR singlet spectrum of the trapped electron in linear Marlex 6050 polyethylene has been investigated by Keyser (1970) over the temperature interval 77–127°K. The results of a typical experiment are shown in Fig. 18. The top curve in this figure pertains to the isothermal decay at 77°K over a 120-min time interval beginning immediately after irradiation of the specimen. After 120 min of decay at 77°K, the temperature was raised to 87°K and the further decay of the trapped-electron singlet was followed for 140 min at this temperature. The data so generated are shown as the second curve in Fig. 18. Subsequent decay curves in Fig. 18 were generated in similar fashion, the temperature in each instance being increased by 10° over that of the preceding series of measurements. All the decay curves are characterized by an initially rapid decay followed by a much slower decay which appears to approach an asymptotic value at long times.

The decay of the trapped electron at the various temperatures in Fig. 18

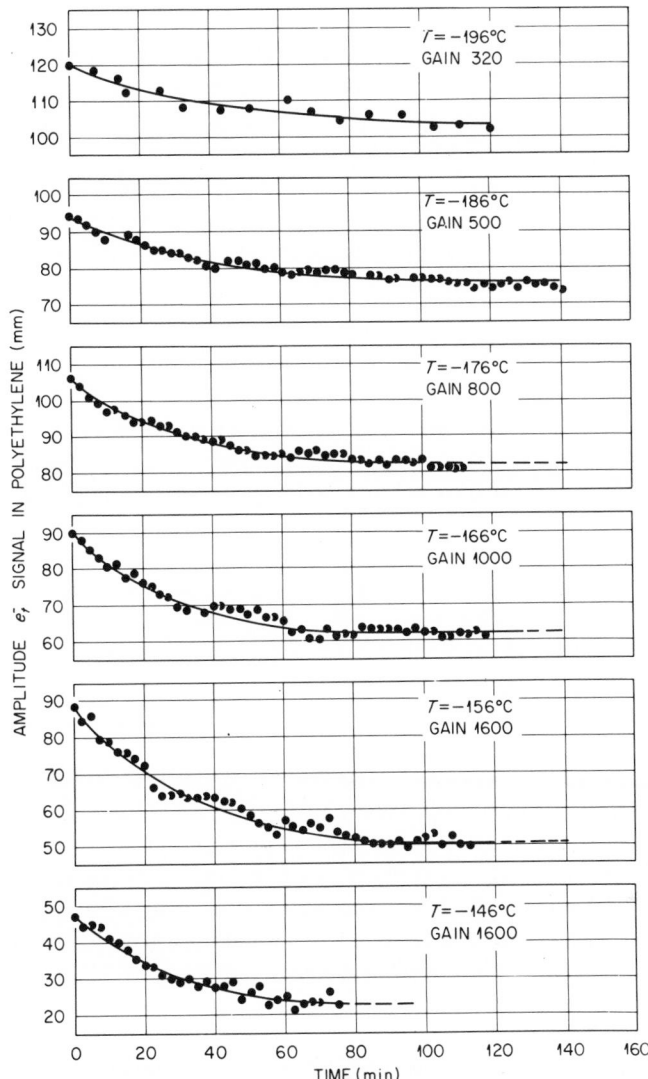

Fig. 18. Stepwise thermal decay of trapped electrons in γ-irradiated Marlex 6050 polyethylene. A comparison of amplitudes at one temperature with those at another does not accurately reflect concentration changes since the temperatures are different.

was tested for first-order kinetics using the Guggenheim method (Frost and Pearson, 1961). According to this method, the relationship between A_t, the trapped-electron singlet amplitude at time t for a first-order process, is given by

$$\ln (A_t - A_{t+\Delta}) = -kt + \ln \left[(A_0 - A_\infty)(1 - e^{-k\Delta}) \right] \qquad (3)$$

where k is the first-order rate constant, A_0 and A_∞ are the singlet amplitudes at zero and infinite times, and Δ is a fixed increment of time. Hence, a plot of the left side of Eq. (3) versus time should be linear for a first-order process with a rate constant given by the slope of the line. Using $\Delta = 40$ min and A_t values from the smooth curves in Fig. 18, the first-order nature of the trapped-electron decay over the time interval involved is demonstrated by the data of Fig. 19. Rate constants obtained from the slopes of the lines are given in Table II.

TABLE II

FIRST-ORDER RATE CONSTANTS FOR THE THERMAL DECAY OF TRAPPED ELECTRONS IN MARLEX 6050 POLYETHYLENE AT VARIOUS TEMPERATURES

Temperature ($°$K)	First-order rate constant, k (min^{-1})
77	0.021
87	0.025
97	0.037
107	0.036
117	0.030
127	0.033

The shape of the decay curves in Fig. 18 implies that the amplitude of the trapped-electron singlet approaches an asymptotic value at long times. The value of the asymptotes, A_∞, for each temperature of decay can be deduced from Eq. (3) using the intercepts at $t = 0$ in Fig. 19. From the values of A_∞ so obtained together with the corresponding value of A_0, it has been determined that if the decay were to be followed to infinite time in each instance, 16% of the trapped electrons initially present after irradiation would decay on standing at 77°K and 16% more would decay each time the temperature was stepped upward by 10°.

These results suggest an interpretation of the thermal decay in polyethylene in terms of a model consisting of various populations of trapped electrons relatively stable toward decay until the temperature is raised to a point where they become vulnerable. Thus, 16% of trapped electrons decay by a first-order process on standing at 77°K, 16% more decay on raising the temperature to 87°K, and so on. (It should be noted that the division into populations each consisting of 16% of the trapped electrons is entirely phenomenological and depends on the rate of temperature increase.) While

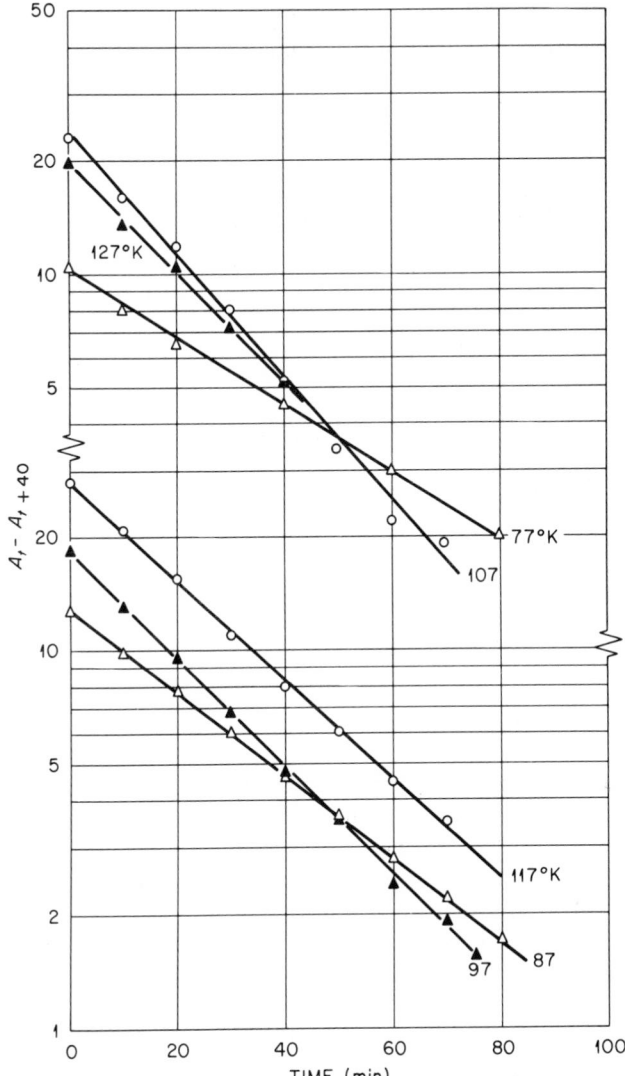

Fig. 19. Test of first-order kinetics for the stepwise thermal decay of trapped electrons in γ-irradiated Marlex 6050 polyethylene by the Guggenheim method.

this type of analysis is attractive, additional thermal decay experiments indicate that it is oversimplified. The results of three such experiments are summarized in Fig. 20. The three curves in this figure represent separate experiments involving three samples for which the thermal decay of the

trapped electron was followed isothermally at the indicated temperatures. All
three samples received an identical irradiation dose of 1.1×10^{19} eV g^{-1}.
The signal amplitudes for each experiment were normalized to the initial
amplitudes observed about 3 min after removal of the sample from the
radiation source. Zero on the time scale refers to these initial measurements.

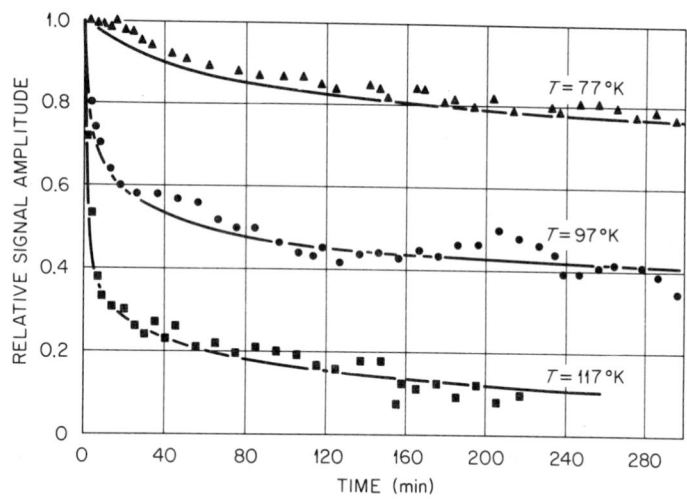

Fig. 20. Isothermal decay of trapped electrons in γ-irradiated Marlex 6050 polyethylene
at various temperatures.

The 77°K curve in Fig. 20 may be compared to the corresponding curve in
Fig. 18, the only difference between the two being that in the former case, the
decay was followed over a longer time period. It is evident that over a long
period of observation, the signal amplitude does not decay to an asymptotic
value. There is a slow decay even out to 350 min. During the first 100 min,
however, the decay is first order as determined from a Guggenheim plot. The
rate constant is 0.019 min^{-1}, virtually the same as that given in Table II at
77°K. For the experiments at 97° and 117°K, there is also a slow decay evident
at long times, but the decay during the first 100 min of observation corresponds
to no simple kinetic order.

It is of interest to note that after about 100 min of decay in Fig. 20, 84% of
the trapped electrons remain at 77°K, 50% remain at 97°K, and lastly, 20%
remain at 117°K. These percentages are very close to those expected on the
basis of a 16% decay for each 10° temperature interval as previously discussed,
and this suggests that a model involving populations of electrons with varying
stabilities toward thermal decay may have some merit. For convenience in
the discussion to follow, the initial trapped-electron concentration will be

divided into six fractions, each consisting of 16% of the initial concentration. That fraction vulnerable to decay at 77°K will be referred to as the 77° population, the fraction decaying at 87°K will be called the 87° population, and so on.

The above observations suggest that the decay curves in Fig. 20 may in fact represent composites of the decay of the 77°–127° trapped-electron populations, these populations decaying not necessarily with the rate constants derived from the stepwise warm-up experiment, but with rate constants characteristic of the actual temperature at which the decay takes place. Proceeding on this assumption, the 77°K curve in Fig. 20 is interpreted as being the sum of the following decay processes: (1) a relatively rapid decay of the 16% of the trapped electrons comprising the 77° population with a first-order rate constant given in Table II and (2) a much slower decay of the 16% of trapped electrons comprising the 87° population. This latter decay is governed, not by the rate constant given in Table II for the 87° fraction, but by a smaller one corresponding to the fact that in this instance the decay occurs at 77°K and not 87°K. On this basis, the first-order contribution from the 77° population can be subtracted from the 77°K curve of Fig. 20. The decay curve remaining after this subtraction process is attributed to the 87° fraction and from it, a half-life of 400 min can be derived for decay of the 87° population at 77°K. This derivation assumes that 16% of the initial concentration of trapped electrons observed immediately after irradiation comprises the 87° population. From the value for the half-life of the 87° population, a first-order rate constant of 1.73×10^{-3} min^{-1} can be calculated for decay of the 87° population at 77°K. This value is to be compared with $k = 2.5 \times 10^{-2}$ min^{-1} observed for the decay of the 87° population at 87°K in the stepwise warm-up experiment (Fig. 18 and Table II).

From these two rate constants, a value of the activation energy for decay of the 87° population may be derived on the assumption that the rate constants are represented by the usual Arrhenius form

$$k = Ae^{-E_a/RT} \tag{4}$$

In Eq. (4), k is the rate constant, A is a frequency factor, E_a is the activation energy, R is the gas constant (1.987 calories per mole per degree Kelvin), and T is the absolute temperature. Since two equations of the form of (4) result for the two values of k at 77° and 87°K, the value $E_a = 3.58$ kcal/mole is readily obtained, assuming that A is temperature independent. Moreover, using this value of E_a for the 87° population, it is also possible to calculate activation energies for decay of the remaining populations of electrons if it is assumed that the frequency factor A is the same for all populations. For example, E_a

for the 77° population may be obtained from the equation

$$\frac{k_{77}}{k_{87}} = \frac{e^{-E_a/77R}}{e^{-3.58 \times 10^3/87R}} \tag{5}$$

where k_{77} and k_{87} are the first-order rate constants for thermal decay of the 77° and 87° populations at these same respective temperatures (Table II). Activation energies for decay of the various populations obtained in this manner are given in Table III.

TABLE III

ACTIVATION ENERGIES FOR THERMAL DECAY OF THE VARIOUS TRAPPED-ELECTRON POPULATIONS

Temperature[a] (°K)	Activation energy (kcal/mole)
77	3.20
87	3.58
97	3.92
107	4.32
117	4.77
127	5.16

[a] These temperatures identify the various trapped-electron populations which are defined in the text.

Knowing the activation energies, it is a simple matter to calculate first-order rate constants for thermal decay of any given trapped-electron population at any temperature of interest from the Arrhenius expression. For example k at 77°K for the 77° population is known to be 2.1×10^{-2} min^{-1} from Table II, and suppose k for this same population is desired at 117°K. This value may be found from the equation

$$k_{\text{at } 117°} = 2.1 \times 10^{-2} \exp\left[\frac{-3.2 \times 10^3}{R}\left(\frac{1}{117°} - \frac{1}{77°}\right)\right] \tag{6}$$

where A has again been taken as temperature independent. Values of k obtained in this fashion for the decay of the various populations at temperatures of interest are given in Table IV.

From the rate constants in Table IV, it is possible to synthesize trapped electron decay curves at any temperature of interest. For example, isothermal decay at 77°K results from the 77° population decaying with a rate constant of 2.1×10^{-2} min^{-1} and the 87° population with a rate constant of 1.7×10^{-3} min^{-1}. The decay of the remaining populations is too slow at 77°K to make a measurable contribution, at least over a 6-hr interval. Likewise, curves for decay at 97° and 117°K can be constructed by simply adding the contributions of the various trapped-electron population as reflected by the

TABLE IV

RATE CONSTANTS FOR THERMAL DECAY OF THE DIFFERENT TRAPPED-ELECTRON POPULATIONS AT VARIOUS TEMPERATURES

Temperature[a] (°K)	First-order rate constant, k (min^{-1}), for decay at					
	77°K	87°K	97°K	107°K	117°K	127°K
77	2.1×10^{-2}	—	1.5	—	25.7	—
87	1.7×10^{-3}	2.5×10^{-2}	2.1×10^{-1}	—	4.9	—
97	2.0×10^{-4}	—	3.7×10^{-2}	—	0.97	—
107	—	—	4.4×10^{-3}	3.6×10^{-2}	0.20	—
117	—	—	4.5×10^{-4}	—	3.0×10^{-2}	—
127	—	—	—	—	5.5×10^{-3}	3.3×10^{-2}

[a] These temperatures identify the various trapped-electron populations which are defined in the text.

rate constant values in Table IV. In each instance, each population is assumed to consist of 16% of the initial number of trapped electrons observed immediately after irradiation. The trapped-electron decay curves calculated on this basis are plotted as the solid lines in Fig. 20. The agreement is seen to be quite satisfactory.

Thus, it is possible to account for the features of the thermal decay of the trapped electron in Marlex 6050 polyethylene in terms of a simple model consisting of various populations of trapped electrons which exhibit varying stability toward decay. This varying stability is governed by the different activation energies for decay. In this connection, it is of interest to note that the optical absorption band maximum at 1650 nm in alkane glasses (Lin et al., 1968) corresponds to an energy of some 17 kcal/mole. Although the band maximum for trapped electrons in poly(4-methylpentene-1) appears to be shifted to somewhat lower energies than this, it does not seem likely that the shift is so extensive that the band maximum corresponds to an energy very much lower than 17 kcal/mole. This value is significantly greater than the activation energies derived for thermal decay of trapped electrons in polyethylene which lie in the range of 3–5 kcal/mole. Thus, the decay of trapped electrons in polyethylene by thermal activation does not appear to involve the upper state reached in the optical transition. An activation energy of some 3–5 kcal/mole is not inconsistent with some type of rotation about a C–C bond. Hence, it is possible that the thermal decay results from some kind of molecular motion involving the surrounding environment of the trapped electron which causes the electron to become mobile. From the previous kinetic analysis, the frequency factor A for the decay of the trapped electron in polyethylene has a value of 10^5 sec^{-1}. Such a value is reasonable for a complicated motion of a polymer chain. Bond vibrations correspond to frequencies of the order of 10^{13} sec^{-1} and rotations to 10^{11} sec^{-1}. Therefore, a more complicated motion, perhaps of a torsional type, would be expected to take place with a lower frequency.

Finally, it should be noted that while the thermal decay of trapped electrons in low-density polyethylene, polypropylene, and poly(4-methylpentene-1) has not been investigated in detail, the shape of the decay curves at 77°K is similar to the Marlex 6050 case in that there is an initial rapid decay of some 30–40% of the trapped electrons in the first 60 min followed by a very much slower rate of decay.

F. THE NATURE OF THE ELECTRON TRAP IN POLYMERS

The earlier sections have dealt with the assignment of the narrow singlet in the ESR spectra of polymers to an electron trapped and physically localized within the polymer matrix as a distinct entity. The question then naturally arises as to what can be said regarding the nature of the trap itself. Such a question is not an easy one to answer since the mechanism of electron trapping in organic materials in general is only poorly understood at present. However, Ekstrom and Willard (1968) have recently shown that many properties of electrons trapped in a variety of matrices including alkane glasses can be correlated in terms of a model whereby the electron is trapped in pre-existing voids or cavities of molecular dimensions which are frozen into the matrix. The electron derives its stability from the lower energy state available to it when localized in a void. In view of the overall similarity between the trapped electron observed in polymers and that in alkane glasses, it seems reasonable to suppose that the environment of the electron in the two cases might exhibit certain similarities as well. It therefore seems advantageous to approach the problem of electron trapping in polymers within the framework of the void concept and to examine the similarities and differences between the behavior of trapped electrons in polymers and alkanes within this context.

In terms of a void model, the results on polymers imply electron trapping in a void that is larger than in alkane glasses. This conclusion follows from three lines of evidence. First, the linewidths of the trapped-electron singlet observed in polymers are in all instances somewhat narrower than the value of 4.0 ± 0.2 G (Lin *et al.*, 1968) found for alkane glasses. If the linewidth is due mainly to unresolved hyperfine interaction between the trapped electron and surrounding protons in the void boundary, then increasing the radius of the void would be expected to weaken the hyperfine interaction and lead to the smaller linewidths observed. A similar conclusion is reached from a comparison of spin–lattice relaxation times. At comparable conditions of radiation dose and modulation frequency, $T_1 = 3 \times 10^{-3}$ sec for the trapped electron in a typical alkane glass (Tsuji and Williams, 1969), whereas in this study, T_1 was found to be $\geq 2 \times 10^{-2}$ sec in Marlex 6050 polyethylene. Thus, T_1 in polyethylene is at least an order of magnitude greater than in alkane glasses. Since T_1 can be regarded as a measure of the strength of the interaction of a trapped electron with its environment, a large T_1 implying a weak interaction, it is evident that the trapped electron in polyethylene is more weakly coupled to its environment than is the case for trapped electrons in alkane glasses so that this result can also be understood naively in terms of a larger void in polymers than in the glasses. Finally, the apparent shift to lower energy of the optical absorption band from the trapped electron in poly(4-methylpentene-1) relative to the corresponding bands in alkane glasses is also of interest in this connection. If a simple "particle-in-a-box" description can be applied to an electron in a void, then the red shift of the absorption band in poly(4-methylpentene-1) implies a larger void in polymers since the optical transition energy decreases with increasing void radius (Ekstrom and Willard, 1968).

TABLE V

EFFECT OF CRYSTALLINITY ON TRAPPED-ELECTRON YIELDS IN γ-IRRADIATED HYDRO-CARBON POLYMERS

Polymer	Percent crystallinity	Initial $G(e_t{}^-)^a$ (electrons/100 eV)	Maximum observable concentration of trapped electrons[b] (electrons/g)
Marlex 6050 polyethylene	$\approx 82^c$	0.46	4.6×10^{16}
Alathon 1414 polyethylene	$\approx 45^c$	0.12	—
Isotactic polypropylene	$\approx 70^d$	0.17	1.5×10^{16}
Atactic polypropylene	0	< 0.02	—
Isotactic poly(4-methylpentene-1)	$\approx 40^d$	0.08	0.5×10^{16}
Polyisobutylene	0	0	—

[a] Determined from plot of trapped-electron concentration versus dose.

[b] Maximum in the concentration-versus-dose curve.

[c] Calculated value based on the densities and specific volume equations of R. Chiang and P. J. Flory, *J. Amer. Chem. Soc.* **83**, 2857 (1961).

[d] Manufacturer's estimate.

It is also of interest to examine the effect of crystallinity on trapped-electron yields in polymers. Table V compares both the initial $G(e_t{}^-)$ and the maximum trapped-electron concentration observed in the various polymers studied in this investigation with the degree of crystallinity of the polymers. Estimates of the degree of crystallinity are based on the conventional assumption that the polymer consists of both crystalline and amorphous phases.

It is evident from the data in Table V that the ability of a polymer to trap electrons is strongly correlated with its crystallinity. Polymers of the highest crystallinity, Marlex 6050 polyethylene and isotactic polypropylene, exhibit the greatest concentration of trapped electrons at the maximum in the concentration versus dose curve and also the highest values of $G(e_t^-)$. Polymers of intermediate crystallinity such as Alathon 1414 and poly(4-methylpentene-1) display intermediate trapped-electron yields. Finally, in noncrystalline polymers such as atactic polypropylene and polyisobutylene, the trapped-electron yield is either vanishingly small or else nonexistent.

This correlation of the trapped-electron yield with crystallinity strongly implies that the trapping site is associated with crystalline regions in the polymer. Such an effect of crystallinity is quite opposite to that observed in the majority of work on electron trapping in low molecular weight systems where in both frozen alcohols (Teply, 1969) and in several glassy alkane systems (Lin *et al.*, 1968), trapped-electron yields are severely diminished by any tendency toward crystallinity in the sample. While at first sight the observation that the trapping site in polymers is associated with crystalline regions would appear to be in conflict with the concept of electron trapping in voids, current views regard the crystalline state of polymers as being characterized by a large number of defects, including those of the void type (Oppenlander, 1968).

Boustead and Charlesby (1970), from studies of thermoluminesence in irradiated polyethylene, have suggested that electron trapping occurs at defects associated with double bonds in the crystalline regions and at physical cavities associated with the folded-chain structure characteristic of the surface of polyethylene crystallites. The first of these possibilities can be discounted, however, since it has been shown that the characteristic trapped-electron ESR singlet is present in a polyethylene sample which had been hydrogenated to remove all unsaturation (Keyser, 1970). Evidence against the second possibility is derived from experiments on irradiated polycrystalline *n*-hexatriacontane (Keyser, 1970). A trapped-electron ESR singlet was observed in this material with linewidth, *g* factor, photobleaching, and thermal decay (at 77°K) properties all identical within experimental error to those for the trapped electron in polyethylene. Therefore the environment of the electron must be similar in both instances. Yet, *n*-hexatriacontane, while possessing the same basic crystal structure as polyethylene, does not possess the folded-chain character at crystallite surfaces as does polyethylene. Thus, it does not appear possible to assign the electron trap in polymers to a specific type of void or defect at the present time. One possibility which should not be overlooked, however, is a void or cavity formed by a chain end which does not extend entirely through the polymer crystal.

The intent of this section has been to explore the possibility that electron trapping in voids, which is thought to be the mode of trapping in low molecular weight glassy systems, also occurs in polymers. The overall similarity in spectral features of the trapped-electron singlet in both polymers and alkane glasses suggests that this may indeed be the case. In terms of the void concept, the experimental results demand that the void be larger in polymers than in alkane glasses and, moreover, that it be associated with crystalline regions in the polymer

On the other hand, one observation which does not seem to be in accord with the idea of trapping in a void is the evident inability of amorphous polymers such as polyisobutylene and atactic polypropylene to trap significant numbers of electrons. High concentrations of voids certainly ought to occur in these materials, owing to their disordered structure, and one would expect a correspondingly high yield of trapped electrons. Such is not the case. It may well be that in order for an electron to be trapped in a void, the molecules forming the void boundary may be required to undergo some type of spatial displacement in order for the electron to be accommodated in the void. If this is the case, then whatever molecular displacements are required to trap the electron would be expected to occur more rapidly and more easily in the case of alkane glasses than in polymers simply on the basis of mobility considerations. The mobility of a segment of a polymer chain is much lower than that of a small molecule.

Of course, if this line of reasoning is adopted to explain the failure of amorphous polymers to trap electrons, then one would need to postulate that voids in crystalline regions of polymers are already preformed, so to speak, such that relatively little if any molecular reorganization is required to trap the electron. The final word on the nature of the electron trap in polymeric systems is obviously yet to come, and clearly, more incisive experiments are needed before definite conclusions can be reached.

REFERENCES

Albrecht, A. C. (1970). *Accounts Chem. Res.* **3**, 238.
Alfimov, M. V., Nikol'skii, V. G., and Buben, N. Ya. (1964). *Kinet. Katal.* **5**, 268.
Bonin, M. A., Lin, J., Tsuji, K., and Williams, F. (1968). *Advan. Chem. Ser.* **82**, 269.
Boustead, I., and Charlesby, A. (1970). *Proc. Roy. Soc. London* **A316**, 291.
Bueche, F. (1962). "Physical Properties of Polymers," p. 97. Wiley, New York.
Campbell, D. (1970). *J. Polymer Sci. B* **8**, 313.
Carpenter, M. R., Davies, D. B., and Matheson, A. J. (1967). *J. Chem. Phys.* **46**, 2451.
Castner, T. G. (1959). *Phys. Res.* **115**, 1506.
Copeland, D. A., Kestner, N. R., and Jortner, J. (1970). *J. Chem. Phys.* **53**, 1189.
Dole, M., and Bodily, D. M. (1967). *Advan. Chem. Ser.* **66**, 31.

Dye, J. L., DeBacker, M. G., and Dorfman, L. M. (1970). *J. Chem. Phys.* **52**, 6251.

Dyne, P. J., and Miller, O. A. (1965). *Can. J. Chem.* **43**, 2696.

Eiben, K. (1970). *Angew. Chem. Int. Ed.* **9**, 619.

Ekstrom, A. (1970). *Radiat. Res. Rev.* **2**, 381.

Ekstrom, A., and Willard, J. E. (1968). *J. Phys. Chem.* **72**, 4599.

Ekstrom, A., Suenram, R., and Willard, J. E. (1970). *J. Phys. Chem.* **74**, 1888.

Ershov, B. G., and Pikaev, A. E. (1969). *Radiat. Res. Rev.* **2**, 1.

Frost, A. A., and Pearson, R. G. (1961). "Kinetics and Mechanism," p. 49. Wiley, New York.

Fowler, J. F. (1956). *Proc. Roy. Soc. London* **A236**, 464.

Gallivan, J. B., and Hamill, W. H. (1966). *J. Chem. Phys.* **44**, 1279.

Hamill, W. H. (1968). *In* "Radical Ions" (E. T. Kaiser and L. Kevan, eds.), p. 321. Wiley, New York.

Johnson, P. M., and Albrecht, A. C. (1966a). *J. Chem. Phys.* **44**, 1845.

Johnson, G. E., and Albrecht, A. C. (1966b). *J. Chem. Phys.* **44**, 3162, 3179.

Jortner, J. (1964). *Radiat. Res. Suppl.* **4**, 24.

Kerr, C. M. L., and Williams, F. (1971). *J. Amer. Chem. Soc.* **93**, 2805.

Kevan, L. (1969). *In* "Actions Chimiques et Biologiques des Radiations" (M. Haissinsky, ed.), Vol. 13. Masson et Cie, Paris.

Keyser, R. M. (1970). An Electron Spin Resonance Study of Trapped Electrons in Gamma-Irradiated Hydrocarbon Polymers, Ph. D. Dissertation, Univ. of Tennessee, Knoxville, Tennessee.

Keyser, R. M., and Williams, F. (1969). *J. Phys. Chem.* **73**, 1623.

Keyser, R. M., Lin, J., Tsuji, K., and Williams, F. (1968a). *Polymer Preprints* **9**, 277.

Keyser, R. M., Tsuji, K., and Williams, F. (1968b). *Macromolecules* **1**, 289.

Lawton, E. J., Balwit, J. S., and Powell, R. S. (1960). *J. Chem. Phys.* **33**, 395, 405.

Leone, J. A., and Hamill, W. H. (1968). *J. Chem. Phys.* **49**, 5294, 5304.

Lewis, G. N. and Lipkin, D. (1942). *J. Amer. Chem. Soc.* **64**, 2801.

Lin, J. (1968). Electron Spin Resonance and Optical Studies of Radiation-Induced Trapped Electrons and Free Radical Intermediates in Organic Glasses, Ph. D. Dissertation, Univ. of Tennessee, Knoxville, Tennessee.

Lin, J., Tsuji, K., and Williams, F. (1967a). *Chem. Phys. Lett.* **1**, 66.

Lin, J., Tsuji, K., and Williams, F. (1967b). *J. Chem. Phys.* **46**, 4982.

Lin, J., Tsuji, K., and Williams, F. (1968). *J. Amer. Chem. Soc.* **90**, 2766.

Lin, J., Lok, G. K.-W., and Williams, F. unpublished work.

Lok, G. K.-W. (1969). Studies on Trapped Electrons in Gamma-Irradiated Hydrocarbons by Electron Spin Resonance, M. S. Thesis, Univ. of Tennessee, Knoxville, Tennessee.

Lombardi, J. R., Raymonda, J. W., and Albrecht, A. C. (1964). *J. Chem. Phys.* **40**, 1148.

Nikol'skii, V. G., Tochin, V. A., and Buben, N. Ya. (1963). *Fiz. Tver. Tela* **5**, 2248 (*English transl.:* (1964). *Sov. Phys. Solid State* **5**, 1636).

Oppenlander, G. C. (1968). *Science* **159**, 1311.

Ormerod, M. G., and Charlesby, A. (1962). *Nature* **195**, 262.

Ormerod, M. G., and Charlesby, A. (1964). *Polymer* **5**, 67.

Partridge, R. H. (1970). *J. Chem. Phys.* **52**, 1277.

Rånby, B., and Carstensen, P. (1967). *Advan. Chem. Ser.* **66**, 256.

Ronayne, M. R., Guarino, J. P., and Hamill, W. H. (1962). *J. Amer. Chem. Soc.* **84**, 4230.

Sauer, M. C., and Mani, I. (1968). *J. Phys. Chem.* **72**, 3857.

Shirom, M., and Willard, J. E. (1968). *J. Amer. Chem. Soc.* **90**, 2184.

Sprague, E. D., Takeda, K., and Williams, F. (1971). *Chem. Phys. Lett.* **10**, 299.

Takeda, K., and Williams, F. (1970). *Abstracts, Symp. Electron Spin Resonance*, No. F4. Div. of Phys. Chem. of the Amer. Chem. Soc. Athens, Georgia.

Teplý, J. (1969). *Radiat. Res. Rev.* **1**, 361.

Tochin, V. A., Nikol'skii, V. G., and Buben, N. Ya. (1966). *Dokl. Akad. Nauk SSR* **163**, 360.

Tsuji, K., and Williams, F. (1967). *J. Amer. Chem. Soc.* **89**, 1526.

Tsuji, K., and Williams, F. (1968). *J. Phys. Chem.* **72**, 3884.

Tsuji, K., and Williams, F. (1969a). *Trans. Faraday Soc.* **65**, 1718.

Tsuji, K., and Williams, F. (1969b). *Int. J. Radiat. Phys. Chem.* **1**, 383.

Whelan, D. J. (1969). *Chem. Rev.* **69**, 179.

Whelan, D. J., and Pinkerton, D. M. (1970). *Aust. J. Chem.* **23**, 391.

Willard, J. E. (1968). *In* "Fundamental Processes in Radiation Chemistry" (P. Ausloos, ed.), Chapter 9, p. 599. Wiley, New York.

Williams, F., and Hayashi, K. (1966). *Nature* **212**, 281.

Williams, M. L., Landel, R. F., and Ferry, J. D. (1955). *J. Amer. Chem. Soc.* **77**, 3701.

10

Thermoluminescence in Polymers

Roger H. Partridge

Division of Materials Applications, National Physical Laboratory, Teddington, Middlesex England

I. Introduction

Thermoluminescence from inorganic materials has been studied for many years, but only in about the last decade has a similar study been made of polymers or, indeed, of any organic solids. It has been found that thermoluminescence is, in fact, a most useful method of locating and studying molecular motion and structural transitions in polymers, and indeed, it is routinely used for this purpose in some laboratories. Furthermore, a detailed study of thermoluminescence can give much information on charge diffusion and trapping in polymers and on the influence on these of additives, impurities, physical and chemical structure, temperature, radiation, and other such factors. However, much remains to be done before a complete understanding of the whole phenomenon can be obtained in even the simplest polymers.

Some general features of polymer thermoluminescence will be discussed in the next five sections, but since few polymers have been examined in any detail, such generalizations are inevitably limited. Thus, in the final and largest section results on individual polymers will be described and discussed.

II. General Mechanism

Thermoluminescence is produced by simply irradiating with ionizing radiation at low temperature (usually liquid nitrogen) and then allowing the sample to warm up. The resulting plot of emission intensity against temperature (or warming time) is called a "glow curve." Samples will usually give considerable emission even if kept for many days between irradiation and warming, so the thermoluminescence cannot be due to simple excitation of particular molecular species since none has excited states with lifetimes of this duration. The only other possible mechanisms are the recombination of trapped ions or of neutral chemical species such as free radicals. Radical recombination can be ruled out in most cases on a number of grounds. First, thermoluminescence is often observed at temperatures well below those at which radical reactions occur (Semenov, 1962; Charlesby and Partridge, 1963b; Nikol'skii *et al.*, 1964; Aulov *et al.*, 1969). Second, no emission is observed when radicals are produced at low temperature by purely mechanical means and the sample is then warmed (Alfimov *et al.*, 1964). Third, the variation of thermoluminescence intensity with dose usually reaches a maximum at a few megarads, but radical production is linear with dose up to more than 50 Mrad (Alfimov *et al.*, 1964). Finally, the thermoluminescence output can usually be reduced to nearly zero by illuminating the sample with visible light prior to warming ("optical bleaching"), whereas this has no effect on the radical concentration (Alfimov *et al.*, 1964). However, all these phenomena are explicable if trapped ions are the reactive species, and there can be no doubt that in most cases thermoluminescence is caused by ion recombination (Charlesby and Partridge 1963b; Nikol'skii *et al.*, 1963a; Alfimov *et al.*, 1964).

The overall thermoluminescence mechanism thus involves initial ionization within the material by the incident radiation, with trapping of at least some of the resulting ions if the temperature is low enough. Subsequent warming (or optical bleaching) promotes recombination of the ions and, thus, formation of neutral excited molecules which, in some cases, return to the ground state by radiative emission. In many cases the polymer does not itself give any significant luminescence and the observed emission is then from impurity or additive molecules lying between the polymer chains. The concentration of these molecules is often extremely low (Partridge, 1966), and thus, most of the initial ionization and excitation will occur in the polymer itself followed by charge and/or excitation energy transfer to the luminescent molecules, causing these to ionize (Partridge, 1970a and Chapter 3). In addition, recombination of electrons with some of the polymer ions on warming may give rise to further migratory excitation that can excite or ionize the luminescent molecules.

III. Electron Traps

The very existence of thermoluminescence in polymers shows that both positive and negative charges must be efficiently trapped at low temperature, so thermoluminescence should be able to give useful information on these traps. The general topic of charge trapping is considered elsewhere in this volume, but some of the evidence on electron trapping from thermoluminescence studies will be noted here. In general, three different types of electron trap might be expected: "dielectric cavities," neutral molecules with a positive electron affinity, and free radicals. Cavity traps, where the cavity is probably best pictured as an irregular space bounded by a particular local arrangement of molecular chains, seem very likely in polymers, particularly since untrapping of electrons from them by molecular chain motion or structural transitions is very reasonable if the chains themselves define the cavities (Section IV). The long wavelength portions of the very broad absorption spectra of irradiated polyethylene (Partridge, 1970c) and poly(4-methylpentene-1) (Keyser and Williams, 1969) and the thermoluminescence optical activation spectrum of polyethylene (Nikol'skii *et al.*, 1963b; Nikol'skii, 1968) are all suggestive of cavity traps. Neutral molecules with positive electron affinities can certainly act as electron traps if in sufficient concentration, as shown, for example, by the extra glow peaks produced in polyethylene by oxygen (Section VII) and by triphenylamine (Partridge, 1970a). However, addition of aromatic or aliphatic molecules to polymers, whether to increase the thermoluminescence emission or modify the initial polymerization process, does not usually have a significant effect on the glow curve shape (Rozman, 1958; Fleming, 1968a); Boustead and Charlesby, 1970c). Thus, it is often true that while the actual luminescence of many polymers is due largely to chemical impurities, the charge trapping is due mainly to the basic polymer structure (including "chemical defects" in the polymer chains and radiation-produced species). Free radicals are quite likely to trap electrons at high doses, but as discussed in Section VI, they may well often not yield up these electrons for thermoluminescence and in such cases are actively competing with the thermoluminescence process.

Some evidence is now accumlating to suggest that a fourth type of electron trap may exist in polymers. This is a molecule, or molecular group, that does *not* have a positive electron affinity but that can nevertheless trap electrons when it is in a polymer matrix, perhaps by distorting the polymer structure in its immediate vicinity to form a sort of cavity. Such traps would be truly intermediate between pure cavity traps and genuine molecular anions. The strongest evidence for this type of trap comes from polytetrafluorethylene in which Mele *et al.* (1968) found a large glow peak that only appeared in the presence of dissolved gases such as oxygen and *helium* and was unobservable in evacuated samples (see also Section VII, H). Since helium could not by itself trap positive or negative charges, it seems that trapping must be a cooperative effort between helium atoms, polymer chains, and electrons. More evidence comes from polyethylene, in which Boustead and Charlesby (1970c) have found that the intensities of the lowest temperature glow peaks are closely correlated with unsaturation in the polymer chains, yet alkenes do not generally trap electrons. [For instance, Gallivan and Hamill (1966) found that there was little or no trapping of electrons by 2-methylpentene in a 3-methylpentane glass even at high alkene concentrations.] Further evidence is that the narrow singlet ESR line of weakly trapped electrons in polyethylene irradiated with low doses is considerably enhanced by the addition of benzene to the polymer (Buben *et al.*, 1962; Campbell, unpublished), and this is certainly not due to formation of benzene anions (as tentatively suggested by Buben *et al.*) since these have a distinctive seven line spectrum (Salem, 1966). Thus, benzene apparently promotes electron trapping in polyethylene without formation of true molecular anions. Finally, the absorption spectrum of irradiated polyethylene (Partridge, 1970c) extends much further into the visible and ultraviolet region than is usually observed for cavity-trapped electrons, yet optical bleaching of the thermoluminescence (Nikol'skii, 1968) indicates a continuous range of optical activation energies that would not be expected from true molecular anions or cations but which would be quite reasonable for the "molecular cavity" type of traps proposed here.

IV. Molecular Motion and Structural Transitions

Measurements on many polymers and other organic materials have shown that nearly all the thermoluminescence glow peaks are associated with some type of molecular motion or with structural changes associated with this motion (Nikol'skii and Buben, 1960; Semenov, 1962; Charlesby and Partridge, 1963a, b; Nikol'skii *et al.*, 1964; Partridge, 1965; Magat, 1966). In particular, glow peaks are almost invariably observed at the glass transition of each material and at its melting point, unless all ion recombination has occurred before the appropriate transition temperature is reached.

In considering thermoluminescence glow curves it seems essential to make a distinction wherever possible between molecular motions which lead to a large scale and fairly sudden change of molecular chain conformation and motions which do not directly cause such changes. The latter types of motion are often observed at quite low temperatures and generally involve rotation of side-chain units, such as methyl or phenyl groups, or of small segments of the main chain [such as the "crankshaft" motion of Schatzki (1962)]. Such rotations will occur over fairly well-defined potential barriers and will thus have a genuine thermal activation energy or, at least, an average activation energy. If the thermoluminescence charge traps are actual polymer groups or alternatively "cavities" formed by a particular local arrangement of molecular chains, then it is reasonable to assume that the rate constant for untrapping will be proportional to the frequency with which molecular motion occurs and, hence, that it will have a Boltzmann-type variation with temperature (Partridge, 1965).

Molecular motions associated with structural transitions are likely to give a considerably different rate of ion recombination. The most important structural transition in the usual thermoluminescence temperature range (up to $\approx 300\,°K$) for amorphous polymers is the glass–rubber transition. Here there is a very considerable increase in molecular chain motion for only a small rise in temperature, where the motion involves quite long sections of the chains but does not give rise to full-scale molecular diffusion. The onset of this more extensive molecular motion must inevitably lead to a more rapid ion recombination and, hence, to the occurrence of a glow peak in the vicinity of the glass-transition temperature; good examples of this can be seen in the glow curves of polytetrafluorethylene (Semenov, 1962) and polybutadiene (Alfimov and Nikol'skii, 1963). The increased molecular motion at the glass transition is due not just to thermal activation but also to the polymer's having obtained, by thermal expansion, a critical volume such that much increased molecular motion is possible. For this reason the molecular relaxation processes at the glass transition have a relaxation time which is not given by a simple Boltzmann factor but is instead found to be quite well described by the Williams, Landel, and Ferry (WLF) equation. A possible application of this equation to glow peaks in the glass-transition region is discussed later.

The other major structural transition for both amorphous and crystalline polymers is usually the melting point although in some crystalline polymers, such as polytetrafluoroethylene, changes in the crystal structure of the solid polymer may occur also at temperatures well below the melting point. The onset of large-scale molecular diffusion will clearly lead to rapid recombination of any remaining ions, with production of another glow peak; such melting-point peaks are common in the glow curves of small alkanes (Section VII, D) since these have low melting points. As with the glass transition, the sudden increase in molecular mobility will not be describable by a Boltzmann-type equation.

V.　Glow-Peak Equations

The thermoluminescence glow curve is essentially a plot of the rate of ion recombination in a material as a function of temperature although it may sometimes be considerably biased by various quenching processes at higher temperatures. However, by no means all ions present will give luminescence on recombination since some types will return to the ground state predominantly by radiationless transitions. The kinetics of ion recombination will usually be either first or second order, depending on whether the rate-determining step is charge untrapping or charge recombination after untrapping. In the only cases studies so far, i.e., polyethylene (Charlesby and Partridge, 1963a), polybutadiene (Tochin and Nikol'skii, 1969), polymethylmethacrylate (Fleming, 1968a), and squalane (Boustead and Charlesby, 1970a), the kinetics appear to be first order. This suggests that most electrons are trapped within the Coulomb field of an ion (probably their parent ion) after ionization and that on warming they recombine only with that ion.

For first-order ion recombination the recombination rate is just

$$dn/dt = -Kn \tag{1}$$

where n is the number of trapped ion pairs at time t and K is the recombination rate constant. If the trapped charges were released by purely thermal activation from fixed traps, as is common in inorganic thermoluminescence, then K could be expected to have the Boltzmann form

$$K = S \exp(-E/kT) \tag{2}$$

where E is the "trap depth" and S is the "frequency constant" (Randall and Wilkins, 1945). But since in polymer thermoluminescence most of the trapped charges seem to be released by molecular motion rather than direct thermal activation, the use of Eq. (2) in this case is *not* automatic, and the basic cause of each particular glow peak should be considered.

A. Glow Peaks not Associated with Structural Changes

As discussed earlier this category is concerned with the very limited motions of side groups or small main chain segments, and it is not unreasonable to assume that the rate of charge release due to these motions will be given by the product of the frequency of this motion and the probability that each individual movement will cause the release of a trapped charge. In this case Eq. (2) should be at least approximately true since the motion is thermally activated and there is no overall structural change to promote charge diffusion, but E now becomes the activation energy of the molecular motion itself and conveys no information at all about the depth of the charge traps (Partridge, 1965). Polymethylmethacrylate (see Section VII, G) is probably an example of a polymer whose main glow peak is due to this type of motion.

The thermoluminescence intensity L at any time t is given by the product of the ion recombination rate and the "luminescence constant" α (i.e., the probability of photon emission from each recombination), so using also Eq. (2)

$$L = -\alpha(dn/dt) = \alpha S n \exp(-E/kT) \tag{3}$$

where

$$n = n_0 \exp\left[S \int_0^t \exp\left(-\frac{E}{kT} \right) dt \right] \tag{4}$$

and n_0 is the number of trapped ion pairs at the start of warming ($t = 0$). The integral in Eq. (4), and thus the glow-peak shape, depends on the warming rate used. For the most common warming rates used experimentally, those with temperature increasing linearly or exponentially with time, it is not possible to obtain a simple analytical expression for the integral, although a solution in terms of exponential integrals is possible (Bonfiglioli *et al.*, 1959; Partridge, 1964). However, a useful expression for the temperature T_M of the glow-peak maximum can be obtained by differentiation of Eq. (3), which for a linear warming rate β gives

$$SkT_M{}^2/E\beta = \exp(E/kTM) \tag{5}$$

or upon further differentiation with respect to warming rate

$$\frac{dT_M}{d\beta} = \frac{T_M{}^2}{2T_M + (E/k)} \tag{6}$$

From Eq. (6) it is clear that an increase in heating rate will always shift the peak maximum towards higher temperatures and vice versa and, furthermore, that this shift will be largest for glow peaks with the lowest activation energy.

Equation (5) provides a good method of measuring E and S for a particular glow peak, so long as Eq. (2) is likely to be applicable, since a plot of $\log_e (T_M{}^2/\beta)$ against $1/T_M$ has slope E/k and intercept k/ES.

Another common method for obtaining E is that of initial rise (Garlick and Gibson, 1948), in which the logarithm of the initial intensity portion of a glow curve is plotted against reciprocal temperature on the assumption that α and n in Eq. (3) are effectively constant. However, this does not give S or the kinetic order.

Many methods have been proposed for obtaining activation energies from glow-peak temperatures and shapes, and an excellent review of these has been given by Fleming (1968a, b). Most of these assume either first- or second-order kinetics from the outset, and most also demand a glow peak which is well separated from its neighboring peaks. Furthermore, it is tacitly assumed that neither the activation energy nor luminescence constant vary during the extent of the peak, and for polymers this is often not true. In fact, α will often decrease rapidly with temperature if a major part of the thermoluminescence is basically phosphorescence emission from excited molecules formed by ion recombination, since such phosphorescence can be quenched by thermally activated collisions between the excited molecules and the polymer chains. Indeed, the same molecular motion may be simultaneously promoting thermoluminescence by causing ion recombination and reducing thermoluminescence by phosphorescence quenching. Such quenching is actually a powerful method of studying molecular motion in polymers (Charlesby and Partridge, 1965b; Boustead, 1970b). The efficiency of such phosphorescence quenching can be expressed (Boustead, 1970b), in terms of α, as

$$\frac{1}{\alpha_T} - \frac{1}{\alpha_0} = a \exp\left(-\frac{E_Q}{kT}\right) \qquad (7a)$$

where α_0 and a are constants and E_Q is the activation energy of the quenching process. Clearly, at very low temperatures quenching will be negligible, but at very high ones Eq. (7a) becomes

$$\alpha_T \simeq (1/a) \exp(E_Q/kT) \qquad (7b)$$

and this will greatly modify the glow curve (see, for instance, Boustead and Charlesby, 1970c) and any kinetic constants calculated from it. For example, in the extreme case of Eq. (7b) the form of Eq. (3) is now

$$L \simeq (Sn/a) \exp[(E_Q - E)/kT]$$

and so the activation energy measured by initial rise would be $E - E_Q$ and not E. Since the same molecular motion will usually both untrap charges and quench phosphorescence, the activation energy measured by initial rise could vary between E and zero, depending upon the temperature of measurement and the phosphorescence content of the thermoluminescence spectrum.

Probably the only method of obtaining E and S that is not affected by phosphorescence quenching is that of luminescence decay at constant temperature (Charlesby and Partridge, 1963a), and this has the further advantage that the kinetic order is determined from direct observation of decay curves before the kinetic constants are evaluated. Thus, for first-order kinetics the decay of luminescence at constant temperature is, from Eq. (1), just

$$L = \alpha K n_0 \, e^{-Kt}$$

and thus K can be obtained for a series of different temperatures by plotting log L against t, whatever the value of α at each temperature. If Eq. (2) holds then a plot of $\log_e K$ against $1/T$ gives both E and S. The main problem with this method is the experimental difficulty of maintaining the sample temperature constant to less than a degree for an hour or more at any temperature down to at least that of liquid nitrogen.

It is important to note that if any of the above methods for obtaining E are applied to a glow peak whose thermoluminescence intensity is *not* given by Eq. (3) (i.e., in particular a peak associated with a large-scale structural transition), then an apparently reasonable "pseudo-activation energy" may often be obtained but the value of S will be impossibly high. Examples of this will be given for polyethylene (Section VII, A) and polybutadiene (see below).

B. Glow Peaks Associated with a Glass Transition

The various structural transitions are too diverse to enable use of a general equation for ion recombination rate as a function of temperature, but this may be possible for the important case of the glass transition. At the glass transition it has been found that the Boltzmann-type equation for mechanical and dielectric relaxation times is replaced by the WLF equation, so by analogy it may be possible to replace Eq. (2) with an equivalent WLF form [McCrum *et al.* (1967) and Chapter 5].

$$K = K_0 \exp\{C_1(T - T_g)/[C_2 + (T - T_g)]\} \tag{8}$$

where T_g is the glass-transition temperature and K_0, C_1, and C_2 are constants. This leads to an equation analogous to Eq. (3) for the thermoluminescence intensity

$$L = \alpha K_0 n \exp\{C_1(T - T_g)/[C_2 + (T - T_g)]\} \tag{9}$$

with an integral expression for n analogous to Eq. (4).

As before, the glow peak maximum T_M can be obtained by differentiation of Eq. (9) which, for a linear warming rate β, gives

$$\frac{\beta C_1 C_2}{K_0[C_2 + (T_M - T_g)]^2} = \exp\left(\frac{C_1(T_M - T_g)}{C_2 + (T_M - T_g)}\right) \tag{10a}$$

Further differentiation, with respect to warming rate, gives an equation analogous to Eq. (6)

$$dT_M/d\beta \simeq C_2/\beta C_1$$

in which it was assumed that $C_2 \gg T_M - T_g$, as is indicated below. This equation shows that, as in the previous case, an increase in the warming rate will always cause an increase in the glow-peak maximum.

A useful test of these equations is provided by the work of Alfimov and Nikol'skii (1963) on the variations of glow-peak temperature with warming rate for the glass-transition glow peaks of some polybutadienes. Application of Eq. (5) to their results gives activation energies of about 2 and 1.5 eV for samples with a high 1,2 and 1,4 structure content, respectively, but the corresponding S values are about 10^{45} and 10^{42} sec^{-1} and so are impossibly high. However, if Eq. (10a) is expressed, again assuming that $C_2 \gg T_M - T_g$, as

$$(C_1/C_2)(T_M - T_g) \simeq \log_e(C_1/K_0 C_2) + \log_e \beta \tag{10b}$$

then T_M plotted against $\log \beta$ gives C_1/C_2. Use of this for the polybutadiene results gave C_2/C_1 as about 2.2 for the 1,4 sample and about 1.7 for the other sample. Now the constants C_1 and C_2 of the WLF equation are "universal constants" with values of about 17 and 51, respectively, so C_2/C_1 should be about 3; the extent of agreement with the experimental results is really quite good, especially since it is known that T_g does itself vary somewhat with heating rate (Shen and Eisenberg, 1967). Since C_2 is 51 and T_M is within a few degrees of T_g for all warming rates (Alfimov and Nikol'skii, 1963) it can be seen that, indeed, $C_2 \gg T_M - T_g$ as assumed. It is not really possible to obtain the frequency constant K_0 because the precise value of T_g is hard to obtain (its practical definition is to some extent a matter of experimental convenience), and thus, the value of $T_M - T_g$ is subject to considerable error. However, it can be seen from Eq. (10a) that, since $T_M - T_g$ will only be a few degrees, the value of K_0 will be quite modest and certainly nothing like the massive values obtained by use of Eq. (5).

Thus, it seems likely that Eq. (9) will provide a much more satisfactory description of glow peaks caused by glass transitions than will Eq. (3).

VI. Thermoluminescence Variation with Dose

The thermoluminescence light output essentially measures the number of trapped ion pairs in the material that are capable of producing thermoluminescence, so the variation of this with dose can indicate effects such as the destruction of luminescent molecules, untrapping of trapped charges by radiation, and production of competing charge traps by radiation. Most dose curves observed so far have been of a few distinct types, so some discussion of these is worthwhile.

The simplest dose curves, apart from pure linear ones, are those of form

$$I = I_0[1 - \exp(-Ar)] \tag{11}$$

where I is the intensity at the glow-peak maximum (assuming peak *shape* is independent of dose), r is the dose, and A is a constant. Such curves, which saturate at high doses, have been observed from some of the glow peaks of n-heptane, squalane, and polyethylene (Boustead 1970d; Boustead and Charlesby, 1970a). The saturation could be due to thermal untrapping (for glow peaks close to the irradiation temperature, particularly at low dose rates) or to untrapping of trapped charges by the radiation itself or to the complete filling of all available charge traps. The latter is probably rather unlikely in most polymers in view of the apparent range of potential traps, noted earlier, unless a few electron-trapping molecules such as oxygen are present. Untrapping by radiation, however, could be quite a general mechanism since most polymer thermoluminescence is activated by molecular chain motion and a considerable amount of the irradiation energy is likely to be ultimately dissipated in molecular vibrations.

A second, less common, type of curve is roughly represented by

$$I = Br(r + C) \tag{12}$$

where B and C are constants, and thus shows a component which increases quadratically with dose. This behavior, found in one glow peak of squalane and one of polyethylene (Boustead, 1970e; Boustead and Charlesby, 1970a), indicates creation of extra charge traps by radiation in addition to the traps already present (creation of extra luminescent molecules is most unlikely). The quadratic increase must reduce to a linear increase at higher doses as only a limited number of charges will be available for trapping after each dose increment [see Eqs. (13) and (16) below].

The most common dose curves increase to a maximum, often at 1–3 Mrad, and then decrease steadily as the dose increases. The actual decrease (as opposed to mere saturation) could be due to destruction of traps, destruction of luminescent molecules, or production by radiation of species which compete for charges or accept excitation energy from excited luminescent molecules. The latter is not very likely since virtually all radiation products will be aliphatic species and so will usually have energy levels which are higher than the predominantly aromatic luminescent molecules. Actual destruction of traps or luminescent molecules is probably not too significant (except at extremely high doses) since samples given a large dose usually "recover" almost completely after warming to room temperature, though some permanent destruction has been observed (Charlesby and Partridge, 1963a, 1965b). However, production of charge traps by radiation, linked also with radiation untrapping, does seem quite possible in that measurements on polybutadiene by Alfimov et al. (1964) strongly suggest that the intensity decrease at high doses is directly related to the production of free radicals by the radiation. Now the rate of radical production in this polymer is linear to more than 50 Mrad while the thermoluminescence maximum occurs at about 1 Mrad. Thus, in this case there may well be competition for charges (probably electrons) between the "shallow" traps of the polymer itself and the free radicals, where the electrons captured by radicals (which have a high electron affinity) are *not* given up for thermoluminescence during subsequent warming. If radiation untrapping only affects electrons in the shallow polymer traps (the traps which give thermoluminescence), then the thermoluminescence intensity will decrease at high doses since the radicals capture most of the new electrons while the electrons already in shallow traps are steadily untrapped by the radiation. It is interesting to express this model in mathematical terms. Let s, f, and i be the concentrations of shallow-trapped electrons, free radicals, and radical ions (electrons trapped on free radicals), respectively, at dose r. Let D be the probability of an electron being captured by a shallow trap (assuming a very high concentration of such traps) and Ef be the probability of its capture by a free radical. Let F be the rate constant for untrapping of shallow-trapped electrons by radiation. Finally, let G and H be the rates of electron trapping (by all traps) and radical production, respectively, with both assumed to be independent of dose. From the above model we then have

$$i + f = Hr \tag{13}$$

$$di/dr = GEf/(D + Ef) \tag{14}$$

$$ds/dr = [GD/(D + Ef)] - Fs \tag{15}$$

Combination of Eq. (13) and Eq. (14) leads to

$$(H - G)r = f - [GD/E(H - G)] \log_e \{1 + [E(H - G)/HD]f\} \tag{16}$$

which shows that the rate of free radical production varies from H at low doses to H-G at high doses. Combination of Eqs. (13) and (16) shows that the concentration of radical ions increases quadratically at low doses but becomes linear at high doses.

Differentiation of Eq. (16) with respect to r and combination with Eq. (15) leads to l, a dimensionless quantity proportional to S and, hence, to the thermoluminescence output, being given as

$$l = S\left(\frac{E(H - G)}{DG}\right) = y^{G\phi} e^{-\phi y} \int_H^y y^{-(1 + G\phi)} e^{\phi y} \, dy \tag{17}$$

where

$$\phi = FD/E(H - G)^2$$

and

$$y = H + [f(E/D)(H - G)]$$

and where the integral can be expressed as the series

$$\left(- \frac{1}{G\phi} y^{-G\phi} + \frac{\phi}{(1 - G\phi)1!} y^{(1 - G\phi)} + \frac{\phi^2}{(2 - G\phi)2!} y^{(2 - G\phi)} + \cdots \right)_H^y$$

Typical graphs of l against r, as given by Eqs. (13) and (17), are shown in Fig. 1, and the form of these is indeed very similar to the high dose curves obtained by Alfimov *et al.* (1964); in particular, their curves for polyethylene and polybutadiene are very similar to curves 2 and 3, respectively. For curves 2, 3 and 4 $G\phi \ll 1$, and in such cases the integral in Eq. (17) is virtually

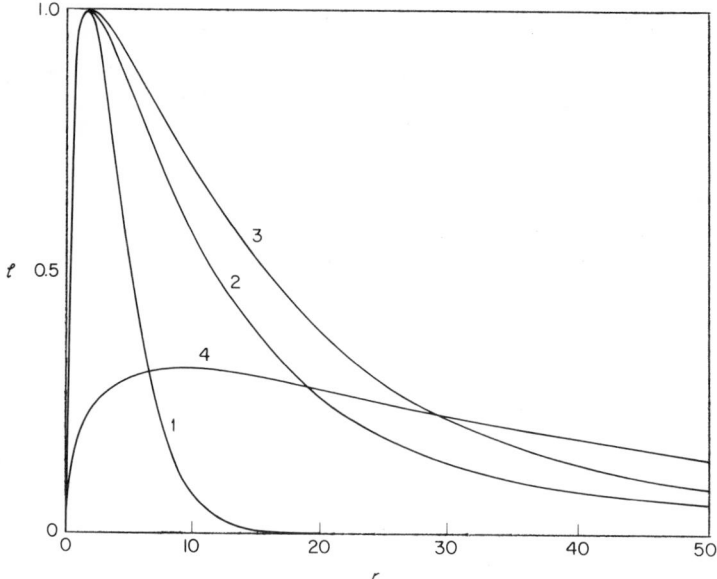

Fig. 1. Typical curves of l against r as calculated from Eqs. (16) and (17). Curves 1, 2, and 3 normalized to unity at their maximum values of l. $H = 1$ and $G = 0.05$ in all cases. Other values are tabulated below.

Curve	F	ϕ	E/D
1	1	4	0.277
2	0.13	0.0025	56.9
3	0.08	0.00013	701.7
4	0.0408	0.0151	3

an exponential integral and can thus be evaluated from the appropriate tables. A useful, though very approximate, relation for the dose r_{MAX} at the maximum of the curve is

$$[r_{MAX}]^4 \simeq \frac{HD}{0.3(H - G)^2 EF^3} \tag{18}$$

which shows that the maximum is principally determined by F and the ratio D/E, i.e., by the rate of radiation-induced untrapping and by the relative electron capture efficiencies of shallow traps and free radicals (or other radiation-produced species). For the same value of r_{MAX} curves will decrease more steeply at high doses if D/E and F are large and vice versa. Equation (18) is valid roughly for values of the dimensionless quantity ϕH in the range about 10^{-5} to 10^2 as long as $G\phi \lesssim 1$

If the above model is appropriate for any particular polymer, then one experimental indication, apart from the shape of the dose curve itself, would be an apparent change in the rate of free radical production from H at low doses to H-G at high doses. Unfortunately, r_{MAX} is often ≈ 2 Mrad, and from the above equations, the change of rate would come at ≈ 0.1 Mrad, which is often barely enough to produce a significant ESR signal; detection of the change also, of course, requires that G is not very much smaller than H. Another experimental indication is that the ratio of shallow-trapped electrons to free radicals is a maximum at low doses but decreases rapidly at high doses when the shallow trapped-electron concentration falls; precisely this behavior has been observed for the trapped electron ESR line in a variety of polymers (Alfimov et al., 1964). It is interesting to note that exposure of low-temperature irradiated samples to visible light removes virtually all thermoluminescence but has no detectable effect on the free radical concentration (Alfimov et al., 1964), which suggests that either the concentration of free radicals is far higher than electrons trapped on free radicals or, more likely, that electrons trapped on free radicals cannot be released by visible radiation.

Equations (13)–(15) are, in fact, fairly general dose curve equations since they will give the exponential behavior of Eq. (11) if $E = 0$ (no trapping on radiation-produced species) and also the "linear plus quadratic" behavior of Eq. (12) at *low* doses if the electrons trapped on radiation-produced species *do* contribute to thermoluminescence along with shallow trapped electrons and if $F = 0$ (no significant untrapping by radiation at these low doses).

VII.   Thermoluminescence in Particular Polymers

A.   Polyethylene

Thermoluminescence has been studied much more in polyethylene than in any other polymer, so it will be considered here at some length. Its glow curve, in samples free of oxygen, contains at least four peaks, though not all these may be resolved under some experimental conditions. These glow peaks, designated for convenience α_1, α_2, β, γ in order of ascending temperature, have peak temperatures of about 100°, 105°, 160°, and 205°K for a 3° min^{-1} warming rate (Boustead, 1970a), but the α peaks can shift by more than 60° for much higher warming rates (Charlesby and Partridge, 1963a). Even for a constant warming rate the peak temperatures can vary considerably between different polyethylene samples, though they are reproducible for samples taken from the same batch. Nikol'skii and Buben (1962) found that the γ peak shifts to lower temperatures with increasing dose for doses above 2 Mrad, and this they ascribed to increased molecular chain mobility due to trapping within the polymer of gaseous irradiation products. The onset of thermoluminescence occurs at about $40°K$ (Aulov *et al.*, 1969) so some luminescence can be seen even at liquid nitrogen temperature.

Dissolved oxygen has a dramatic effect on the glow curve (Charlesby and Partridge, 1963b) since it completely removes the β and γ peaks (for doses below about 0.1 Mrad) but produces one extra peak (ε) at a temperature, for $3°$ $\min^{-1}$ warming rate, of about $145°K$ (Boustead, 1970c). The rate of oxygen diffusion into a previously evacuated sample can be measured using this effect, and indeed, the dependence of this rate on sample thickness is the simplest proof that thermoluminescence is a bulk and not a surface phenomenon. Two basic mechanisms have been suggested for the production of the ε peak, one (Charlesby and Partridge, 1965a) that the O_2 acts as an efficient electron trap and the other (Nikol'skii *et al.*, 1964) that the O_2 combines with the electron traps, assigned as alkyl radicals, when it becomes mobile at higher temperatures and in doing so releases any trapped electrons. The latter mechanism was proposed mainly because the onset of the ε peak was observed to occur at about the same temperature as O_2 molecules started to become mobile and combine with alkyl radicals to form peroxide radicals [this O_2 mobility is likely to be due to the onset of a particular type of molecular chain motion, though not the one associated with the glass transition (Nikol'skii *et al.*, 1964)]. Also, the peak maximum roughly coincided in temperature with the maximum rate of peroxide formation. However, it is also found that virtually all peroxide radical formation finishes soon after the ε peak maximum is reached (Charlesby and Partridge, 1963a; Nikol'skii *et al.*, 1964), so most of the high-temperature portion of the peak cannot be a by-product of peroxide radical formation. Furthermore, trapping of radiation-produced charges on radiation-produced radicals would be expected to give a quadratic dependence of output on dose at low doses, such as has been observed for some other glow peaks (Section VI), whereas it is found that the ε peak is linear with dose down to at least 0.8 rad (Charlesby and Partridge, 1963a,b). Another argument against this mechanism, and in support of O_2 acting as an electron trap, is that if an irradiated sample containing O_2 is left for several days at liquid nitrogen temperature before warming then the ε peak intensity is found to be unchanged but the other peaks (α and γ) have decayed by a factor of around 3 (Partridge, 1964). This suggests the presence of a new and "deep" electron trap, presumably O_2 itself, which can efficiently capture electrons at $77°K$. [O_2 would not capture positive charges from the polyethylene chains as its ionization potential is about 2 eV higher than that of a long-chain alkane molecule (Watanabe *et al.*, 1962).] This is also likely in that formation of O_2^- has often been observed in radiation chemical reactions.

Electron release from O_2^- could be either by direct thermal activation or by migration of the O_2^- until it found a suitable reaction site. For the first of these to be significant the electron affinity of O_2 would probably have to be reduced somewhat below its vacuum value by interaction with the surrounding polymer matrix, and indeed, Boustead (1970c) has shown that this may well occur. Equally, however, the appearance of the ε peak in the region of O_2 mobility suggests the second process; indeed, the reaction site could well be an alkyl radical and the electron be released during formation of a peroxide radical, which is similar to the proposal of Nikol'skii *et al.* (1964) except that initial trapping is on O_2 rather than an alkyl radical. It is interesting to note that a counterpart of the ε peak, appearing only when O_2 is present, has been observed in measurements of electrical conductivity during warming after irradiation (Sichkar *et al.*, 1969).

Also difficult to explain at present is the complete quenching of the β and γ peaks by O_2 at low doses. This is not just due to the thermoluminescence having been released at lower temperature (as the ε peak) because the total luminescence output in low-density polyethylene containing O_2 is about a factor of 5 less than in samples free of O_2 (Charlesby and Partridge, 1963b). It is significant that at these *low* doses there are still many free O_2 molecules in the polymer even after peroxide radical formation, since the O_2 molecules outnumber the radiation-produced alkyl radicals. These O_2 molecules probably do not quench by accepting energy from the excited luminescent molecules, though this is well known to occur at higher O_2 concentrations in other systems (see, for example, Berlman and Walter, 1962), since no significant oxygen quenching of the UV-induced phosphorescence from polyethylene is found (Charlesby and Partridge, 1965b). The quenching probably involves capture of trapped electrons from other electron traps since the broad absorption spectrum of these (Partridge, 1970c) fades as the ε peak becomes exhausted, whereas in samples free of oxygen the color persists to much higher temperature (Charlesby and Partridge, 1963b). It may well be that the now mobile O_2 molecules capture electrons from their traps and deliver them to the various cations and that the instantaneous complex formed during this exchange promotes radiationless dissipation of the excitation energy. Another possible mechanism for quenching by O_2 molecules is by their acting as "excitation traps" in competition with the luminescent molecules during irradiation, thus reducing the number of ionized luminescent molecules and, hence, the ultimate thermoluminescence emission. Such transfer of excitation energy from alkanes to O_2 molecules was predicted on the "exciton model" of alkanes (Partridge, 1970b, and

Chapter 3) and has recently been observed experimentally for a number of *n*-alkanes (Hirayama and Lipsky, 1969). However, the concentration of O_2 in polyethylene due to air diffusion may not be large enough to make this mechanism very significant.

The glow curve shape of polyethylene is little dependent on the type of ionizing radiation used (except for near-UV radiation, which interacts by a different mechanism, as noted below) but it is sensitive to crystallinity. The oxygen sensitivity of the β and γ peaks is sufficient to show that they arise from, or at least on the edge of, the amorphous regions since the crystalline regions are virtually impermeable to gases (Michaels and Parker, 1959). The α peaks are relatively much larger than the β and γ peaks in samples of high crystallinity and thus almost certainly arise mainly from the crystalline regions (Partridge and Charlesby, 1963), although recent work by Boustead and Charlesby (1970c) suggests that there is some contribution also from amorphous regions.

The luminescence output of all the glow peaks (except β and γ in the presence of oxygen) is linear with dose up to at least 0.2 Mrad (Charlesby and Partridge, 1963a,b). At 0.1–5 Mrad doses the behavior differs from peak to peak, and also depends on the oxygen content and general chemical purity of the sample; indeed Boustead (1970e) has observed from different peaks all three types of behavior discussed in Section VI. At still higher doses, 5–50 Mrad, the output tends to decrease steadily with increasing dose, as discussed in Section VI. The thermoluminescence output is generally independent of dose rate (except for irradiations of long duration when the finite decay at $77°K$ may become significant) for true ionizing radiations. However, it was found to be *linear* with dose rate for UV radiation (Charlesby and Partridge, 1965b) because of a two photon ionization process in which the first photon excites a suitable molecule to (indirectly) its lowest triplet state and absorption of a second photon while in that state is sufficient to cause ionization. This process will occur particularly in molecules with a high singlet-triplet crossover efficiency and a long triplet state lifetime; these are usually efficient phosphorescence emitters, and so the thermo-luminescence spectrum will probably be mainly a phosphorescence one. The relative intensities of the glow peaks are considerably different under UV irradiation (Charlesby and Partridge, 1965b), and this is probably because the peaks are associated with different concentrations of phosphores-cence-emitting molecules, plus the fact that thermal quenching of phos-phorescence from the high-temperature peaks will be significant.

The thermoluminescence emission spectrum varies considerably between different batches of polyethylene but usually shows two broad and irregular bands centered around 4700 and 3500 Å. These comprise the phosphorescence and fluorescence bands, respectively, of a number of different impurity molecules. Spectra so far partially or fully identified are those of phenanthrene (Partridge, 1966), benzoic acid, naphthalene, anthracene, and an alkylarylketone (Boustead and Charlesby, 1967). Identification of the other spectra is difficult since they consist of featureless bands, but they may include emission from carbonyl groups attached to the polymer chains (Charlesby and Partridge, 1965a). Emission from polyethylene itself is not likely to be significant although a weak fluorescence has recently been observed at much shorter wavelengths ($\approx$2100 Å) from a variety of alkanes (Hirayama *et al.*, 1970). Observation of various aromatic molecules in polyethylene is not too surprising since contact with lubricating oils during manufacture is to be expected (Boustead and Charlesby, 1967). More surprising, however, is that some of the spectra increase in intensity with time (Charlesby and Partridge, 1965a), and a corresponding build-up in absorption can also be seen even after initial purification of the sample (Partridge, 1966). Simple oxidation could hardly produce such complex aromatic molecules, but it is possible that they result from atmospheric pollution. Air pollution studies have revealed many luminescent aromatic molecules (Altshuller 1965), and these molecules could well get trapped within polymer samples after having diffused into them with the air. Another important feature of the thermoluminescence emission spectrum is that it changes somewhat with *temperature* (Boustead and Charlesby, 1967). One reason for this is that, as shown later, different luminescent molecules tend to be found in different structural regions of the polymer and thus become associated with different glow peaks. A second important reason is that the phosphorescence component is susceptible to quenching by thermally activated molecular collisions while the fluorescence component is not, and thus the latter predominates at high temperatures. The quenching of UV-induced phosphorescence from additives in polyethylene has been studied as a function of temperature by Charlesby and Partridge (1965b) and by Boustead (1970b). The quenching is so strong that phosphorescence is unobservable above about 0°C.

Activation energies for polyethylene thermoluminescence have been measured by three of the methods mentioned in Section IV, warming rate variation, initial rise, and decay at constant temperature. Nikol'skii and Buben (1960) and Tochin and Nikol'skii (1969) used the warming-rate method, the former obtaining values of 1.28 eV for the γ peak and 0.3 eV for the α peak, while from the graph of the latter 2.8 eV can be calculated for the γ peak. However, the frequency constant of the γ peak is found, using Eq. (5), to be about 10^{27} sec^{-1} from the earlier work and 10^{61} sec^{-1} from the latter work. These values are far too high to be associated with any real physical process, and it is therefore likely that use of Eq. (5) is invalid for the γ peak because it is associated with a structural transition, as discussed in Section V. The polyethylene transition in this temperature region is sometimes called the glass transition although there is some doubt as to whether it is really meaningful to assign a glass transition to polyethylene (McCrum *et al.*, 1967). The γ peak seems to shift more with warming rate than does the associated structural transition and is thus rather intermediate in behavior between a pure "molecular motion peak" and a pure "glass-transition peak." Thomas *et al.* (1968) used a modified initial rise method to obtain a continuous range of activation energies from 0.12 eV at low temperatures up to 0.59 eV at higher temperatures although, unfortunately, the associated temperature range was not given. The reason for obtaining here a continuum of activation energies rather than a few discrete values corresponding with molecular motion activation energies may be linked with the structural transition effects which complicate the warming-rate measurements. Another factor is that the thermoluminescence was induced here by UV irradiation and is therefore likely to suffer from the phosphorescence quenching error discussed in Section V, although this by itself would lead to activation energies which decreased rather than increased with temperature. The decay of luminescence at constant temperature was measured by Charlesby and Partridge (1963a), who found three simultaneous exponential decays (in the absence of oxygen); these all had approximately the same activation energy of 0.027 eV at low temperatures but different frequency constants, ranging from about 4×10^{-3} to 10^{-1} sec^{-1} (the molecular chain frequency below the glass transition is usually $\approx 10^{-1}$ sec^{-1}). At higher temperature, around 175° and 230°K, the activation energies increased to about 0.21 eV for the fluorescence emission component and 0.44 eV for the phosphorescence component. These changes in decay rate correlate quite well in temperature and activation energy with the onset of various molecular motions in the polymer as observed by NMR and electrical conductivity (Partridge, 1965; Sichkar *et al.*, 1967) and by phosphorescence quenching (Boustead, 1970b), while the frequency constants are certainly reasonable. The activation energy difference should not be due to thermal quenching (Section V) and may instead indicate that the majority of the phosphorescence-emitting molecules in this particular sample are in different regions from the majority of the fluorescence emitters. The low-temperature activation energy is so low that it could correspond to direct thermal activation from fixed "cavity" traps, though the frequency constants suggest molecular motion; Boustead (1970b) did observe some phosphorescence quenching at around 100°K that was presumably caused by molecular motion, but its activation energy was much higher (≈ 0.15 eV). Initial rise measurements on samples containing O_2 suggest for the ε peak an activation energy of 0.28 eV and frequency constant of 10^{10} sec^{-1} (Partridge, 1964).

The thermoluminescence output of irradiated polyethylene, and many other polymers, is reduced if it has been exposed to near-ultraviolet, visible, or infrared radiation before warming, and thus, samples should always be irradiated, stored and measured in the dark. The reduced output is due to optical untrapping of the trapped charges (Alfimov *et al.*, 1964), and recombination luminescence can be seen during the optical irradiation (Nikol'skii *et al.*, 1963b). Such optical bleaching also bleaches a broad absorption spectrum which must be due to the trapped charges (Nikol'skii *et al.*, 1963b; Partridge, 1970c). The bleaching efficiency steadily increases with decreasing wavelength from zero at about 15,500 Å (Thomas *et al.*, 1968) to at least 4000 Å (Nikol'skii, 1968). Furthermore, the fraction of the total thermoluminescence output that can be bleached by an "infinite" irradiation at a given wavelength *also* increases with decreasing wavelength (Nikol'skii, 1968), which strongly suggests that there is a continuous range of trap depths in the polymer with optical activation energies extending from 0.8 to at least 3 eV. Nikol'skii (1968) also found that the glow-curve *shape* was independent of the extent of optical bleaching, though Partridge (1964) and Nakai and Matsuda (1965) found the α peak to bleach faster than the others, while Aulov *et al.* (1969) found it to bleach slower. This relative insensitivity of the glow-curve shape to selective bleaching of the shallower (optical) traps indicates that the bulk of the thermal untrapping can occur from all types of trap at all temperatures, not just only from shallow traps at low temperatures and deep ones at high temperatures, and so supports the view that untrapping is predominantly due to molecular chain motion and not to thermal escape from fixed traps of different depths. However, Charlesby and Partridge (1963a) found that, if during warming a sample is suddenly quenched back to the irradiation temperature, then on subsequent warming little light is emitted until just below the quenching temperature, and this seems to indicate that shallow traps *are* emptied before deep ones. A likely explanation of this situation is that the traps, of all types, are distributed at different distances from the polymer chains. Thus, at low temperatures only charges trapped close to a chain (or near a particularly mobile segment of a chain) are released, while at higher temperatures charges further away can get untrapped as the frequency and amplitude of the molecular motion increases. Thus, the different glow peaks are associated more with different frequency constants than with different activation energies (Partridge, 1965). Aulov *et al.* (1969) carried this temperature quenching a stage further by irradiating at 4°K, warming to 180°K (sufficient to exhaust the α and β peaks), quenching back to 4°K, and then bleaching with visible light. On subsequent warming some luminescence did appear in the 4°–180°K range, which indicates that while the untrapping caused by optical bleaching usually results in ion recombination some of the released charges are trapped again in other traps.

Immersion of polyethylene in suitable liquids, particularly alkanes, has proved a powerful method of studying its thermoluminescence because the liquid penetrates into the amorphous (but not the crystalline) regions and leaches out most foreign molecules not actually attached to the polymer chains (Partridge, 1966). The thermoluminescence, fluorescence, and phosphorescence spectra observed before and after such leaching (Boustead and Charlesby, 1967) indicate that of the various identifiable luminescent molecules phenanthrene and anthracene are only in the amorphous regions, benzoic acid and an alkylarylketone are in the crystalline regions, and a naphthalene derivative is in the amorphous regions attached to the molecular chains. A further advantage of this technique is that a range of additives can be put into polyethylene by first dissolving them in a liquid alkane (Partridge, 1970a). In this way Boustead (1970a) showed that the extractable aromatic molecules act only as luminescence centers and not as charge traps, though electron trapping by diphenyl and triphenylamine molecules when in high concentration has been observed (Partridge, 1970a,c).

B. POLYPROPYLENE

Thermoluminescence commences at about 40°K for irradiations at 4°K (Aulov *et al.*, 1969) and rises to a peak at around 125°K for a 4–10° min^{-1} warming rate (Mozisek, 1967; Aulov *et al.*, 1969; Boustead, 1970a). Boustead also found a smaller peak at about 95°K, while Charlesby and Partridge (1963a) found another weak one at much higher temperature. The thermoluminescence spectrum is predominantly due to phosphorescence and is quite similar to the polyethylene spectrum (Charlesby and Partridge, 1965a; Boustead, 1969). Thus, the various impurities noted earlier in polyethylene may often also be present in polypropylene, and for the same reason. Polypropylene may also contain an *m*-cresol derivative (Partridge, 1968). Alfimov *et al.* (1964) noted that the production of free radicals in polypropylene by mechanical treatment of the sample at low temperature did not lead to any luminescence emission on warming.

C. POLYISOBUTYLENE

The glow curve for samples irradiated at 4°K starts around 50°K and rises slowly to a peak at about 150°K for a 9° min^{-1} warming rate (Aulov *et al.*, 1968). Charlesby and Partridge (1963a) observed only this peak, possibly because their sample contained oxygen, but Aulov *et al.* (1968) found a second peak at 180°K and Nikol'skii and Buben (1960) a second peak at 228°K. The latter workers correlated their second peak with the glass transition occurring at about the same temperature, as measured by mechanical deformation of the same sample. The lower second peak temperature of Aulov *et al.* may be due to some difference in chemical structure, similar to that in polybutadiene when the 1,4 and 1,2 ratio is altered (see below).

D. SHORT-CHAIN ALKANES

Since polyethylene can be roughly regarded as an infinite linear alkane, and polypropylene and polyisobutylene as infinite highly branched alkanes, the thermoluminescence of much smaller alkanes is of great interest. Various workers have studied linear alkanes but the most comprehensive results are those of Boustead (1970c,d) on most of the *n*-alkanes from pentane to eicosane. It was found that the glow curves of these alkanes (doped with some anthracene to enhance the emission) could be divided into two groups, those from alkanes with an even number of C atoms and those with an odd number. Even *n*-alkanes generally gave two close peaks in the 90–120°K range, a single peak due to oxygen between 186° and 145°K (whose temperature steadily decreased with increasing chain length), and one or two peaks close to the melting point; all peaks were of comparable intensity. By contrast the odd alkanes usually gave only one large peak, plus several smaller ones, where the large peak was in the 90°K region for C_{11}, C_{13}, and C_{15} alkanes but at 140°K for C_5 and 180°K for C_7. The other odd alkane, C_9, gave six peaks between 91° and 208°K with comparable intensities. None of the odd alkanes had oxygen peaks. Such a division of glow-curve properties is most interesting because a similar division exists between their crystal symmetry properties, with most odd alkanes having hexagonal symmetry and most even alkanes (below C_{20}) having triclinic symmetry (McClure, 1968). Such a division is also clearly seen in measurements of phosphorescence quenching in these alkanes (Boustead and Charlesby, 1970b). These crystal structure differences probably affect both the type of charge traps in the material and the ease with which molecular motion can accomplish untrapping from these traps and, also perhaps, the mobility of free charges within the matrix.

The variation of glow-peak output with dose seems to differ very widely between different alkanes and between the different peaks of each alkane. For instance, Boustead (1970d) found that in *n*-heptane the three lowest temperature peaks increase up to about 1 Mrad and are then constant up to 3 Mrad, whereas the dominant peak at 180°K reaches a maximum at

1 Mrad but thereafter decreases so fast that it is down to a fifth of its maximum value at 3 Mrad. In complete contrast Alfimov *et al.* (1964) found that emission from the next odd alkane, *n*-nonane, was *linear* with dose up to at least 50–70 Mrads.

Kustanovich *et al.* (1962) found the thermoluminescence emission spectrum of purified *n*-heptane to extend from 5700 to 4100 Å and include two peaks. They assigned the thermoluminescence to radical recombination, but this seems most unlikely since later work (Alfimov *et al.*, 1964) showed that optical bleaching could remove all thermoluminescence from alkanes yet leave the radical concentration unchanged.

Nikol'skii *et al.* (1963b) studied the optically activated luminescence from irradiated alkanes and found spectra similar to those in polyethylene, with a "shallow trap" absorption at low doses extending from more than 10,000 to less than 4000 Å and a "deep trap" absorption at higher doses from 6000 Å to shorter wavelengths. These spectra, like the associated thermoluminescence, bleached easily under visible light irradiation.

Branched alkanes form amorphous glasses rather than crystals on cooling, and one consequence of this is to greatly increase the extent of oxygen diffusion into the sample as compared with linear alkanes and, hence, the size of the oxygen effect (Nikol'skii *et al.*, 1964; Boustead, 1970d; Boustead and Charlesby, 1970a). Squalane (hexamethyl tetracosane) was studied in detail by Boustead and Charlesby (1970a). This gave four glow peaks, three of which were closely associated with phase transitions (one at the melting point) while the fourth was due to oxygen and reached a maximum at a temperature intermediate between the two lowest phase transitions. The thermoluminescence spectrum, variation of glow intensity with dose, possible electron traps, and phosphorescence quenching were all discussed.

E. DIENE POLYMERS

The thermoluminescence of natural and synthetic rubbers has been studied considerably by Nikol'skii, Buben, and other Russian workers. Most samples gave two glow peaks (Nikol'skii and Buben, 1960) in the 130–160°K and 180–230°K ranges, and some a third peak around 325°K. For polybutadiene the central glow peak is closely related to the chemical structure of the polymer (Alfimov and Nikol'skii, 1963) since its maximum varies from 180°K for a polybutadiene with a 95% content of 1,4 structure to 230°K for one with only a 34% 1,4 content. This central peak temperature, measured at 5° min^{-1} warming rate, was found in all cases to be close to the glass-transition temperature of the polymer (see the discussion of this in Section V). The glow peaks on either side of the central peak are not sensitive to the

1,2 to 1,4 structure ratio. The lowest temperature peak (often two close together) was associated by Alfimov and Nikol'skii with the onset of methylene group motion in the 120–160°K region, as observed by NMR. The highest glow peak at about 325°K occurs at a transition to a viscous liquid state (Nikol'skii and Buben, 1960). The central peak shifts to higher temperatures with increase in warming rate, as expected, but only by about 6° for a warming rate increase of 30 times; this shift was ascribed to a shift in the glass-transition temperature itself (Alfimov and Nikol'skii, 1963).

The thermoluminescence output of polybutadiene, studied by Alfimov *et al.* (1964), increases with dose up to ≈ 1 Mrad and then steadily decreases, with the increase most marked for the central peak and the highest temperature portion of the lowest glow-peak "doublet." But if a sample that has been given a high dose is then warmed to room temperature, it will give a "low dose" glow curve on subsequent recooling and low dose irradiation. Furthermore, Alfimov *et al.* found in parallel ESR measurements that the extent to which the "low dose" glow curve could be recovered by warming after a high dose correlated very closely with the extent to which free radicals produced by the high dose were able to decay during warming. This strongly supports claims that free radical can act as efficient electron traps (Nikol'skii *et al.*, 1964), but since the radical concentration is linear with dose up to at least 70 Mrad while the thermoluminescence output falls above 1 Mrad it seems that any electrons trapped by radicals do *not* give rise to thermoluminescence on subsequent warming. The observed variation of glow-peak output with dose can then be explained as due to competition for electrons between "shallow" traps which do release electrons for thermoluminescence emission on subsequent warming and radiation-produced free radicals which do not, as described in Section VI.

Russian workers have made considerable routine use of thermoluminescence in studying glass-transition temperatures, the homogeneity of physical mixtures of different rubbers, and the effects of vulcanization by various additives (Buben *et al.*, 1965, 1967). The glow curves were found to give a very clear indication of the efficiency with which two rubbers had been mixed together, and also the physical state of the mixture and its stability with time.

F. POLYSTYRENE

Thermoluminescence in polystyrene (containing fluorescent additives) was first observed by Pannell and Manning (1955), and a prominent optical absorption was also seen which must have been due to various trapped ionic species. The polystyrene glow curve has just one glow peak (Rozman, 1958; Charlesby and Partridge, 1963a) with a maximum at about 120°K for an $8°$ min^{-1} warming rate. Rozman found the glow-peak shape largely independent of the nature of the luminescence additives, though the intensity was dependent on these, so charge trapping probably occurs in the polymer matrix itself. The glow peak is probably associated with the mechanical loss maximum at $130°K$ which may be due to limited rotational motion of small main chain segments (Shen and Eisenberg, 1967). Initial rise measurements by Partridge (1964) gave an activation energy of 0.21 eV for the glow peak and a frequency constant, using Eq. (5), of 8×10^5 sec^{-1}.

G. POLYMETHYLMETHACRYLATE

Usually, only one glow peak, of very low intensity, is observed from this polymer (Charlesby and Partridge, 1963a; Fleming, 1968a) at about $162°K$ for a $20°$ min^{-1} warming rate. Fleming also observed a second peak on a few occasions at $239°K$, though its appearance was unpredictable; the glow-curve shape was unaffected by oxygen, residual monomer, or a variety of polymerization initiators and chain-transfer additives. Fleming measured the glow-peak activation energies by a number of different methods and obtained a value of about 0.084 eV for the main peak and 0.43 eV for the $239°K$ peak; the main peak almost certainly followed first-order kinetics, and it had, from Fleming's warming rate graph and Eq. (5), a frequency constant of about 1.6 sec^{-1}. The temperature of the main peak maximum shifted by more than $50°$ when the warming rate was varied from $3°$ to $40°$ min^{-1}, and this, together with the low frequency constant, indicates that this is a "molecular motion" peak rather than a "structural transition" peak. Fleming noted that the temperature and activation energy of the main peak are close to the values for rotation of the main chain methyl groups (as measured by other methods). The glow-peak output of this polymer seems to saturate at doses of 1–2 Mrad (Alfimov *et al.*, 1964; Fleming, 1968a).

H. POLYTETRAFLUOROETHYLENE

Nikol'skii and Buben (1960) observed two glow peaks from this polymer at about $148°$ and $295°K$ for a $15°$ min^{-1} warming rate and noted that it has structural transitions around $170°$ and $295°K$ [for their glow curve, see Semenov (1962)]. Mozisek (1967) found only one peak, at about $130°K$, for a $4°$ min^{-1} warming rate. But the most surprising observation was that of Mele *et al.* (1968), who found that the low-temperature peak appears *only* in the presence of dissolved gases such as oxygen and *helium*. Samples irradiated *in vacuo* gave just one weak peak at $308°K$ for a $10°$ min^{-1} warming rate and very low irradiation dose, or two peaks at $292°$ and

327°K at higher doses (≈ 0.1 Mrad). Another small peak at about 190°K is apparent from some of Mele's glow curves of samples containing dissolved gas. The two low-temperature peaks produced by dissolved gas must come from the amorphous region of the polymer as significant gas diffusion into the crystalline regions is unlikely. Addition of gas *after* irradiation does *not* produce the low-temperature peaks. The thermoluminescence emission spectrum seems similar whether the sample contains air (Charlesby and Partridge, 1965a) or oxygen or helium (Mele *et al.*, 1968), and consists of four prominent maxima between about 3500 and 6000 Å. Since this spectrum is apparently independent of the type of gas, it must be concluded that the gas acts by creating shallow charge traps rather than luminescent molecular groups. It may be that the gas molecules distort the polymer matrix in their immediate vicinity and so create local electron "cavity" traps (Section III).

Differential thermal analysis by Mele *et al.* indicated that molecular motion, probably rotational motion of small chain segments in the amorphous region, starts at ≈ 130°K and continues up to 250°K, and this is probably responsible for the low-temperature glow peaks. An activation energy of about 0.1 eV for these peaks was deduced by initial rise measurements, with 0.5 eV for the high-temperature peaks. The high-temperature peaks must be closely connected with two known crystalline transitions of this polymer at about 292° and 303°K (McCrum *et al.*, 1967), and thus their associated activation energies are probably not "real" (Section V).

Mele *et al.* made generally similar observations for the thermoluminescence of polytetrafluoroethylene oxide.

I. OTHER SYNTHETIC POLYMERS

Thermoluminescence has been observed from a variety of other polymers although detailed study is at present lacking. These include polyoxymethylene and its monomer trioxane (Charlesby and Gupta, 1968), polydimethyl-siloxane (Nikol'skii and Buben, 1960), polyvinylpyrrolidone (Charlesby and Partridge, 1963a), and various chloro- and fluoro-substituted olefin polymers (Kozlov *et al.*, 1967). The work on trioxane is of special interest since this material also showed the rather rare phenomenon of triboluminescence (luminescence activated by mechanical pressure) due to thermal stresses in the sample during warming.

J. BIOLOGICAL MACROMOLECULES

The thermoluminescence of a considerable range of biological materials has been investigated, but their complexity has so far prevented any real understanding of the detailed processes involved. Published studies include those of Augenstein *et al.* (1960), Weinberg *et al.* (1962), Lehman and Wallace (1964), Singh and Charlesby (1965), Nelson *et al.* (1967), Lillicrap and Fielden (1969), and Sharpatyi *et al.* (1969).

REFERENCES

Alfimov, M. V., and Nikol'skii, V. G. (1963). *Vysokomol. soed.* **5**, 1388 (*English trans.*: *Polymer Sci. USSR* **5**, 477).
Alfimov, M. V., Nikol'skii, V. G., and Buben, N.Ya. (1964). *Kinet. Katet.* **5**, 268 (*English trans.*: *Kinet. Catal.* **5**, 238).
Altshuller, A. P. (1965). *Anal. Chem.* **37**, 11R.
Augenstein, L. G., Carter, J. G. Nelson, D. R., and Yockey, H. P. (1960). *Radiat. Res Suppl.* **2**, 19.
Aulov, V. A., Sukhov, F. F., Chernyak, I. V., and Slovokhotova, N. A. (1968). *Khim. Vys. Energ.* **2**, 191 (*English trans.*: *High Energy Chem.* **2**, 165).
Aulov, V. A., Sukhov, F. F., Slovokhotova, N. A., and Chernyak, I. V. (1969). *Khim. Vys. Energ.* **3**, 452 (*English trans.*: *High Energy Chem.* **3**, 407).
Berlman, I. B., and Walter, T. A. (1962). *J. Chem. Phys.* **37**, 1888.
Bonfiglioli, G., Brovetto, P., and Cortese, C. (1959). *Phys. Rev.* **114**, 951.
Boustead, I. (1969). Thesis. University of London.
Boustead, I. (1970a). *J. Polym. Sci.* A-2 **8**, 143.
Boustead, I. (1970b). *Eur. Polym. J.* **6**, 731.
Boustead, I. (1970c). *Nature* **225**, 846.
Boustead, I. (1970d). *Proc. Roy. Soc.* **A319**, 237.
Boustead, I. (1970e). *Proc. Roy. Soc.* **A318**, 459.
Boustead, I., and Charlesby A. (1967). *Eur. Polym. J.* **3**, 459.
Boustead, I., and Charlesby, A. (1970a). *Proc. Roy. Soc. A* **315**, 271.
Boustead, I., and Charlesby, A. (1970b). *Proc. Roy. Soc. A* **315**, 419.
Boustead, I., and Charlesby, A. (1970c). *Proc. Roy. Soc. A.* **316**, 291.
Buben, N.Ya., Koritskii, A. T., and Shamshev, V. N. (1962). *Proc. All-Union Conf. Radiat. Chem. 2nd* (*English trans.*: (1964). *Israel Sci. Transl. Serv.* p. 576).
Buben, N.Ya., Gol'danskii, V. I., Zlatkevich, L.Yu., Nikol'skii, V. G., and Raevskii, V. G. (1965). *Dokl. Akad. Nauk. SSSR* **162**, 370 (*English transl.*: *Proc. Acad. Sci. USSR Phys. Chem.* **162**, 386).
Buben, N.Ya., Gol'danskii, V. I., Zlatkevich, L.Yu., Nikol'skii, V. G., and Raevskii, V. G. (1967). *Vysokomol. soed.* **A9**, 2275 (*English transl.*: *Poly. Sci. USSR* **9**, 2575).
Charlesby, A., and Partridge, R. H. (1963a). *Proc. Roy. Soc.* **A271**, 170.
Charlesby, A., and Partridge, R. H. (1963b). *Proc. Roy. Soc.* **A271**, 188.
Charlesby, A., and Partridge, R. H. (1965a). *Proc. Roy. Soc.* **A283**, 312.
Charlesby, A., and Partridge, R. H. (1965b). *Proc. Roy Soc.* **A283**, 329.
Charlesby, A., and Gupta, P. S. (1968). *J. Mater. Sci.* **3**, 70.
Fleming, R. J. (1968a). *J. Polym. Sci.* A-2 **6**, 1283.
Fleming, R. J. (1968b). *Can. J. Phys.* **46**, 1569.
Gallivan, J. B., and Hamill, W. H. (1966). *J. Chem. Phys.* **44**, 2378.
Garlick, G. F. J., and Gibson, A. F. (1948). *Proc. Phys. Soc.* **60**, 574.
Hirayama, F., and Lipsky, S. (1969). *J. Chem. Phys.* **51**, 3616.
Hirayama, F., Rothman, W., and Lipsky, S. (1970). *Chem. Phys. Lett.* **5**, 296.

Keyser, R. M., and Williams, F. (1969). *J. Phys. Chem.* **73**, 1623.
Kozlov, V. T., Ivanov, S. I., and Smagin, E. N. (1967). *Khim. Vys. Energ.* **1**, 400 (*English transl.: High Energy Chem.* **1**, 350).
Kustanovich, I. M., Polak, L. S. and Rytova, N. M. (1962). *Proc. All-Union Conf. Radiat. Chem. 2nd* (*English transl.*: (1964). *Israel Sci. Transl. Serv.* p. 341).
Lehman, R. L., and Wallace, R. (1964). *In* "Electronic Aspects of Biochemistry" (B. Pullman, ed.), p. 43. Academic Press, New York.
Lillicrap, S. G., and Fielden, E. M. (1969). *J. Chem. Phys.* **51**, 3503.
McCrum, N. G., Read, B. E., and Williams, G. (1967). "Anelastic and Dielectric Effects in Polymeric Solids." Wiley, New York.
McClure, D. W. (1968). *J. Chem. Phys.* **49**. 1830.
Magat, M. (1966). *J. Chim. Phys.* **63**, 142.
Mele, A., Delle Site, A., Bettinali, C., and Di Domenico, A. (1968) *J. Chem. Phys.* **49**, 3297.
Michaels, A. S., and Parker R. B. (1959). *J. Poly. Sci.* **41**, 53.
Mozisek, M. (1967). *Proc. Tihany Symp. Radiat. Chem. Akad. Kiado, Budapest* p. 785.
Nakai, Y., and Matsuda, K. (1965). *Jap. J. Appl. Phys.* **4**, 264.
Nelson, D. R., Carter J. G., Birkhoff, R. D., Hamm, R. N., and Augenstein, L. G. (1967). *Radiat. Res.* **32**, 723.
Nikol'skii, V. G. (1968). *Khim. Vys. Energ.* **2**, 271 (*English transl.: High Energy Chem.* **2**, 233).
Nikol'skii, V. G., and Buben, N. Ya. (1960). *Dokl. Akad. Nauk. SSSR* **134**, 134 (*English transl.: Proc. Acad. Sci. USSR Phys. Chem.* **134**, 827).
Nikol'skii, V. G., and Buben, N.Ya. (1962). *Dokl. Akad. Nauk. SSSR* **147**, 1406 (*English transl.: Proc. Acad. Sci. USSR Phys. Chem.* **147**, 896).
Nikol'skii, V. G., Alfimov, M. V., and Buben, N.Ya. (1963a). *Izv. Akad. Nauk. SSSR Otd. Khim. Nauk.* No. 5, 955 (*English transl.: Bull. Acad. Sci. USSR Chem. Sci.* No. 5, 870).
Nikol'skii, V. G., Tochin, V. A., and Buben, N. Ya. (1963b). *Fiz. Tver. Tela.* **5**, 2248 (*English transl.*: (1964). *Sov. Phys. Solid State* **5**, 1636).
Nikol'skii, V. G., Chkheidze, I. I., and Buben, N.Ya. (1964). *Kinet. Katet.* **5**, 82 (*English transl.: Kinet. Catal.* **5**, 69).
Pannell, J. H., and Manning, B. (1955). *J. Chem. Phys.* **23**, 1368.
Partridge, R. H. (1964). Thesis, Univ. of London.
Partridge, R. H. (1965). *J. Polym. Sci.* **A3**, 2817.
Partridge, R. H. (1966). *J. Chem. Phys.* **45**, 1679.
Partridge, R. H. (1968). *J. Chem. Phys.* **49**, 3656.
Partridge, R. H. (1970a). *J. Chem. Phys.* **52**, 2491.
Partridge, R. H. (1970b). *J. Chem. Phys.* **52**, 2501.
Partridge, R. H., (1970c). *J. Chem. Phys.* **52**, 1277.
Partridge, R. H., and Charlesby, A. (1963). *J. Polym. Sci. B* **1**, 439.
Randall, J. T., and Wilkins, M. H. F. (1945). *Proc. Roy. Soc.* **A184**, 366.
Rozman, I. M. (1958). *Izv. Akad. SSSR Ser. Fiz.* **22**, 50 (*English transl.: Bull. Acad. Sci. USSR Phys. Ser.* **22**, 48).
Salem, L. (1966). "The Molecular Orbital Theory of Conjugated Systems." Benjamin, New York.
Schatzki, T. F. (1962). *J. Polym. Sci.* **57**, 496.
Semenov, N. N. (1962). *Pure Appl. Chem.* **5**, 353.
Sharpatyi, V. A., Nadzhimiddinova, M. T., Tochin, V. A., and Nikol'skii (1969). *Khim. Vys. Energ.* **3**, 469 (*English transl.: High Energy Chem.* **3**, 427).
Shen, M. C., and Eisenberg, A. (1967). *Progr. Solid State Chem.* **3**, 407.

Sichkar, V. P., Vaisberg, S. E., and Karpov, V. L. (1967). *Khim. Vys. Energ.* **1**, 561 (*English transl.: High Energy Chem.* **1**, 493).
Sichkar, V. P., Vaisberg, S. E., and Karpov, V. L. (1969). *Khim. Vys. Energ.* **3**, 438 (*English transl.: High Energy Chem.* **3**, 394).
Singh, B. B., and Charlesby, A. (1965). *Int. J. Radiat. Biol.* **9**, 157.
Thomas, B., Houston, E. and Weeks, J. C. (1968). *In* "Energetics and Mechanisms in Radiation Biology" (G. O. Phillips, ed.), p. 493. Academic Press, New York.
Tochin, V. A., and Nikol'skii, V. G. (1969). *Khim. Vys. Energ.* **3**, 281 (*English transl.: High Energy Chem.* **3**, 256).
Watanabe, K., Nakayama, T., and Mottl, J. R. (1962). *J. Quant. Spectrosc. Radiat. Trans.* **2**, 369.
Weinberg, C. J., Nelson, D. R., Carter, J. G., and Augenstein, L. G. (1962). *J. Chem. Phys.* **36**, 2869.

11

Statistical Theories of Cross-Linking

Osamu Saito

Department of Physics, Chuo University, Tokyo, Japan

I. Degradation of Polymers

A. Introduction

When a polymer substance is exposed to high-energy radiation, the main effects which the substance undergoes are the scission of main chains and the creation of free radicals, double bonds, cross-links, end-links, etc. Among these effects, those which change the molecular size distribution are main chain scission, cross-linking, and end-linking.

When a main chain scission occurs, a polymer molecule is divided into two smaller molecules, so that the molecular size distribution of the polymer is changed. It is not clear that the probability of a main chain scission is independent of its position in the molecule, but it may be reasonable to suppose that the scission probability of a bond which belongs to a structural unit near the molecular end will be different from one in the remaining part of molecule. If a main chain scission is more probable in the vicinity of a molecular end than in any other part of it, smaller molecules will be more easily degraded than larger ones. Then, the smaller the molecular weight of polymer, the greater will be the number of main chain scissions. However, we have no such evidence for most polymers (Saeman *et al.*, 1952; Charlesby, 1955b). This might presumably be due to the following reason: As most polymer substances contain molecules whose degree of polymerization is very large, even if a few structural units at the molecular end are more susceptible to main chain scission, such an end-effect would not be observed because it would be buried under the uniform effect of the remaining part of molecule.

The constancy of the probability of main chain scission irrespective of its position in the polymer molecule is the basic assumption of our theoretical considerations. This assumption could be easily removed, if required (Simha, 1941). However, this assumption will give information on the standard effect of main chain scissions. Of many theoretical treatments concerning this problem most are based on the combinatorial theory which counts the number of main chain scissions (Kuhn, 1930; Montroll and Simha, 1940; Matsumoto, 1949; Charlesby, 1954). In these theories different treatments are needed for different initial molecular weight distributions of polymers. Main chain scissions produce directly the change in molecular weight distribution. Therefore, the formulation of a law governing the change in molecular weight distribution due to main chain scissions will be fundamental for many purposes (Saito, 1958), and many quantities concerning the molecular size distribution will be derived from the solution of this fundamental equation (Saito, 1958, 1959). It will also be possible in many ways to extend the theory from this basic consideration, for instance, to the consideration of the end effect, impurity effect, degradation of copolymers, etc. (Saito, 1959; Inokuti, 1960).

B. MOLECULAR SIZE DISTRIBUTION

Polymer substances are usually based on molecules having a dispersed molecular weight system. Uniform molecular size distribution in which all molecules have the same size is a special case which rarely occurs. If we cut at random a circular polymer molecule of infinite molecular weight, we have the following molecular size distribution (Charlesby, 1960):

$$w(p) = (p/u^2) \exp(-p/u) \tag{1}$$

where p is the degree of polymerization of a molecule, u is the number average degree of polymerization, and $w(p)$ is the weight fraction of molecules having p structural units. The weight-average degree of polymerization of this distribution is $2u$. When u is large, Eq. (1) is equivalent to

$$w(p) = p\alpha^{p-1}(1 - \alpha)^2 \tag{2}$$

where $1 - \alpha$ corresponds to $1/u$. Such a distribution as expressed by Eqs. (1) or (2) is called the most probable distribution or the random distribution. The number- and the weight-average degrees of polymerization of Eq. (2) are $1/(1-\alpha)$ and $(1 + \alpha)/(1-\alpha)$, respectively. Random distribution is observed in condensation polymers (Flory, 1953), radical polymerization polymers associated with disproportionation termination (Flory, 1953), and cation polymerization polymers (Flory, 1953; Jordan and Mathieson, 1952). Conversely, the following molecular size distribution appears in the radical polymerization of polymers associated with recombination (Gee and Melville, 1944)

$$w(p) = (4p^2/u^3) \exp(-2p/u) \tag{3}$$

As a general molecular size distribution which implies Eqs. (1) and (3) we introduce

$$w(p) = [\sigma^\sigma/u\Gamma(\sigma)](p/u)^\sigma \exp(-\sigma p/u) \tag{4}$$

which is called the Schulz–Zimm distribution (Schulz, 1939; Zimm, 1948), where u is the number-average degree of polymerization, σ is a parameter indicating the narrowness of the distribution breadth, and $\Gamma(\sigma)$ is the gamma function. When σ increases infinitely, the Schulz–Zimm distribution tends to the uniform distribution $w(p) = \delta(p-u)$, where $\delta(x)$ is the Dirac delta function, which satisfies

$$\delta(x) = 0 \qquad \text{for } x \neq 0$$

and

$$\int_{-\infty}^{\infty} \delta(x)\,dx = 1$$

The weight-average degree of polymerization of the Schulz–Zimm distribution is equal to $u(\sigma + 1)/\sigma$. Equation (4) is the same as the Poisson distribution in statistics. In addition, the following Tung distribution is often useful for expressing experimental values of molecular weight distributions (Tung, 1956):

$$w(p) = (\sigma p^{\sigma-1}/p_0^\sigma)\exp[-(p/p_0)^\sigma] \tag{5}$$

where σ is a parameter to indicate the narrowness of the distribution breadth and p_0 is related to the number-average degree of polymerization u by $p_0 = u\Gamma(1-1/\sigma)$. The weight-average degree of polymerization is equal to $u\pi/\sigma \sin(\pi/\sigma)$. The Tung distribution is similar in its expression to the Schulz–Zimm distribution. When σ is nearly equal to unity, both distributions are almost the same. When σ is large, the breadth of the Tung distribution is smaller than that of the Schulz–Zimm distribution. When σ is less than unity, the weight fraction $w(p)$ of the Tung distribution becomes infinite at $p = 0$. Thus, the Tung distribution loses its meaning for σ smaller than unity. Therefore, this distribution is only useful for expressing distributions that are narrower than the most probable distribution. On the other hand, the Schulz–Zimm distribution is useful for expressing those distributions which are broader as well as narrower than the most probable distribution.

Another distribution which is compared with experimental values of molecular size distributions is the Wesslau distribution or the logarithmic regular distribution (Wesslau, 1956):

$$w(p) = \left[\frac{\exp(-\beta^2/4)}{\beta\sqrt{\pi}p_0}\right]\exp\left(\frac{-[\ln(p/p_0)]^2}{\beta^2}\right) \tag{6}$$

where β is a parameter indicating the broadness of the distribution, p_0 is related to the number-average degree of polymerization u with $p_0 = u\exp(-\beta^2/4)$. When $\beta = 0$, this distribution is the same as the uniform distribution $w(p) = \delta(p - u)$. When β is nearly equal to unity, the Wesslau distribution is similar to the Schulz–Zimm distribution having $\sigma = 1$. When β is larger than unity, that is, the distribution breadth is larger than that of the most probable distribution, the Wesslau distribution is different from the Schulz–Zimm distribution. When the number- and the weight-average degrees of polymerization of both distributions are equal, respectively, the peak of the Wesslau distribution is higher than that of the Schulz–Zimm distribution, the abscissa of the peak of the former is smaller than that of the latter, and the weight fraction of the former at sufficiently large degrees of polymerization is larger than that of the latter. The distinction between both distributions is clear when they are broad.

Molecular size distribution is characterized by the average degrees of polymerization, which are given by

$$P_j = \int_0^\infty p^{j-1} w(p)\, dp \Big/ \int_0^\infty p^{j-2} w(p)\, dp, \qquad j = 1, 2, \cdots \qquad (7)$$

The quantity P_1 is the number-average degree of polymerization, which is often written as P_n or $(DP)_n$. P_2 is the weight-average degree of polymerization and is often denoted by P_w or $(DP)_w$, and P_3 is the z-average degree of polymerization and often denoted by P_z or $(DP)_z$. In general, P_{k+3} is the $(z + k)$-average degree of polymerization and often denoted by P_{z+k} or $(DP)_{z+k}$. We also use M_n, M_w, M_z, and M_{z+k} for the number-, weight-, z-, and $(z+k)$-average molecular weight, respectively. The ratio of corresponding average molecular weight and degree of polymerization is equal to the molecular weight of a structural unit.

We have also the following relationships for the Schulz–Zimm distribution

$$\sigma = M_n/(M_w - M_n) = M_n/(M_z - M_w) = \cdots$$
$$\cdots = M_n/(M_{z+k+1} - M_{z+k}) = \cdots \qquad (8)$$

and for the Wesslau distribution (Saito et al., 1967)

$$\exp(\beta^2/2) = M_w/M_n = M_z/M_w = \cdots$$
$$\cdots = M_{z+k+1}/M_{z+k} = \cdots \qquad (9)$$

These equations will be useful for classifying molecular size distributions.

C. Basic Equation of Degradation

As discussed already, even if main chain scission was more probable in the structural units near the molecular end than in other parts, it would be hard to detect the difference for sufficiently large molecular weight of polymers. Thus, many theories of polymer degradation assume that every structural unit is equally fractured by irradiation. It is not easy to estimate the change in molecular size distribution of branched polymers undergoing main chain scissions. Such difficulty disappears in the fracture of linear polymers. In order to clarify the behavior of main chain scissions, the change in molecular size distribution of fractured polymers should be studied. We will establish a statistical theory of polymer degradation under the following assumptions (Saito, 1958): (1) All polymer molecules are linear; (2) every structural unit is fractured with equal probability; (3) average molecular weight is sufficiently large; and (4) the total number of main chain scissions is sufficiently smaller than the total number of structural units.

The basic equation which expresses the change in molecular size distribution of linear polymer molecules undergoing main chain scissions is

$$\frac{\partial w(p, y)}{\partial y} = -pw(p, y) + 2p \int_p^\infty \frac{w(l, y)}{l} \, dl \qquad (10)$$

where

$$y = \int_0^t r \, dt \qquad (11)$$

t is time, p is the degree of polymerization of a polymer molecule, r is the probability that a structural unit is fractured in unit time, and $w(p,y)$ is the weight fraction of polymer molecules having p structural units (Saito, 1958). An integration with respect to p is introduced in Eq. (10) instead of summation. It will assure us sufficient accuracy as well as ease in calculation when the average molecular weight is sufficiently large. The first term of the right-hand side of Eq. (10) corresponds to the decrease of the molecules having p structural units due to main chain scission, and the last term corresponds to the increase of the molecules having p structural units due to the scissions of those molecules having l structural units. The quantity y is the number of main chain scissions per structural unit, and it will be called the density of main chain scissions. We may put y equal to rt, because r is usually independent of t. Equation (10) has the solution

$$w(p, y) = \left[w(p, 0) + py \int_p^\infty \frac{(2 + yl - yp)}{l} w(l, 0) \, dl \right] \exp(-py) \qquad (12)$$

where $w(p,0)$ stands for the initial weight fraction (Saito, 1958). It is possible to derive much information about polymer degradation from this solution. The following expression is useful for the calculation of average molecular weights.

$$f_j(y) = \int_0^\infty p^{j-1} w(p, y) \, dp, \qquad j = 0, 1, 2, \cdots \qquad (13)$$

which is called the jth moment of molecular size distribution.

D. AVERAGE MOLECULAR WEIGHTS

Main chain scission reduces the average molecular weights of polymers. We know how to get the number-, weight-, and z-average molecular weights from the measurement of osmotic pressure, light scattering, and sedimentation, respectively (Flory, 1953). The change in these average molecular weights due to main chain scission is calculated from Eqs. (12) and (13). The number-average degree of polymerization P_n is given in the absence of cross-linking as

$$1/P_n = 1/u + y \qquad (14)$$

where u is the number-average degree of polymerization prior to irradiation. This relation holds for any initial molecular size distribution. If we use $M_n(0)$ and $M_n(y)$ for expressing the number-average molecular weight before and after irradiation, respectively, we get

$$uy = [M_n(0) - M_n(y)]/M_n(y) \qquad (15)$$

from which the value uy can be obtained from the measurement of the number-average molecular weight.

The weight-average degree of polymerization P_w is calculated from $f_2(y)/f_1(y)$, giving

$$P_w(y) = \frac{2}{y} - \frac{2}{y^2} \int_0^\infty \frac{(1 - e^{-py})}{p} w(p,0)\, dp$$

$$= 2 \sum_{k=0}^{\infty} \frac{(-y)^k}{(k+2)!} f_{k+2}(0) \qquad (16)$$

which depends on the initial molecular size distribution. If the initial distribution is the uniform distribution, then we have

$$P_w(y)/P_w(0) = 2(uy - 1 + e^{-uy})/(uy)^2$$

$$= 2 \sum_{k=0}^{\infty} \frac{(-1)^k}{(k+2)!} (uy)^k \qquad (17)$$

If the initial distribution is the most probable distribution, we have

$$P_w(y)/P_w(0) = 1/(1 + uy) \qquad (18)$$

which shows the linear relationship between $P_w(0)/P_w(y)$ and y. For the initial Schulz–Zimm distribution, we have

$$\frac{P_w(y)}{P_w(0)} = \left[uy - 1 + \left(\frac{1 + uy}{\sigma} \right)^{-\sigma} \right] \frac{2\sigma}{(1 + \sigma)(uy)^2}$$

$$= \sum_{k=0}^{\infty} (\sigma + 2)_k \left(\frac{-uy}{\sigma} \right)^k \bigg/ (3)_k \qquad (19)$$

where $(z)_k = z(z + 1) \cdots (z + k - 1)$ and $(z)_0 = 1$. For the initial Wesslau distribution, we have

$$\frac{P_w(y)}{P_w(0)} = 2 \frac{uy - 1 + I_\beta(\eta)}{(uy)^2 \exp(\beta^2/2)} \qquad (20)$$

where

$$I_\beta(\eta) = \frac{1}{\sqrt{\pi}} \int_{-\infty}^{\infty} \exp[-\xi^2 - \eta \exp(\beta\xi)] \, d\xi$$

and

$$\eta = uy \exp(-\beta^2/4) \tag{21}$$

Equation (20) cannot be expanded in the series of uy because it does not converge for any small value of uy except for zero. The ratio of the average degree of polymerization is equal to the ratio of the corresponding average molecular weights. Therefore, the above-mentioned relationships also indicate the ratios between the number- and the weight-average molecular weights, which are measured experimentally. It is known how to get the value y from osmotic pressure measurements. The effect of radiation on polymers is often expressed by the yield, which is assumed as the number of events per 100 eV of energy absorbed in the polymer. The yield of main chain scission $G(S)$ is calculated by the relationship

$$G(S) = 100 \, N_A y/R \tag{22}$$

where N_A is Avogadro's number and R is the radiation dose in electron volts which has been absorbed in a mole of structural units.

E. INTRINSIC VISCOSITY

Intrinsic viscosity of linear homopolymers is expressed by the Mark–Houwink relationship

$$[\eta] = KM^a \tag{23}$$

where M is molecular weight and K and a are constants which depend on the system of polymer and solvent. Intrinsic viscosity of a polydispersed system is given by

$$[\eta] = K \int_0^{\infty} (wp)^a w(p) \, dp \tag{24}$$

where w is the molecular weight of a structural unit and $w(p)$ is the weight fraction of molecules having p structural units. When the polymer substance has the random molecular weight distribution, we have

$$[\eta] = K\Gamma(a + 2)M_n^a \tag{25}$$

and when the polymer molecules have the Schulz–Zimm distribution, we have

$$[\eta] = K[\Gamma(\sigma + a + 1)/\sigma^{a+1}\Gamma(\sigma)]M_n^a \tag{26}$$

where M_n is the number-average molecular weight of polymers. If the polymer molecules have the Wesslau distribution, we have

$$[\eta] = K \exp[a(a + 1)\beta^2/4]M_n^a \tag{27}$$

When polymer molecules undergo main chain scission, intrinsic viscosity changes with respect to the number of scissions. It will be calculated from Eqs. (12) and (24). If the initial molecular size distribution is the random distribution, the intrinsic viscosity is

$$[\eta]_0/[\eta] = (1 + uy)^a \tag{28}$$

where $[\eta]_0$ is the value of the intrinsic viscosity prior to undergoing scission. This relationship means that the plot of log $[\eta]_0/[\eta]$ vs uy is linear, where uy is given by the measurement of osmotic pressure. When the initial molecular sizes follow the Schulz–Zimm distribution, we have

$$[\eta]/[\eta]_0 = \sum_{k=0}^{\infty} \{(a)_k(a + \sigma + 1)_k/k!(a + 2)_k\}(-uy/\sigma)^k \tag{29}$$

Since the density of main chain scissions y is usually proportional to the radiation dose R, we may put $y = y_0R$, where y_0 is a constant and means the density of main chain scissions per unit dose. If we plot $([\eta]_0 - [\eta])/[\eta]_0R$ from experimental values at low dose, we will obtain the coefficients A and B of the relationship

$$([\eta]_0 - [\eta])/[\eta]_0R = A - BR \tag{30}$$

Theoretical calculation gives

$$A = a(a + \sigma + 1)uy_0/\sigma(a + 2) \tag{31}$$

$$B = a(a + 1)(a + \sigma + 1)(a + \sigma + 2)(uy_0)^2/2\sigma^2(a + 2)(a + 3) \tag{32}$$

and

$$B/A^2 = (a + 1)(a + 2)(a + \sigma + 2)/2a(a + 3)(a + \sigma + 1) \tag{33}$$

Thus we get the value of σ from Eq. (33) and then the value of uy_0 from Eq. (31).

F. CHANGE IN MOLECULAR WEIGHT DISTRIBUTION

The change in molecular size distribution due to main chain scissions is expressed in Eq. (10). When molecular sizes are initially predicted by the random distribution function (1), the weight fraction of molecules having p structural units, undergoing main chain scissions whose density is y, is calculated by Eq. (12), giving

$$w(p, y) = (1/u + y)^2 p \exp[-(1/u + y)p] \tag{34}$$

This expression is quite similar to that of Eq. (1), which gives the random distribution. Thus, it is said that when a polymer substance has initially the random molecular size distribution, the shape of molecular size distribution is conserved, however large the amount of chain scissions, and the change caused by main chain scissions is only the reduction of the average degree of polymerization. Equation (34) shows us that the height of the peak in $w(p, y)$ vs p curve increases with the density of scissions y, while the abscissa of the peak decreases. When the initial distribution is uniform, the molecular size distribution at an arbitrary density of chain scissions is

$$
\begin{aligned}
w(p, y) &= (py/u)[2 + (u - p)y] \exp(-py) & p < u \\
&= \exp(-uy) & p = u \\
&= 0 & p > u
\end{aligned} \tag{35}
$$

When the initial distribution is given as the Schulz–Zimm distribution (4), we have

$$
uw(\xi, \eta) = \left(2\xi\eta + (1 - \xi)\xi\eta^2 + \frac{1}{\Gamma(\sigma)} \sum_{k=0}^{\infty} \frac{(-1)^k}{k!} (\sigma\xi)^{\sigma+k} C_k \right) e^{-\xi\eta} \tag{36}
$$

where $\xi = p/u$, $\eta = uy$, and

$$
C_k = \frac{(\sigma + k)(\sigma + k + 1) - 2(\sigma + k + 1)\xi\eta + (\xi\eta)^2}{(\sigma + k)(\sigma + k + 1)}
$$

Figure 1 shows the case when $\sigma = 20$.

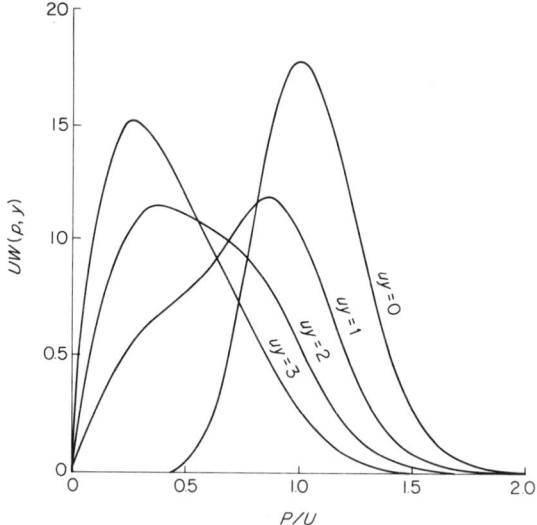

Fig. 1. The change in molecular size distribution due to main chain scissions, initial molecular size distribution being the Schulz–Zimm distribution for $\sigma = 20$.

II. Cross-Linking of Polymers

A. INTRODUCTION

There are many polymer substances in which cross-links are produced by irradiation. It is usually assumed that cross-links are produced randomly by radiations, and the number of cross-links is proportional to the radiation dose. It will be supposed that the random creation of cross-links will be supported experimentally, when such quantities as average molecular weights and solubility, which are averaged over whole polymer substance, are taken into consideration, even if ionization and excitation due to irradiation are located in the spurs of radiations. The proportionality of the number of cross-links to radiation dose is a reasonable assumption, so long as the number of cross-links is sufficiently small in comparison to the total number of structural units, and so that the probability that radiations attack cross-links is negligible. A structural unit in the vicinity of a chain end might differ in the ability of making a cross-link from that in the other part of chain. However, such an end effect will be negligible in the polymers whose average molecular weight is sufficiently large.

A statistical theory of cross-linking was first given by Flory (1941, 1942). He has shown theoretically that polymers of uniform molecules change the shape of molecular size distribution owing to the formation of cross-links and an insoluble part is created after a certain number of cross-links is produced. The change in molecular size distribution due to cross-linking of polymers which have initially had an arbitrary distribution of molecular sizes has been given by Stockmayer (1943, 1944). However, it is not always simple to calculate the extent of cross-linking according to Stockmayer's theory. Charlesby (1953, 1954) has solved by probability theorem the problem of cross-linked polymers which have an initial arbitrary molecular size distribution. His theory is preferable to those of Flory and Stockmayer in the ease of calculation. However, even his theory is not successful in solving the problem of cross-linking as a whole. Because it needs different theories to consider average molecular weights and solubility. And also, viscosity is difficult to solve in his theory. The author has proposed a basic equation which governs the change in molecular size distribution of polymers due to cross-linking and from which many properties of cross-linked polymers are derived easily (Saito, 1958). The results derived from this theory are the same as those from other theories. It is also not difficult to extend this theory to the complicated problems, such as simultaneous cross-linking and degradation, simultaneous cross-linking and end-linking, cross-linking of copolymers, etc. (Saito, 1958, 1959; Inokuti, 1960).

As is discussed later, the effect of initial molecular size distribution is quite important. For example, the behavior of viscosity and solubility depends greatly on it. Therefore, much care should be paid in order to obtain information of cross-linking from measurements of intrinsic viscosity and solubility.

B. Basic Equation of Cross-Linking

For establishing a mathematical theory of cross-linking, it is important to settle the basic assumptions. Many theories which attempt to explain the experimental behavior of cross-linked polymers require the following assumptions (Charlesby, 1954; Saito, 1958): (1) Cross-links are produced at random; (2) every structural unit crosslinks with the same probability regardless of its position in the polymer molecule; (3) the number of cross-links is sufficiently small in comparison to the total number of structural units; and (4) intramolecular linkings in molecules of finite sizes are negligible.

The fundamental effect of cross-links on polymers is to change the molecular size distribution. Many properties will be derived from the formalism of this change. Let us denote the degree of polymerization of a molecule by p, the total number of cross-links by X, the total number of structural units by N. Then, the density of cross-links is equal to $x = X/N$. Since N is extremely large, x will be assumed to increase continuously. As p takes on large values on an average, it will be sufficiently accurate to take it as a continuous variable. When we make $w(p, x)$ the weight fraction of molecules having p structural units, the change in $w(p, x)$ with respect to the density of cross-links x satisfies the following equation (Saito, 1858):

$$\frac{1}{p}\frac{\partial w(p,x)}{\partial x} = -2w(p,x)\int_0^\infty w(l,x)\,dl + \int_0^p w(l,x)w(p-l,x)\,dl \tag{37}$$

The left-hand side expresses the rate of increase of the number of molecules having p structural units, the first term in the right-hand side gives the rate with which the molecules of p structural units change to larger molecules by cross-linking with other molecules, and the second term gives the rate with which the molecules of p structural units are produced by cross-linking the molecules of l units with those of $p - l$ units. Equation (37) is to be solved with a given initial value $w(p,0)$. If we put

$$f_n(x) = \int_0^\infty p^{n-1}w(p,x)\,dp, \qquad n = 0,1,2,\cdots \tag{38}$$

we have from Eq. (37)

$$\frac{df_n}{dx} = -2f_1 f_{n+1} + \sum_{k=0}^n \binom{n}{k} f_{k+1} f_{n-k+1} \tag{39}$$

For $n = 1$, it gives

$$df_1/dx = 0$$

This shows that f_1 is constant, which is equal to unity according to the definition. For $n = 0$ and 2, we have

$$\frac{df_0}{dx} = -f_1^2 = -1, \quad \text{and} \quad \frac{df_2}{dx} = 2f_2^2$$

which give

$$f_0(x) = f_0(0) - x, \quad \text{and} \quad f_2(x) = f_2(0)/[1 - 2xf_2(0)] \tag{40}$$

The second moment $f_2(x)$ increases with x, until x reaches $\frac{1}{2}f_2(0)$, where it becomes infinite. It loses the meaning for x larger than $\frac{1}{2}f_2(0)$, because $f_2(x)$ is negative there. The nth moment $f_n(x)$ is given by solving Eq. (39) successively. From these the average degree of polymerization P is given as follows:

$$P_j(x) = f_j(x)/f_{j-1}(x) \tag{41}$$

Another method of solving the basic Eq. (37) refers the Laplace transform

$$f(\zeta, x) = \int_0^\infty w(p, x)\, e^{-p\zeta} dp \tag{42}$$

giving

$$(\partial f/\partial x) + 2(f - 1)(\partial f/\partial \zeta) = 0 \tag{43}$$

The solution which satisfies an initial condition $f(\zeta, 0) = F(\zeta)$ is

$$f(\zeta, x) = F\{\zeta + 2x[1 - f(\zeta, x)]\} \tag{44}$$

The relationship between $f_n(x)$ and $f(\zeta, x)$ is

$$f_n(x) = (-1)^{n-1}[\partial^{n-1} f(\zeta, x)/\partial \zeta^{n-1}]_{\zeta = 0} \tag{45}$$

When cross-links and main chain scissions are produced simultaneously by irradiation, it may usually be permitted to assume that they are independent of each other, and we are able to obtain the change in molecular size distribution by considering that cross-linking occurs after all main chain scissions have taken place.

When a polymer substance, whose molecular size distribution is given by a weight fraction function $w(p)$, undergoes cross-linking, whose number is X, and main chain scissions, whose number is Y, we calculate owing to Eq. (12) the modified weight fraction $w(p, y)$ due to scissions, where y is the density of main chain scissions defined in the preceding section. And then, considering the cross-linking effect, we obtain the last weight fraction $w(p, x, y)$ by solving Eq. (37) under the initial condition $w(p, 0, y) = w(p, y)$.

C. Average Molecular Weights

When cross-linking and main chain scission occur simultaneously, the number-average degree of polymerization $P_n(x, y)$ is given

$$1/P_n(x, y) = 1/u + y - x \qquad (46)$$

where u is the initial number-average degree of polymerization. The number-average molecular weight M_n is given by multiplying $P_n(x, y)$ with the molecular weight of a structural unit w. Equation (46) holds for any arbitrary shape of the initial molecular size distribution. It gives the way in which we get the difference of the density of cross-links and that of scissions x-y from osmotic measurement. The weight-average degree of polymerization $P_w(x, y)$ depends on the shape of molecular size distribution prior to undergoing cross-linking and main chain scission. For the initial random distribution, we have

$$\frac{1}{P_w(x, y)} = \frac{1}{2u} + \frac{y}{2} - 2x \qquad (47)$$

Since x and y are usually proportional to radiation dose, the plot of the reciprocal weight-average degree of polymerization versus radiation dose is linear. This linear relationship is particular to the initial random distribution, while the plots for other molecular size distributions are not linear, although the difference from linearity is usually not distinct. When the initial distribution is uniform, we have

$$1/P_w(x, y) = uy^2/2(uy - 1 + e^{-uy}) - 2x$$

$$= \frac{1}{u} + \frac{y}{3} - 2x + \frac{uy^2}{36} + 0(u^2 y^3) \qquad (48)$$

and for the initial Schulz–Zimm distribution, we get

$$u/P_w(x, y) = (uy)^2/2[uy - 1 + (1 + uy/\sigma)^{-\sigma}] - 2ux$$

$$= \frac{\sigma}{(\sigma + 1)} + \frac{(\sigma + 2)uy}{3(\sigma + 1)} - 2ux$$

$$+ \frac{(\sigma + 2)(\sigma - 1)(uy)^2}{36\sigma(\sigma + 1)} + 0(u^3 y^3) \qquad (49)$$

These equation are useful for calculating the number of cross-links and main chain scissions. When the density of main chain scissions y is supposed sufficiently small, those terms in Eq. (49) containing higher powers than y will be negligible. If we put

$$A = [P_n(0,0) - P_n(x,y)]/P_n(0,0)P_n(x,y)$$

$$B = [P_w(0,0) - P_w(x,y)]/P_w(0,0)P_w(x,y) \tag{50}$$

which are calculated from the experimental values of osmotic pressure and light scattering, the number of cross-links X and that of main chain scissions Y are given by the relation:

$$X = [(\sigma + 2)A - 3(\sigma + 1)B]N/(5\sigma + 4)$$

$$Y = 3(\sigma + 1)(2A - B)N/(5\sigma + 4) \tag{51}$$

where N denotes the total number of structural units. The case of the random and the uniform initial distribution will be given by putting $\sigma = 1$ and $\sigma \to \infty$, respectively.

D. GELATION

It is well known that when a polymer substance undergoes radiation cross-linking, the weight-average molecular weight and the intrinsic viscosity increase with radiation dose, until at last they tend to infinitely large values. If radiation dose increases further, a part which is insoluble in any solvent grows in the polymer substance. It is called the gel and this phenomenon is called gelation; the instant when an incipient gel is created is called the gel point. The density of cross-links at the gel point is given from the value of x which makes the denominater of the last of Eq. (40) vanish, namely,

$$x_g = 1/2f_2(0) = 1/2P_w(0) \tag{52}$$

This shows that the number of cross-links X at gel point is equal to half the ratio of the total number of structural units N to the weight-average degree of polymerization P_w of the polymer prior to undergoing cross-linking.

When cross-linking and main chain scission occur simultaneously, $P_w(0)$ of the above Eq. (52) must be replaced with $P_w(y)$ which is given by Eq. (16). Since $P_w(y)$ in general decreases, the density of cross-links at a gel point in the case when main chain scissions are accompanied with cross-linking is larger than that when no scission occurs. For instance, when the initial distribution is the random one (1), we have

$$x_g = y/4 + 1/2P_w(0) \tag{53}$$

Since the insoluble part grows with cross-links in the post-gel-point irradiation and the structural units in the gel part also take part in cross-linking, the basic equation (37) will be rewritten as follows:

$$\frac{1}{p}\frac{\partial w(p,x)}{\partial x} = -2w(p,x) + \int_0^p w(l,x)w(p-l,x)\,dl \tag{54}$$

where $w(p, x)$ is the weight fraction of those molecules which are contained in the soluble or sol part and have p structural units. The solution of this equation is also given by Eq. (44). Since we have the following relation

$$1 - g = \int_0^\infty w(p,x)\,dp = f(0,x) \tag{55}$$

where g stands for gel fraction, we have

$$1 - g = F(2gx) \tag{56}$$

$$F(z) = \int_0^\infty w(p,0)\,e^{-pz}\,dp \tag{57}$$

where Eq. (56) is the relationship between the gel fraction g and the density of cross-links x. This relationship depends on the initial molecular size distribution. For instance, the initial random distribution gives

$$s + s^{1/2} = 1/xP_w(0) \tag{58}$$

which is the Charlesby–Pinner relationship, where s is the sol fraction and equal to $1 - g$ (Charlesby and Pinner, 1959). When cross-linking and main chain scission occur simultaneously, $P_w(0)$ in the above equation is replaced with Eq. (18), giving

$$s + s^{1/2} = y/2x + 1/xP_w(0) \tag{59}$$

Since the density of cross-links x and that of main chain scissions y are usually proportional to radiation dose, the plot of $s + s^{1/2}$ versus reciprocal radiation dose gives a straight line. The intersection of this line with the ordinate axis gives the ratio $y/2x$. The ratio of the yields of cross-linking and main chain scission is equal to y/x.

When the initial molecular size distribution is not the random distribution, the plot of $s + s^{1/2}$ versus reciprocal radiation dose is not linear. The relationship between gel fraction g and the density of cross-links x and main chain scissions y is generally given as follows:

$$(1 - g)(2gx + y)^3 = y^2(2gx + y) + 4xyg/u$$

$$- 4xyg\int_{2gx+y}^\infty F(z)\,dz + 4y^2g^2(2gx + y)F(2gx + y) \tag{60}$$

where $F(z)$ is given by Eq. (57). When the densities of cross-links and main chain scissions are proportional to radiation dose, the ratio y/x is equal to a constant, say λ, and Eq. (60) becomes (Inokuti, 1963)

$$(1 - g)x(2g + \lambda)^3 = x\lambda^2(2g + \lambda) + 4\lambda g/u$$

$$- 4\lambda g \int_{(2g + \lambda)x}^{\infty} F(z)\,dz + 4\lambda^2 xg^2(2g + \lambda)F[(2g + \lambda)x] \qquad (61)$$

The ratio of the yields of cross-links and main chain scissions is equal to λ. The gel fraction versus the density of cross-links relationship for the initial Schulz–Zimm distribution is given in Table I (Inokuti, 1963). In this case the Charlesby–Pinner plot is convex upward for $\sigma < 1$, concave upward for $\sigma > 1$, and linear for $\sigma = 1$. However, even when σ is not equal to unity, the difference of the plot from a straight line is not remarkable. It is known that some of the Charlesby–Pinner plots of polyethylene deviate distinctly from a straight line and show the convex upward behavior (Saito *et al.*, 1967; Kang *et al.*, 1967). From this it was concluded that the initial molecular size distribution was not the Schulz–Zimm type. The relationship between gel fraction g and density of cross-links x is given in Table II in the case when the initial molecular size distribution is assumed to be the Wesslau distribution (Satio *et al.*, 1967). These theoretical values are compared with the experimental values of irradiated polyethylene in Fig. 2 (Saito *et al.*, 1967). It shows that the Wesslau distribution describes well the initial molecular size distribution of polyethylene.

E. Molecular Weight Distribution

It is not always easy to measure accurately the molecular weight distribution, although it is important for estimating radiation effects on polymers. The method of fractionation and gel permeation chromatography will be available for the determination of molecular weight distribution. However, it is not powerful for polymers which have branches. Since cross-linking necessarily makes branches, it is difficult to measure the change in molecular weight distribution due to cross-linking. Thus, theoretical calculation becomes necessary for knowing the behavior. The change in molecular size distribution due to cross-linking has been given theoretically by Flory and Stockmayer. Flory (1941) has derived the change in molecular size distribution of cross-linked polymers which have the uniform distribution prior to undergoing cross-linking. Stockmayer (1944) has solved the same problem for an arbitrary initial molecular size distribution. However, the resulting formulation is not always simple for calculation. Charlesby (1954) has given the change in molecular size distribution in an infinite series. We will deal only with a theory which derives the change in molecular size distribution from the basic equation of cross-linked polymers (37).

TABLE I

THE GEL POINT AND THE RELATION BETWEEN THE GEL FRACTION AND THE DENSITY OF CROSS-LINKS [a]

σ	λ	x_g	x_g/x_1	x_g/x_2	x_g/x_3	x_g/x_4	x_g/x_5	x_g/x_6	x_g/x_7	x_g/x_8	x_g/x_9
0.01	0.0	0.004950	0.9005	0.8010	0.7014	0.6017	0.5020	0.4021	0.3021	0.2020	0.1016
	0.5	0.005908	0.8914	0.7835	0.6764	0.5702	0.4654	0.3623	0.2617	0.1647	0.0742
	1.0	0.007249	0.8860	0.7732	0.6620	0.5528	0.4461	0.3425	0.2432	0.1497	0.0652
	1.5	0.009227	0.8820	0.7660	0.6524	0.5416	0.4342	0.3309	0.2330	0.1420	0.0612
	2.0	0.012383	0.8793	0.7611	0.6460	0.5343	0.4267	0.3240	0.2272	0.1380	0.0593
	2.5	0.01805	0.8775	0.7581	0.6420	0.5300	0.4226	0.3203	0.2244	0.1363	0.0588
	3.0	0.03057	0.8767	0.7567	0.6406	0.5387	0.4216	0.3199	0.2245	0.1369	0.0595
	3.5	0.07403	0.8775	0.7584	0.6431	0.5321	0.4256	0.3243	0.2289	0.1408	0.0621
0.02	0.0	0.009804	0.9010	0.8018	0.7026	0.6033	0.5038	0.4042	0.3043	0.2040	0.1031
	0.5	0.011692	0.8919	0.7844	0.6777	0.5719	0.4673	0.3643	0.2636	0.1664	0.0752
	1.0	0.014330	0.8864	0.7741	0.6633	0.5544	0.4479	0.3444	0.2449	0.1511	0.0661
	1.5	0.018218	0.8826	0.7670	0.6538	0.5433	0.4360	0.3327	0.2347	0.1434	0.0620
	2.0	0.024415	0.8798	0.7621	0.6474	0.5359	0.4285	0.3258	0.2288	0.1393	0.0601
	2.5	0.03554	0.8782	0.7591	0.6435	0.5317	0.4243	0.3221	0.2261	0.1377	0.0596
	3.0	0.06001	0.8772	0.7577	0.6420	0.5303	0.4233	0.3216	0.2261	0.1382	0.0603
	3.5	0.14473	0.8780	0.7594	0.6445	0.5336	0.4273	0.3260	0.2305	0.1421	0.0629
0.03	0.0	0.0145663	0.9015	0.8027	0.7039	0.6049	0.5057	0.4062	0.3063	0.2060	0.1046
	0.5	0.017355	0.8925	0.7855	0.6791	0.5736	0.4692	0.3663	0.2655	0.1680	0.0763
	1.0	0.021248	0.8870	0.7751	0.6647	0.5561	0.4497	0.3462	0.2467	0.1526	0.0670
	1.5	0.026981	0.8831	0.7680	0.6551	0.5449	0.4377	0.3345	0.2363	0.1448	0.0629
	2.0	0.036109	0.8803	0.7631	0.6487	0.5375	0.4302	0.3275	0.2304	0.1406	0.0609
	2.5	0.05246	0.8786	0.7600	0.6447	0.5333	0.4260	0.3238	0.2276	0.1389	0.0604
	3.0	0.08839	0.8778	0.7587	0.6433	0.5319	0.4250	0.3233	0.2277	0.1395	0.0611
	3.5	0.21228	0.8785	0.7603	0.6458	0.5352	0.4289	0.3277	0.2321	0.1434	0.0637

0.05	0.0	0.023810	0.9023	0.8044	0.7063	0.6080	0.5093	0.4101	0.3104	0.2098	0.1077
	0.5	0.028328	0.8934	0.7873	0.6817	0.5768	0.4728	0.3701	0.2692	0.1712	0.0785
	1.0	0.034616	0.8879	0.7769	0.6673	0.5592	0.4532	0.3499	0.2501	0.1554	0.0688
	1.5	0.043859	0.8841	0.7699	0.6577	0.5480	0.4412	0.3380	0.2396	0.1474	0.0645
	2.0	0.058538	0.8814	0.7650	0.6513	0.5407	0.4336	0.3309	0.2336	0.1432	0.0625
	2.5	0.08476	0.8796	0.7619	0.6474	0.5363	0.4294	0.3272	0.2307	0.1415	0.0619
	3.0	0.14211	0.8787	0.7606	0.6458	0.5349	0.4283	0.3266	0.2307	0.1419	0.0626
	3.5	0.33864	0.8795	0.7622	0.6483	0.5382	0.4322	0.3309	0.2351	0.1458	0.0652
0.10	0.0	0.045455	0.9044	0.8084	0.7121	0.6153	0.5178	0.4195	0.3201	0.2191	0.1151
	0.5	0.053873	0.8957	0.7916	0.6877	0.5842	0.4812	0.3790	0.2780	0.1790	0.0836
	1.0	0.065540	0.8903	0.7814	0.6734	0.5667	0.4615	0.3584	0.2583	0.1623	0.0732
	1.5	0.082608	0.8865	0.7744	0.6639	0.5554	0.4493	0.3464	0.2474	0.1539	0.0686
	2.0	0.109560	0.8838	0.7695	0.6575	0.5480	0.4416	0.3390	0.2411	0.1494	0.0664
	2.5	0.15736	0.8820	0.7664	0.6534	0.5435	0.4372	0.3351	0.2381	0.1475	0.0656
	3.0	0.26095	0.8812	0.7650	0.6519	0.4521	0.4360	0.3344	0.2379	0.1478	0.0663
	3.5	0.61055	0.8819	0.7665	0.6541	0.5451	0.4397	0.3385	0.2422	0.1517	0.0689
0.020	0.0	0.083333	0.9081	0.8155	0.7223	0.6282	0.5330	0.4363	0.3377	0.2361	0.1290
	0.5	0.098103	0.8997	0.7991	0.6984	0.5975	0.4964	0.3952	0.2940	0.1931	0.0934
	1.0	0.118427	0.8944	0.7891	0.6842		0.4764	0.3739	0.2731	0.1750	0.0815
	1.5	0.147913	0.8907	0.7823	0.6748	0.5686	0.4639	0.3613	0.2615	0.1657	0.0761
	2.0	0.194018	0.8881	0.7774	0.6683	0.5610	0.4559	0.3536	0.2547	0.1606	0.0734
	2.5	0.27481	0.8863	0.7742	0.6641	0.5563	0.4512	0.3492	0.2512	0.1583	0.0725
	3.0	0.44704	0.8854	0.7727	0.6623	0.5545	0.4496	0.3481	0.2507	0.1584	0.0729
	3.5	1.01379	0.8859	0.7738	0.6641	0.5570	0.4527	0.3516	0.2544	0.1619	0.0755
0.30	0.0	0.115385	0.9112	0.8216	0.7310	0.6393	0.5461	0.4509	0.3532	0.2513	0.1419
	0.5	0.135065	0.9030	0.8056	0.7075	0.6089	0.5095	0.4093	0.3081	0.2058	0.1024
	1.0	0.161977	0.8979	0.7957	0.6935	0.5913	0.4892	0.3875	0.2863	0.1863	0.0890
	1.5	0.200752	0.8943	0.7890	0.6841	0.5799	0.4765	0.3743	0.2739	0.1762	0.0829
	2.0	0.260888	0.8917	0.7841	0.6775	0.5721	0.4682	0.3662	0.2666	0.1706	0.0798
	2.5	0.36520	0.8898	0.7808	0.6732	0.5672	0.4631	0.3614	0.2627	0.1678	0.0786
	3.0	0.58465	0.8888	0.7791	0.6711	0.5650	0.4611	0.3598	0.2616	0.1674	0.0787
	3.5	1.29221	0.8891	0.7798	0.6723	0.5667	0.4634	0.3625	0.2646	0.1704	0.0810

TABLE I (*continued*)

σ	λ	x_g	x_g/x_1	x_g/x_2	x_g/x_3	x_g/x_4	x_g/x_5	x_g/x_6	x_g/x_7	x_g/x_8	x_g/x_9
0.50	0.0	0.166667	0.9162	0.8313	0.7451	0.6573	0.5675	0.4751	0.3790	0.2772	0.1648
	0.5	0.193321	0.9084	0.8159	0.7224	0.6275	0.5311	0.4327	0.3319	0.2277	0.1185
	1.0	0.229421	0.9035	0.8064	0.7086	0.6100	0.5105	0.4100	0.3084	0.2058	0.1022
	1.5	0.280861	0.9000	0.7998	0.6992	0.5983	0.4972	0.3960	0.2948	0.1941	0.0947
	2.0	0.359612	0.8975	0.7949	0.6925	0.5902	0.4883	0.3870	0.2864	0.1873	0.0907
	2.5	0.49408	0.8955	0.7914	0.6878	0.5848	0.4825	0.3813	0.2815	0.1836	0.0887
	3.0	0.77128	0.8943	0.7892	0.6850	0.5816	0.4794	0.3785	0.2792	0.1822	0.0882
	3.5	1.63841	0.8941	0.7890	0.6848	0.5817	0.4798	0.3792	0.2804	0.1835	0.0895
2.0	0.0	0.333333	0.9326	0.8634	0.7921	0.7183	0.6412	0.5599	0.4725	0.3756	0.2599
	0.5	0.375345	0.9262	0.8504	0.7722	0.06912	0.6912	0.5170	0.4206	0.3135	0.1861
	1.0	0.430500	0.9220	0.8420	0.7595	0.6741	0.5850	0.4912	0.3908	0.2810	0.1554
	1.5	1.506394	0.9188	0.8357	0.7501	0.6616	0.5697	0.4734	0.3713	0.2614	0.1398
	2.0	0.618034	0.9162	0.8304	0.7423	0.6516	0.5576	0.4598	0.3570	0.2479	0.1302
	2.5	0.80000	0.9137	0.8255	0.7353	0.6426	0.5471	0.4482	0.3453	0.2374	0.1230
	3.0	1.15470	0.9110	0.8204	0.7279	0.6333	0.5364	0.4368	0.3340	0.2276	0.1166
	3.5	2.18767	0.9074	0.8135	0.7183	0.6215	0.5231	0.4230	0.3208	0.2164	0.1096
5.0	0.0	0.416667	0.9408	0.8797	0.8164	0.7502	0.6805	0.6061	0.5250	0.4334	0.3207
	0.5	0.462273	0.9351	0.8680	0.7982	0.7250	0.6474	0.5640	0.4720	0.3660	0.2303
	1.0	0.521242	0.9314	0.8603	0.7862	0.7083	0.6258	0.5368	0.4387	0.3262	0.1872
	1.5	0.601022	0.9284	0.8541	0.7766	0.6953	0.6091	0.5164	0.4147	0.2999	0.1646
	2.0	0.716232	0.9256	0.8484	0.7680	0.6837	0.5944	0.4989	0.3952	0.2802	0.1498
	2.5	0.90035	0.9226	0.8424	0.7589	0.6716	0.5795	0.4817	0.3768	0.2628	0.1378
	3.0	1.25243	0.9186	0.8344	0.7472	0.6563	0.5614	0.4616	0.3564	0.2448	0.1262
	3.5	2.26595	0.9118	0.8214	0.7286	0.6333	0.5354	0.4345	0.3307	0.2237	0.1135

λ	σ	χ_g	χ_g/χ_1	χ_g/χ_2	χ_g/χ_3	χ_g/χ_4	χ_g/χ_5	χ_g/χ_6	χ_g/χ_7	χ_g/χ_8	χ_g/χ_9
10.0	0.0	0.454545	0.9446	0.8872	0.8275	0.7650	0.6989	0.6279	0.5502	0.4616	0.3514
	0.5	0.50091	0.9393	0.8762	0.8102	0.7408	0.6668	0.5865	0.4971	0.3922	0.2529
	1.0	0.56045	0.9357	0.8687	0.7985	0.7244	0.6451	0.5588	0.4622	0.3485	0.2022
	1.5	0.64039	0.9327	0.8626	0.7880	0.7111	0.6278	0.5370	0.4356	0.3182	0.1755
	2.0	0.75492	0.9299	0.8567	0.7799	0.6986	0.6116	0.5172	0.4128	0.2945	0.1579
	2.5	0.93653	0.9266	0.8500	0.7696	0.6846	0.5940	0.4964	0.3901	0.2730	0.1433
	3.0	1.28183	0.9218	0.8403	0.7551	0.6656	0.5711	0.4709	0.3642	0.2505	0.1291
	3.5	2.27923	0.9131	0.8235	0.7313	0.6362	0.5381	0.4369	0.3326	0.2250	0.1141
20.0	0.0	0.476190	0.9467	0.8915	0.8340	0.7736	0.7095	0.6406	0.5649	0.4783	0.3698
	0.5	0.52273	0.9416	0.8808	0.8172	0.7499	0.6780	0.5997	0.5119	0.4079	0.2666
	1.0	0.58228	0.9381	0.8735	0.8057	0.7337	0.6564	0.5717	0.4760	0.3618	0.2108
	1.5	0.66190	0.9352	0.8675	0.7962	0.7203	0.6387	0.5490	0.4478	0.3287	0.1815
	2.0	0.77547	0.9324	0.8615	0.7868	0.7073	0.6216	0.5277	0.4228	0.3024	0.1622
	2.5	0.95479	0.9289	0.8543	0.7756	0.6920	0.6021	0.5046	0.3973	0.2783	0.1461
	3.0	1.29505	0.9235	0.8434	0.7593	0.6704	0.5760	0.4754	0.3679	0.2530	0.1305
	3.5	2.28296	0.9135	0.8243	0.7322	0.6371	0.5390	0.4376	0.3331	0.2254	0.1143

[a] The first three columns give σ, λ, and χ_g. The remaining nine columns give χ_g/χ_n, where χ_n denotes the value of χ corresponding to $g = (n/10)g_{max}$ ($n = 1,2,\ldots,9$).

TABLE II

THE RELATION BETWEEN THE GEL FRACTION AND THE NUMBER OF CROSS-LINKS FOR THE INITIAL WESSLAU DISTRIBUTION[a]

β	x_g	x_1	x_2	x_3	x_4	x_5	x_6	x_7	x_8	x_9
					$\lambda = 0.0$	$g_{max}(\lambda) = 1.0$				
0.5	0.4413	0.4682	0.4996	0.5370	0.5826	0.6398	0.7148	0.8195	0.9824	1.3037
1.0	0.3033	0.3303	0.3623	0.4012	0.4490	0.5109	0.5944	0.7158	0.9145̄	1.3369
1.5	0.1623	0.1882	0.2181	0.2543	0.3001	0.3605̄	0.4450	0.5734	0.7969	1.3184
1.8	0.09895̄	0.1225	0.1487	0.1806	0.2214	0.2765	0.3555	0.4794	0.7044	1.2617
2.1	0.05513	0.0750	0.0961	0.1221	0.1562	0.2034	0.2731	0.3862	0.6911	1.1694
2.4	0.02807	0.0433	0.0590	0.0788	0.1054	0.1434	0.2014	0.2993	0.4942	1.0457
2.6	0.01702	0.0291	0.0415̄	0.0573	0.0791	0.1109	0.1606	0.2469	0.4245̄	0.9506
2.8	0.009921	0.0191	0.0285	0.0409	0.0582	0.0841	0.1257	0.2000	0.3581	0.8494
3.0	0.005554	0.0123	0.0192	0.0285	0.0420	0.0626	0.966	0.1591	0.2967	0.7456
					$\lambda = 0.5$	$g_{max}(\lambda) = 0.9571$				
0.5	0.4889	0.5208	0.5596	0.6968	0.6656	0.7419	0.8466	1.0032	1.2779	1.9902
1.0	0.3499	0.3829	0.4226	0.4715̄	0.5336	0.6160	0.7316	0.9092	1.2287	2.0589
1.5	0.2058	0.2355	0.2716	0.3169	0.3757	0.4556	0.5711	0.7548	1.0985	2.0255
1.8	0.1373	0.1629	0.1942	0.2340	0.2866	0.3595̄	0.4672	0.6431	0.9825	1.9326
2.1	0.08635̄	0.1067	0.1320	0.1648	0.2090	0.2717	0.3669	0.5270	0.8472	1.7848
2.4	0.05138	0.0664	0.0855̄	6.1107	0.1457	0.1967	0.2764	0.4149	0.7029	1.5912
2.6	0.03525	0.0469	0.0621	0.0827	0.1115	0.1547	0.2216	0.3457	0.6072	1.4439
2.8	0.02362	0.0325	0.0444	0.0606	0.0839	0.1194	0.1773	0.2828	0.5152	1.2885̄
3.0	0.01546	0.0221	0.0310	0.0435	0.0619	0.0904	0.1381	0.2270	0.4292	1.1298

$\lambda = 1.0$ $\quad g_{max}(\lambda) = 0.8660$

0.5	0.5482	0.5872	0.6339	0.6913	0.7642	0.8607	0.9969	1.2096	1.6091	2.7745
1.0	0.4104	0.4500	0.4981	0.5580	0.6351	0.7389	0.8876	1.1226	1.5644	2.8037
1.5	0.2599	0.2956	0.3397	0.3956	0.4690	0.5699	0.7180	0.9582	1.4207	2.7331
1.8	0.1837	0.2146	0.2532	0.3029	0.3693	0.4620	0.6010	0.8312	1.2854	2.6078
2.1	0.1233	0.1483	0.1800	0.2216	0.2783	0.3592	0.4832	0.6940	1.1226	2.4157
2.4	0.07866	0.0976	0.1222	0.1550	0.2007	0.2677	0.3729	0.5571	0.9444	2.1649
2.6	0.05420	0.0719	0.0919	0.1191	0.1575	0.2148	0.3065	0.4704	0.8238	1.9739
2.8	0.04004	0.0519	0.0678	0.0897	0.1213	0.1691	0.2473	0.3901	0.7060	1.7703
3.0	0.02766	0.0367	0.0489	0.0662	0.0915	0.1306	0.1958	0.3175	0.5943	1.5617

$\lambda = 1.5$ $\quad g_{max}(\lambda) = 0.7500$

0.5	0.6290	0.6758	0.7323	0.8024	0.8924	1.0136	1.1881	1.4680	2.0131	3.6499
1.0	0.4926	0.5406	0.5993	0.6730	0.7686	0.8985	1.0869	1.3895	1.9711	3.6526
1.5	0.3341	0.3782	0.4329	0.5027	0.5949	0.7224	0.9108	1.2191	1.8205	3.5621
1.8	0.2480	0.2869	0.3357	0.3988	0.4833	0.6020	0.7806	1.0780	1.6696	3.4165
2.1	0.1756	0.2078	0.2487	0.3029	0.3765	0.4819	0.6436	0.9190	1.4809	3.1902
2.4	0.1186	0.1438	0.1765	0.2203	0.2812	0.3701	0.5098	0.7539	1.2673	2.8887
2.6	0.08898	0.1097	0.1369	0.1739	0.2260	0.3033	0.4267	0.6464	1.1187	2.6540
2.8	0.06539	0.0820	0.1042	0.1347	0.1782	0.2439	0.3506	0.5444	0.9709	2.4008
3.0	0.04707	0.0601	0.0777	0.1022	0.1378	0.1924	0.2828	0.4503	0.8279	2.1369

[a] g is the gel fraction; χ is the mean number of cross-links per initial number average molecule; λ is the ratio of the yield of scissions to the yield of cross-links; g_{max} is the gel fraction at infinite dose; χ_g is the value of χ at the gel point; χ_n is the value of χ corresponding to $g = (n/10)g_{max}$, where $n = 1, 2, \ldots, 9$; β is the parameter of Wesslau distribution, satisfying the relation $M_w/M_n = \exp(\beta^2/2)$.

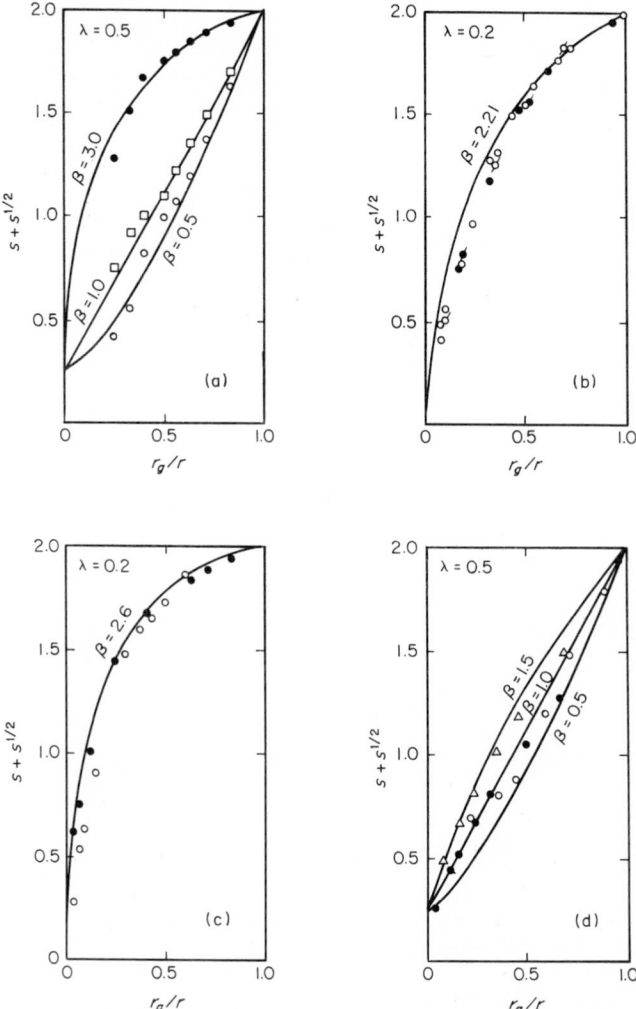

Fig. 2. Experimental tests of the Charlesby–Pinner functions based on the Wesslau distribution: (a) data of Kitamaru *et al.* (1964) on fractionated samples of polyethylene at 133°, where (○) $M = 1.82 \times 10^5$, (□) $M = 4.25 \times 10^5$, (●) $M = 1.04 \times 10^6$; (b) data of Kang *et al.* (1967) on bulk Marlex-50 polyethylene at the four temperatures, 35°, 80°, 100°, and 120°, of which two are shown, (○) 35° and (●) 120°; (c) data of Lawton *et al.* (1958), with (○) Marlex-50 polyethylene and (●) Plax film; (d) data of Kawai *et al.* (1965), with (○) bulk Marlex-50 polyethylene and others polyethylene single crystals.

When an initial distribution of weight fraction $w(p, 0)$ is given we have obtained by Eq. (44) the Laplace transform $f(\zeta, x)$, from which weight fraction $w(p, x)$ is given by calculating the inverse transform. Amemiya (1967) has derived the weight fraction of such polymers as have uniform molecular size distribution prior to undergoing cross-linking. It is

$$w(p,x) = (2px)^{p-1} e^{-2px}/p! \tag{62}$$

where the degree of polymerization of the initial uniform distribution has been chosen to be unity. This result agrees with that of Flory (1941). Amemiya (1967) has also obtained the change in weight fraction of polymers which have initially the random size distribution. We will now instead refer to an infinite series of solution derived by Kimura (1962, 1964). He has given the weight fraction distribution of cross-linked polymers which have initially the random size distribution as follows:

$$w(p,x) = \frac{p}{u^2} e^{-(2x+1)p/u} \sum_{k=0}^{\infty} \frac{2^k}{(k+1)!(2k+1)!} \left(\frac{p}{u}\right)^{3k} (ux)^k \tag{63}$$

which are plotted in Fig. 3. The more the density of cross-links increases, the lower becomes the height of the peak of the weight fraction distribution and the more the abscissa of the peak decreases. Kimura (1962, 1964) has also given the weight fraction of cross-linked polymers which have initially the Schulz–Zimm molecular size distribution.

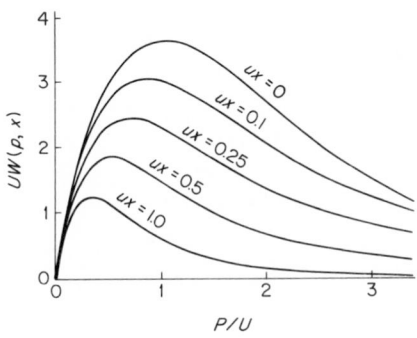

Fig. 3. The change in molecular size distribution due to cross-linking, initial molecular size distribution being the random distribution. From Kimura (1962).

F. Intrinsic Viscosity

Intrinsic viscosity or limiting viscosity number depends not only on the molecular weight of polymer, but also the number of branches. Denoting by M molecular weight, by b the number of branches, and by $R(b)$ the radius of gyration of a molecule which has b branches, the intrinsic viscosity $[\eta]$ is given by

$$[\eta] = Kg_b{}^v M^a \tag{64}$$

where K, a, and v are constants, and $g_b = R^2(b)/R^2(0)$. Zimm and Stockmayer (1949) have proposed $v = 2 - a$, and Zimm and Kilb (1959) have proposed $v = \frac{1}{2}$. The values of g_b for small values of b are

$$g_0 = 1, \quad g_1 = 0.800, \quad g_2 = 0.690, \quad \text{and} \quad g_3 = 0618$$

Since, in general, g_b is less than unity, the intrinsic viscosity of a branched polymer is less than that of a linear polymer, so long as their molecular weights are equal.

When a polymer substance, which has had a certain molecular size distribution prior to irradiation, undergoes radiation-induced cross-linking and main chain scission simultaneously, the intrinsic viscosity is given by

$$[\eta] = Kw^a \sum_{b=0}^{\infty} g_b{}^v \int_0^{\infty} p^a w_b(p, x, y)\, dp \tag{65}$$

where w is the molecular weight of a structural unit and $w_b(p, x, y)$ is the weight fraction of polymer molecules which have p structural units and b branches, x and y being the densities of cross-links and main chain scissions. Katsuura (1960) has calculated the intrinsic viscosity of a polymer which has the random molecular size distribution prior to undergoing radiation-induced cross-linking and main chain scission, where he assumes $v = 2 - a$. Inokuti and Katsuura (1959) have also calculated the intrinsic viscosity of polymers which have initially the Schulz–Zimm molecular size distribution.

Let us consider the case when linear polymers change their weight fraction from $w_0 (p, 0, 0)$ to $w_0(p, 0, y)$ owing to main chain scissions and then change it to $w_b(p, x, y)$ owing to cross-linking. Then, we have

$$\frac{\partial w_0(p, 0, y)}{\partial y} = -p w_0(p, 0, y) + 2p \int_p^{\infty} \frac{w_0(l, 0, y)}{l}\, dl \tag{66}$$

and

$$\frac{\partial w_0(p, x, y)}{\partial x} = -2p w_0(p, x, y) \tag{67}$$

$$\frac{\partial w_b(p, x, y)}{\partial x} = -2p w_b(p, x, y)$$

$$+ p \sum_{k=0}^{b-1} \int_0^p w_k(l, x, y) w_{b-k-1}(p - l, x, y)\, dl, \qquad b \geq 1 \tag{68}$$

When a polymer has initially the random molecular size distribution (1), we have

$$w_b(p, x, y) = \frac{C_b}{u\Gamma(3b + 1)} x_n{}^b(1 + y_n)^{2(b+1)} \left(\frac{p}{u}\right)^{3b+1}$$

$$\times \exp\left(-(1 + 2x_n + y_n)\frac{p}{u}\right) \qquad (69)$$

where $x_n = ux$ and $y_n = uy$ and C_b is given by the recurrence formula:

$$bC_b = \sum_{k=0}^{b-1} (3k + 1)(3b - 3k - 2)C_k C_b - k - 1, \qquad b \geq 1 \qquad (70)$$

Intrinsic viscosity is calculated by

$$[\eta]/[\eta]_0 = \sum_{b=0}^{\infty} g_b{}^v \frac{\Gamma(a + 3b + 2)C_b x_n{}^b(1 + y_n)^{2(b+1)}}{\Gamma(a + 2)\Gamma(3b + 1)(1 + 2x_n + y_n)^{a+3b+2}} \qquad (71)$$

where $[\eta]_0$ is the intrinsic viscosity prior to undergoing cross-linking and main chain scission. The values of g_b have been given by Zimm and Stockmayer (1949) and Kataoka (1953).

G. Effects of Impurity

Radiation effects on polymers depend on the conditions of irradiation. It is known that the existence of oxygen reduces radiation-induced cross-links. Impurities or additives change the density of cross-links. For instance, oxygen in polyethylene hinders the cross-linking reaction of polymers (St. Pierre and Dewhurst, 1958), while N_2O promotes it (Okada and Amemiya, 1961, Okada, 1964). Thus, impurities or additives are divided into two groups, one promotes cross-linking and the other hinders it. A statistical theory of analyzing quantitatively these effects will be given in this section.

When an impurity hinders radiation-induced cross-linking, the impurity molecule will deactivate an active center created on a polymer molecule, which otherwise would make a cross-link. If we denote by k the ratio of the rate coefficient of the reaction between an active center and an impurity molecule to that of the cross-linking reaction between an active center and a structural unit, and by m the ratio of the number of impurity molecules to the total number of structural units, the probability that cross-link will be made between an active center and a structural unit is equal to $1/(1 + km)$. If we use x as the virtual density of cross-links and x_e as true density of them, $dx/(1 + km)$ is equal to dx_e. Thus, we have the following equation (Saito, 1959)

$$\frac{\partial w(p, x)}{\partial x} = \frac{p}{1 + km}\left[-2w(p, x) + \int_0^p w(l, x)w(p - l, x)\, dl\right] \qquad (72)$$

which is the same as Eq. (37) if the true density of cross-links x_e is replaced with the virtual density of them x. The change in the number of impurity molecules is expressed by

$$dm = -km\ dx/(1 + km) = -km\ dx_e \qquad (73)$$

from which the relationship between the virtual density of cross-links x and the true density x_e is derived as

$$x = x_e + m(0)(1 - e^{-kx_e}) \qquad (74)$$

The virtual density of cross-links x is proportional to the radiation dose, while the true density x_e is not. The above equation shows the following behavior. Since the true density of cross-links of the system containing an impurity is equal to the density of cross-links of the system containing no impurity, the above equation gives the relationship between the virtual density of cross-links of the system containing impurity and the density of cross-links of the system containing no impurity. Density of cross-links is usually proportional to radiation dose R. Putting the proportionality coefficient to be c which is equal to the probability that a structural unit will take part in cross-linking under the irradiation of unit dose, we have

$$R = R_e + \frac{m(0)}{c}(1 - e^{-kcR_e}) \qquad (75)$$

which expresses the relationship between the radiation dose R which is necessary to produce a certain number of cross-links in a polymer containing an impurity and the dose R_e which is necessary to produce the same number of cross-links in the polymer containing no impurity. In other words, the above relationship will be satisfied if we measure radiation doses necessary for making the same amount of radiation effect for both cases when impurity and no impurity is contained. In the plot of R vs R_e, the slope in the region of small R_e gives $km(0)$ and for large R_e the plot becomes a straight line, whose slope is equal to unity and whose extension will intersect with the ordinate axis at the ordinate $m(0)/c$. When the initial amount of impurity is known, the constants k and c are obtained, from the last of which we calculate the yield of cross-links as follows:

$$G(X) = 0.96 \times 10^6 c/w$$

where w is the molecular weight of a structural unit and the unit radiation dose is chosen as 10^6 rad.

Let us now consider the case where the cross linking reaction is promoted by some additive substance or impurity, as is the case of polyethylene irradiated in N_2O vapor (Okada, 1964). In such cases impurity molecules will be activated by radiation and the active center will help polymer molecules to combine into one with a cross-link or the existence of an impurity molecule will help an active center of a polymer molecule combine with another polymer molecule through a cross-link. In the former case the formulation of the reaction will be given by the following reasoning. Let us denote by m the ratio of the number of impurity molecules to that of structural units and by k the ratio of the probability that an impurity molecule will have an active center to the probability that a structural unit will have an active center. Then, the change in weight fraction of polymer molecules will satisfy the following equation (Saito, 1959):

$$\frac{\partial w(p, x)}{\partial x} = (1 + km)p\left(-2w(p, x) + \int_0^p w(l, x)w(p - l, x)\, dl\right) \quad (76)$$

The change in the number of impurity molecules will be given by

$$dm/dx = -km \quad (77)$$

If we put

$$x_e = \int_0^x (1 + km)\, dx = x + m(0)(1 - e^{-kx}) \quad (78)$$

x_e gives the density of cross-links, which is the sum of cross-links produced directly by radiation and those produced through impurity reactions. If we use the radiation dose R_e in the case when no impurity is contained and the dose R which induces the same effect in polymer containing the impurity, we have

$$R_e = R + \frac{m(0)}{c}(1 - e^{-kcR}) \quad (79)$$

where $x = cR$ and c is the density of cross-links for the polymer which is irradiated by unit dose and contains no impurity. Equation (79) is quite similar to Eq. (75) except that R and R_e are exchanged with each other.

When an impurity molecule helps an active center, created by radiation on a polymer molecule, combine with a structural unit of another molecule to make a cross-link, the same equation as Eq. (76) is derived, where k is assumed to be the ratio of the rate constant of the reaction, in which an impurity molecule helps an active center combine with a structural unit of polymer molecule, to the rate constant of the reaction in which an active center creates a cross-link without the help of impurity molecule. Similar other results can also be derived (Saito, 1959).

III. End-Linking of Polymers

A. END-LINKING OF TERMINAL DOUBLE BONDS

Charlesby (1955a) has studied the case when molecular ends newly produced by main chain scissions of polymers are incorporated in end-links and shown that gelation also occurs in this case. Since it is difficult to distinguish the effect of end-linking and that of cross-linking when they are superimposed on each other, the study of the former effect has not much been developed. Graessley (1967) has recently suggested that a certain amount of end-links which are supposed to be made by irradiated terminal double bonds of polyethylene is produced along with the usual cross-links. The effect of such an end-linking process will be discussed in this section.

Let us denote the degree of polymerization by p, time by t, the total number of structural units by N, the ratio of the number of polymer molecules having no terminal double bond to N by $m_0(p, t)$, the ratio of the number of polymer molecules having a terminal double bond to N by $m_1(p, t)$, and the probability that a terminal double bond will be incorporated into an end-link per unit time by e. Then, the changes in molecular size distribution accompanied with the increase of end-links are given by the following equations:

$$\frac{\partial m_0(p, z)}{dz} = -pm_0(p, z) \int_0^\infty m_1(p, z) \, dp$$

$$+ \int_0^p (p - l)m_0(p - l, z)m_1(l, z) \, dl \tag{80}$$

$$\frac{\partial m_1(p, z)}{dz} = -\left(1 + p \int_0^\infty m_1(l, z) \, dl\right) m_1(p, z)$$

$$+ \int_0^p (p - l)m_1(p - l, z)m_1(l, z) \, dl \tag{81}$$

where

$$z = \int_0^t e \, dt \tag{82}$$

and z is equal to the ratio of the number of end-links to N, and expresses the density of end-links. Since $m_0(p, z)$ is the ratio of the number of polymer molecules which have p structural units and no terminal double bond to the total number of structural units, let us call it the number fraction of such polymer molecules. Then $m_1(p, z)$ is the number fraction of polymer molecules which have p structural units and a terminal double bond. These number fractions give the weight fractions by multiplying themselves with p. The left-hand side of Eq. (80) gives the rate of increase of $m_0(p, z)$ with respect to z, the first term of the right-hand side gives the rate of creating end-links by combining backbone chains of molecules having p structural units with terminal double bonds of other molecules, and the second term expresses the rate of creating those molecules having p structural units by combining the terminal double bonds of molecules having l structural units with the backbone chains of molecules having $p-l$ structural units. Equation (81) is likewise derived. If we put

$$f_{i,j}(z) = \int_0^\infty p^j m_i(p, z)\, dp \tag{83}$$

following equations are derived from Eqs. (80) and (81)

$$\frac{df_{0,j}}{dz} = -f_{0,j+1}f_{1,0} + \sum_{k=0}^{j}\binom{j}{k}f_{0,k+1}f_{1,j-k} \tag{84}$$

$$\frac{df_{1,j}}{dz} = -f_{1,j} - f_{1,j+1}f_{1,0} + \sum_{k=0}^{j}\binom{j}{k}f_{1,k+1}f_{1,j-k} \tag{85}$$

By solving these equations, we have

$$f_0(z) = f_{0,0}(0) + f_{1,0}(0)\, e^{-z} \tag{86}$$

$$f_1(z) = 1 \tag{87}$$

$$f_2(z) = f_2(0)/[1 - f_{1,1}(0)(1 - e^{-z})]^2 \tag{88}$$

where

$$f_0(z) = f_{0,0}(z) + f_{1,0}(z) \tag{89}$$

$$f_1(z) = f_{0,1}(z) + f_{1,1}(z) \tag{90}$$

$$f_2(z) = f_{0,2}(z) + f_{1,2}(z) \tag{91}$$

Since the number- and the weight-average degrees of polymerization satisfy $P_n = f_1/f_0$ and $P_w = f_2/f_1$, respectively, we have

$$P_n(0)/P_n(z) = 1 - \rho(1 - e^{-z}) \tag{92}$$

$$P_w(0)/P_w(z) = [1 - \alpha(1 - e^{-z})]^2 \tag{93}$$

where

$$\rho = f_{1,0}(0)/[f_{0,0}(0) + f_{1,0}(0)] \quad \text{and} \quad \alpha = f_{1,1}(0) \tag{94}$$

and ρ is equal to the ratio of the number of molecules, which have terminal double bonds, to the total number of molecules prior to undergoing end-linking, and α denotes the weight fraction of molecules which have terminal double bonds prior to undergoing end-linking. As α is less than unity, we know from Eq. (93) that the weight-average degree of polymerization stays finite. This shows that end-linking in which the terminal double bond is incorporated into another molecule does not make any gel. Equations (92) and (93) give the relation

$$\sqrt{P_w(0)/P_w(z)} = (\alpha/\rho)P_n(0)/P_n(z) + 1 - \alpha/\rho \tag{95}$$

This shows that the plot of root reciprocal weight-average degree of polymerization vs reciprocal number-average degree of polymerization is linear.

B. END-LINKING ASSOCIATED WITH MAIN CHAIN SCISSION

Charlesby (1955) has shown that when molecular ends newly produced by main chain scissions create end-links by combining with backbone chains of other polymer molecules, the molecular size distribution changes as end-links increase and in a certain period gelation occurs. Though main chain scissions and end-linkings occur successively in this case, we will consider them as follows: When main chain scissions and end-links occur independently, exchange of their reaction order will not have any influence on the overall change in molecular size distribution. Then, even if we assume that after all main chain scissions have taken place end-linking occurs, we will get the correct molecular size distribution. Assume that μ is the ratio of those newly produced ends which take part in end-linking to all ends which are produced by main chain scissions, and that all polymer molecules are initially linear, the number-average degree of polymerization is u, and the total number of structural units is N. When the polymers are irradiated and Y chain scissions are produced, the molecular size distribution will be given by $w(p, y)$ in Eq. (12), where $y = Y/N$. Those $2\mu Y$ ends in $2Y$ newly produced ends are incorporated in end-linking. Let us call the ratio of the number of end-links to N the density of end-links and denote it by z. If we denote the ratio of the number of molecules, which have k ends and p structural units, to N by $m_k(p, z)$, we have (Saito, 1958)

$$\frac{\partial m_k(p, z)}{\partial z} + km_k(p, z) + pm_k(p, z)\psi(z)$$

$$= \sum_{i=0}^{k-1} i \int_0^p (p - l)m_i(l, z)m_{k-i+1}(p - l, z)\, dl \tag{96}$$

where

$$\psi(z) = \sum_{i=2}^{\infty} i \int_0^\infty m_i(p, z)\, dp \tag{97}$$

The rate that the ends of those molecules having p structural units and k ends are incorporated into end-links is expressed by the second term of the left-hand side of Eq. (96), the rate that the backbone chains of those molecules are incorporated into end-links is expressed by the third term of the left-hand side, and the rate that the ends of molecules having l structural units and i ends are combined with the backbone chains of molecules having $p-l$ structural units and $k-i+1$ ends so that those molecules having p structural units and k ends are produced is expressed by the right-hand side of Eq. (96). Solving Eq. (96) will give us average molecular weights, gel fraction, molecular size distribution, etc. However, only the average degree of polymerization will be given here.

When a polymer substance having initially the random molecular size distribution undergoes irradiation, and main chain scissions and end-links are produced, the number- and the weight-average degree of polymerization are given as

$$P_n(y, z) = u/(1 + uy - uz) \qquad (98)$$

$$P_w(y, z) = 2u/(1 + uy - 2uz) \qquad (99)$$

where u is the number-average degree of polymerization prior to irradiation, y is the density of main chain scissions, and z is the density of end-links. It is clear from Eq. (99) that the weight-average degree of polymerization becomes infinite at a finite value of $2z - y$. This shows that gelation occurs. As is shown in the preceding section, when only those molecular ends which have existed before irradiation produce endlinks gelation never occurs, but when those ends produced by main chain scissions take part in end-linking, gelation occurs. This difference is explained as follows: In the former case the ends which take part in end-linking are only those ends which existed prior to irradiation and, on the contrary, in the latter case every structural unit has the probability of taking part in end-linking.

IV. Cross-Linking of Copolymers

A. INTRODUCTION

It is well known that copolymers subjected to high-energy radiations become cross-linked and/or degraded, so that the molecular size distribution is changed. There have been few theories dealing with such behavior of copolymers. The change in molecular size distribution due to cross-linking and main chain scission of copolymers is very similar to that of mixed polymers, which are composed of different species of homopolymers. Inokuti (1960) has developed a theory of the molecular size distribution and gelation of irradiated copolymers composed of two different kinds of structural units. Amemiya (1962) has discussed theoretically those cases when irradiated copolymers which are composed of many kinds of structural units undergo main chain scissions. We will consider the effect of radiation on the molecular size distribution along Inokuti's theoretical work.

B. BASIC EQUATION

Let us consider a sytem of copolymers composed of two kinds of structural units, each of which will be referred to as the first and the second, respectively. We will assume the following for the cross-linking of copolymers as well as for homopolymers: (1) cross-links are produced at random, (2) every structural unit of each kind equally likely takes part in cross-linking, respectively, (3) the number of cross-links is sufficiently small in comparison to the total number of structural units, and (4) intramolecular linkage is neglected.

Let us denote by $i - j$ such cross-links as are made between a structural unit of the ith kind and one of the jth kind, where i and j equal 1 or 2. And also let us assume c_{ij} as the probability of the occurrence of an $i - j$ linkage per unit dose and per a structural unit taking part in the $i-j$ linkage. If we denote by $m(p_1, p_2, R)$ the ratio of the number of copolymer molecules, which have p_1 units of the first kind and p_2 units of the second kind when the system has been irradiated up to radiation dose R, and the total number of both structural units by N, we have the following integrodifferential equation:

$$\frac{\partial m(p_1, p_2, R)}{\partial R}$$

$$= -\{2c_{11}v_1p_1 + 2c_{22}v_2p_2 + 2c_{12}(v_2p_1 + v_1p_2)\}m(p_1, p_2, R)$$

$$+ \int_0^{p_1} dq_1 \int_0^{p_2} dq_2 [c_{11}q_1(p_1 - q_1) + c_{22}q_2(p_2 - q_2)$$

$$+ 2c_{12}q_1(p_2 - q_2)]m(q_1, q_2, R)m(p_1 - q_1, p_2 - q_2, R) \quad (100)$$

where $v_1 = N_1/(N_1 + N_2)$, $v_2 = N_2/(N_1 + N_2)$, and N_1 and N_2 are the number of structural units of the first kind and that of the second kind, respectively. The first term of the right-hand side expresses the rate of decrease of the molecules p_1, p_2 due to the formation of linkages, and the second term gives the increase of the molecules p_1, p_2 due to the combination of the molecule q_1, q_2 and the molecule $p_1 - q_1, p_2 - q_2$ through $i - j$ linkages, where the molecule p_1, p_2 means the molecule which has p_1 units of the first kind and p_2 units of the second kind. Using the notation

$$f^{(k,l)}(R) = \int_0^\infty dp_1 \int_0^\infty dp_2 p_1{}^k p_2{}^l m(p_1, p_2, R) \tag{101}$$

Eq. (99) yields

$$f^{(0,0)}(R) = f^{(0,0)}(0) - (c_{11}v_1{}^2 + 2c_{12}v_1v_2 + c_{22}v_2{}^2)R \tag{102}$$

$$f^{(1,0)}(R) = v_1, f^{(0,1)}(R) = v_2 \tag{103}$$

$$f^{(2,0)}(R) = [A_1 - 2(A_1A_2 - B^2)c_{22}R]/\Delta \tag{104}$$

$$f^{(1,1)}(R) = [B + 2(A_1A_2 - B^2)c_{12}R]/\Delta \tag{105}$$

$$f^{(0,2)}(R) = [A_2 - 2(A_1A_2 - B^2)c_{11}R]/\Delta \tag{106}$$

where

$$\Delta = 1 - 2R(A_1c_{11} + 2Bc_{12} + A_2c_{22})$$
$$+ 4R^2(A_1A_2 - B^2)(c_{11}c_{22} - c_{12}^2) \tag{107}$$

and

$$A_1 = f^{(2,0)}(0), A_2 = f^{(0,2)}(0), B = f^{(1,1)}(0) \tag{108}$$

Considering that the number-average and the weight-average degree of polymerization are given, respectively, by

$$P_n(R) = \{f^{(1,0)}(R) + f^{(0,1)}(R)\}/f^{(0,0)}(R) \tag{109}$$

$$P_w(R) = f^{(2,0)}(R) + 2f^{(1,1)}(R) + f^{(0,2)}(R) \tag{110}$$

we get

$$P_n(0)/P_n(R) = 1 - (c_{11}v_1{}^2 + 2c_{12}v_1v_2 + c_{22}v_2{}^2)P_n(0)R \tag{111}$$

$$P_w(R)/P_w(0) = (1 - \rho_0 R)/(1 - \rho_1 R)(1 - \rho_2 R) \tag{112}$$

where $P_n(0)$ and $P_w(0)$ are the number- and the weight-average degree of polymerization prior to irradiation, respectively, and

$$\rho_0 = 2(A_1A_2 - B^2)(c_{11} - 2c_{12} + c_{22})/(A_1 + A_2 + 2B) \tag{113}$$

$$\rho_1, \rho_2 = A_1c_{11} + A_2c_{22} + 2Bc_{12} \pm D^{1/2} \tag{114}$$

$$D = (A_1c_{11} + A_2c_{22} + 2Bc_{12})^2 - 4(A_1A_2 - B^2)(c_{11}c_{22} - c_{12}^2) \tag{115}$$

Since D is positive, then ρ_1 is larger than ρ_2 and the gelation dose R_g, with which the weight-average degree of polymerization becomes infinite, is given by

$$1/R_g = \rho_1 = A_1c_{11} + A_2c_{22} + 2Bc_{12} + D^{1/2} \tag{116}$$

C. EXAMPLES

1. *Mixture of Two Kinds of Homopolmyers*

Let us consider the case when we have a mixture of two kinds of polymers, in each of which the molecular size distribution is uniform and the number-average degree of polymerization is denoted by $u_i (i = 1, 2)$. The initial distribution is then expressed as

$$m(p_1, p_2, 0) = \left(\frac{v_1}{u_1} \delta(p_1 - u_1) \delta(p_2) + \frac{v_2}{u_2} \delta(p_1) \delta(p_2 - u_2) \right) \quad (117)$$

from which we find

$$f^{(k,l)}(0) = u_1^{\ k} u_2^{\ l} \{(v_1/u_1) \delta_{l0} + (v_2/u_2) \delta_{k0}\} \quad (118)$$

where $\delta(x)$ is the Dirac delta function and δ_{ij} is the Kronecker symbol, that is

$$\delta_{ij} = 1, \quad i = j$$
$$= 0, \quad i \neq j$$

The number-average degree of polymerization is calculated from Eq. (111) as

$$1/P_n(R) = (v_1 u_2 + v_2 u_1)/u_1 u_2 - (c_{11} v_1^{\ 2} + 2c_{12} v_1 v_2 + c_{22} v_2^{\ 2})R \quad (119)$$

If we plot $1/P_n(R)$ with respect to dose R, we will have a straight line, whose slope K will be

$$K = c_{11} v_1^{\ 2} + 2c_{12} v_1 v_2 + c_{22} v_2^{\ 2} \quad (120)$$

This is independent of the number-average degree of polymerization of both kinds of polymers prior to irradiation. If we plot K with respect to v_1 or v_2, we will have a parabola, which is convex upward for $c_{11} - 2c_{12} + c_{22} > 0$, and concave for $c_{11} - 2c_{12} + c_{22} < 0$, and for $c_{11} - 2c_{12} + c_{22} = 0$ it reduces to a straight line. From these curves c_{11}, c_{12}, and c_{22} can all be determined.

The gelation dose is calculated from Eq. (117) as

$$1/R_g = v_1 c_{11} u_1 + v_2 c_{22} u_2 + [(v_1 c_{11} u_1 - v_2 c_{22} u_2)^2 + 4v_1 v_2 c_{12} u_1 u_2]^{1/2} \quad (121)$$

If we plot $1/R_g$ with respect to v_1, the ordinates at $v_1 = 0$ and 1 will give $2c_{11} u_1$ and $2c_{22} u_2$, respectively, and the slope of the curve at the points corresponding to $v_1 = 0$ and 1 will be given by

$$2(c_{12}^2 u_1 - c_{22}^2 u_2)/c_{22} \quad \text{for } v_1 = 0$$
$$2(c_{12}^2 u_2 - c_{11}^2 u_1)/c_{11} \quad \text{for } v_1 = 1$$

These results can be compared with experimental values.

2. Mixture of Two Kinds of Polymers, Each of Which Has the Random Molecular Size Distribution

Suppose that a mixture is composed of two kinds of polymers, each has the random molecular size distribution, in which the number-average degree of polymerization is equal to $u_i(i = 1, 2)$. Then, the molecular size distribution is specified by

$$m(p_1, p_2, 0) = 2\left[\frac{v_1}{u_1}\exp\left(-\frac{p_1}{u_1}\right)\delta(p_2 - u_2)\right.$$

$$\left. + \frac{v_2}{u_2}\exp\left(-\frac{p_2}{u_2}\right)\delta(p_1 - u_1)\right] \qquad (122)$$

Now we have

$$f^{(k,l)}(0) = u_1{}^k u_2{}^l[(v_1/u_1)\Gamma(k + 1)\,\delta_{l0} + (v_2/u_2)\Gamma(l + 1)\,\delta_{k0}] \qquad (123)$$

The number-average degree of polymerization is given by the same equation as (119), so that the same discussion along Eq. (120) is also developed.

The gelation dose is then calculated from Eq. (117) as

$$1/R_g = 2\{v_1 c_{11} u_1 + v_2 c_{22} u_2 + [(v_1 c_{11} u_1 - v_2 c_{22} u_2)^2 + 4v_1 v_2 c_{12}^2 u_1 u_2]^{1/2}\} \qquad (124)$$

This is quite similar to Eq. (121). Thus, the behavior in the plot of $1/R_g$ versus v_1 differs by a factor 2 from the above-mentioned.

3. Copolymers of Uniform Molecular Size Distribution

Let us consider a system of copolymers which have uniform size, and the number-average degree of polymerization is given by u. We assume further that a polymer molecule is composed of $v_1 u$ units of the first kind and $v_2 u$ units of the second one, where $v_1 + v_2 = 1$. Then, we have

$$m(p_1, p_2, 0) = u^{-1}\,\delta(p_1 - v_1 u)\,\delta(p_2 - v_2 u) \qquad (125)$$

from which

$$f^{(k,l)}(0) = v_1{}^k v_2{}^l u^{k+l-1} \qquad (126)$$

When such a system of copolymers is irradiated, the number-average degree of polymerization is calculated from Eq. (119), whereas the gelation dose is given by

$$1/R_g = 2u(c_{11} v_1{}^2 + 2c_{12} v_1 v_2 + c_{22} v_2{}^2) \qquad (127)$$

which is similiar to Eq. (120).

4. Copolymers of Random Size Distribution

Here we consider a system of copolymers, which has the random molecular size distribution, the number-average degree of polymerization being u.

We assume further that the ratio of the number of structural units of the first kind to that of the second one in every polymer molecule is constant and equal to $v_1 : v_2$, then we have

$$m(p_1, p_2, 0) = u^{-2} \exp[-(p_1 + p_2)/u] \tag{128}$$

also

$$f^{(k,l)}(0) = u^{k+l-1}\Gamma(k + 1)\Gamma(l + 1) \tag{129}$$

The number-average degree of polymerization under irradiation is given by Eq. (119) and the gelation dose is calculated from

$$1/R_g = u\{2(c_{11}v_1^2 + 2c_{12}v_1v_2 + c_{22}v_2^2)$$
$$+ [(c_{11}v_1^2 + 4c_{12}v_1v_2 + c_{22}v_2^2)^2 + 3(v_1^2c_{11} - v_2^2c_{22})^2]^{1/2}\} \tag{130}$$

If we plot $1/R_g$ with respect to v_1, the ordinates at $v_1 = 0$ and 1 are $4c_{11}u$ and $4c_{22}u$, respectively, and the slopes of the plot at $v_1 = 0$ and 1 are, respectively,

$$2u(3c_{12} - 4c_{22}), \quad \text{and} \quad 2u(4c_{11} - 3c_{12})$$

These can be compared with experimental values.

REFERENCES

Amemiya, A. (1962). *J. Phys. Soc. Japan* **17**, 1245, 1694.
Amemiya, A. (1967). *J. Phys. Soc. Japan* **23**, 1394, 1402.
Charlesby, A. (1953). *J. Polym. Sci.* **11**, 513.
Charlesby, A. (1954). *Proc. Roy. Soc.* **A222**, 60, 542; **A224**, 120.
Charlesby, A. (1955a). *Proc. Roy. Soc.* **A231**, 521.
Charlesby, A. (1955b). *J. Polym. Sci.* **15**, 263.
Charlesby, A. (1960). "Atomic Radiation and Polymers." Pergamon Press, New York.
Charlesby, A., and Pinner, S. H. (1959). *Proc. Roy. Soc.* **A249**, 367.
Flory, P. J. (1941). *J. Amer. Chem. Soc.* **63**, 3083, 3091, 3096.
Flory, P. J. (1942). *J. Phys. Chem.* **46**, 132.
Flory, P. J. (1953). "Principles of Polymer Chemistry." Cornell Univ. Press, Ithaca, New York.
Gee, G. V., and Melville, H. W. (1944). *Trans. Faraday Soc.* **40**, 240.
Graessley, W. W. (1967). Private communication.
Inokuti, M. (1960). *J. Chem. Phys.* **33**, 1607.
Inokuti, M. (1963). *J. Chem. Phys.* **38**, 2999.
Inokuti, M., and Katsuura, K. (1959). *J. Phys. Soc. Japan* **14**, 1379.
Jordan, D. O., and Mathieson, A. R. (1952). *J. Chem. Soc.* 2358.
Kang, H. Y., Saito, O., and Dole, M. (1967). *J. Amer. Chem. Soc.* **89**, 1980.
Kataoka, S. (1953). *Busseiron-Kenkyu* **66**, 402.
Katsuura, K. (1960). *J. Phys. Soc. Japan* **15**, 2310.
Kawai, T., Keller, A., Charlesby, A., and Ormerod, M. G. (1965). *Phil. Mag.* **12**, 657.
Kimura, T. (1962). *J. Phys. Soc. Japan* **17**, 1884.
Kimura, T. (1964). *J. Phys. Soc. Japan* **19**, 777.

Kitamaru, R., Mandelkern, L., and Fatou, J. (1964). *J. Polym. Sci.* **B2**, 511.
Kuhn, W. (1930). *Ber.* **63**, 1503.
Lawton, E. J., Balwit, J. S., and Powell, R. S. (1958). *J. Polym. Sci.* **32**, 257.
Matsumoto, S. (1949). *Kobunshi-Kagaku* **6**, 40.
Montroll, E. W., and Simha, R. (1940). *J. Chem. Phys.* **8**, 721.
Okada, Y. (1964). *J. Phys. Chem.* **68**, 2120.
Okada, Y., and Amemiya, A. (1961). *J. Polym. Sci.* **50**, 22.
Saeman, J. F., Millett, M., and Lawton, E. J. (1952) *Ind. Eng. Chem.* **44**, 2848.
Saito, O. (1958). *J. Phys. Soc. Japan* **13**, 198, 1451, 1465.
Saito, O. (1959). *J. Phys. Soc. Japan* **14**, 798.
Saito, O., Kang, H. Y., and Dole, M. (1967). *J. Chem. Phys.* **46**, 3607.
Schulz, G. V. (1939). *Z. Phys. Chem.* **B43**, 25; **B44**, 227.
Simha, R. (1941). *J. Appl. Phys.* **12**, 569.
Stockmayer, W. H. (1943). *J. Chem. Phys.* **11**, 45.
Stockmayer, W. H. (1944). *J. Chem. Phys.* **12**, 125.
St. Pierre, L. E., and Dewhurst, H. A. (1958). *J. Chem. Phys.* **26**, 241.
Tung, L. H. (1956). *J. Polym. Sci.* **20**, 495.
Wesslau, H. (1956). *Makromol. Chem.* **20**, 111.
Zimm, B. H., and Kilb, R. W. (1959). *J. Polym. Sci.* **37**, 19.
Zimm, B. H., and Stockmayer, W. H. (1949). *J. Chem. Phys.* **17**, 1301.

Experimental Techniques and Applications
to Polyethylene

12

Experimental Techniques

Malcolm Dole
Department of Chemistry, Baylor University, Waco, Texas

* By K. C. Humpherys, Nuclear Science and Instrumentation Labs, EG & G, Goleta, California.

I. Radiation Cells and Techniques

A. INTRODUCTION

The development of the atomic pile during the Second World War made available to scientists large sources of very intense ionizing radiation and stimulated some of the earliest work on the radiation chemistry of macromolecules. Thus, Davidson and Geib (1948) (natural rubber, polyisobutylene, and butyl rubber), Dole (1948) (polyethylene), and Charlesby (1952) (polyethylene) all used pile irradiations in their initial studies. However, pile radiations are a mixture of γ-rays and neutrons; from a purely scientific purpose and for ease of interpretation of the observed data, pure γ-rays or high-speed electrons represent a better type of radiation to use. Electron beams have the advantage of being able to be turned on or off at will; ^{60}Co-gamma rays have the advantage of high penetrability. Both types of radiation sources are available commercially. We shall now consider types of irradiation cells to use at room, above room, and below room temperatures.

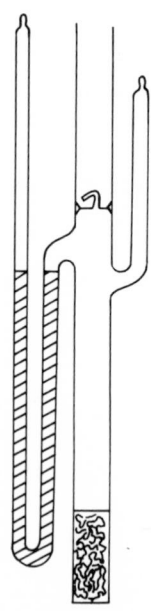

Fig. 1. Polymer irradiation cell with break-off seal for gas analysis.

B. ROOM-TEMPERATURE IRRADIATIONS

Most frequently radiation chemical studies of high polymers have been made at room temperature. The first observation to make is that of gas evolution, because the amount of the latter is a good indication of the extent of radiation chemical reactions resulting from the irradiation. The type of irradiation cell to use depends on a number of factors, such as the shape of the irradiation zone in the source, the amount of material to be irradiated, and temperature of irradiation. A simple cell used by the author and his students is shown in Fig. 1. The side arm enabled the cell to be evacuated before sealing off. The gas pressure was measured in the mercury manometer at different doses. The cell was removed from the irradiation zone and placed in a constant temperature bath at 25°C. Mercury levels were read with a cathetometer, and after determining the void volume of the cell, the total moles of gas liberated could be calculated. The latter, of course, did not include the moles of gas dissolved in the polymer at the time of making the manometer readings. By pouring liquid nitrogen into the space above the break-off seal, gases that condense at 77°K were frozen out and the residual pressure measured. Suitable corrections determined in a control experiment had to be applied to allow for the temperature gradient at the junction between the constant temperature bath and the liquid nitrogen. After measuring the pressures as a function of dose, the cell was connected to a vacuum line, the break-off seal opened, and moles of gas measured in a Toepler gauge. The 77°K uncondensable gases were measured first by passing the gas through a trap at 77°K and then the moles of condensable gases were measured. Analysis of the gases could be carried out in an analytical mass spectrometer or in a gas chromatograph.

Figure 2 illustrates a cell especially designed for the irradiation of polymer films held in a metallic frame that could be inserted into an infrared spectrometer for infrared analysis. Dole *et al.* (1958) used a similar cell with a Pirani gauge attached for low-pressure measurements and were able to demonstrate no increase in gas pressure for irradiations of polyethylene at 77°K. The hydrogen that was produced by the irradiation was trapped in the solid and was not evolved until the solid was heated to about 150°K and above.

In the case of electron spin resonance (ESR) studies of free radicals produced by the irradiation, it is customary to irradiate a small plug of the polymer, about 15–20 mg in special quartz tubing of the size to fit into the ESR spectrometer cavity (4.2-mm external diameter in our work). The tubes were made of Vitreosil pure fused quartz which exhibits no ESR signal before irradiation, but does after irradiation, so that the empty end of

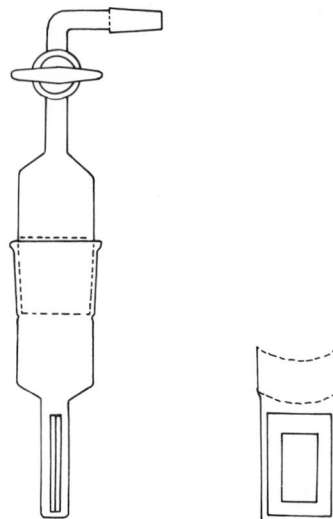

Fig. 2. Polymer irradiation cell containing a frame for holding a polymer film.

the tube had to be heated to dull redness to remove trapped free radicals while the other end, containing the sample, was immersed in liquid nitrogen. After annealing, the tube was inverted, allowing the sample plug to slide to the annealed end of the quartz tube. It is this end which is inserted into the ESR cavity. In the case of powders they were formed into a plug of sufficient integrity for the sliding operation by putting the powder into an open-ended glass tube held vertically on a glass plate and compressing the powder together with a glass rod. The rather fragile cylinder is then easily slipped into the quartz irradiation tube. The quartz tubes can be attached to a vacuum line with rubber tubing sealed on with black wax, Apiezon C.

In all irradiation work exact position reproducibility both horizontal and vertical in the source and a knowledge of the radiation intensity at that position are essential. Methods of calibrating the source are described below.

C. IRRADIATIONS AT ELEVATED TEMPERATURES

A simple method of maintaining an elevated temperature during an irradiation is to place a tube-shaped irradiation cell into a cylindrical cavity drilled out of a copper block. Heating wires can be wrapped around the outside of the block and the temperature measured and controlled by means of a Speedomax temperature controller through a thermocouple attached to the copper block. Such a device can be only used if highly penetrating γ-rays are the radiation source. In this connection it should be pointed out that without any heating, the intensity of the γ-ray energy absorption in the copper block may be sufficient to raise it several degrees in temperature. With a radiation intensity of about 0.7 Mrad hr^{-1} we have found in our work that the temperature of the copper block rose to about 35°C.

In many cases irradiation with electron beams is to be preferred (dose rates are much greater and irradiation time correspondingly reduced), and the above copper block holder for the radiation cells cannot be used. For the irradiation of thin polymer films at elevated temperature in a vacuum we have used the metal cell illustrated in Fig. 3. The body of the cell (A) is

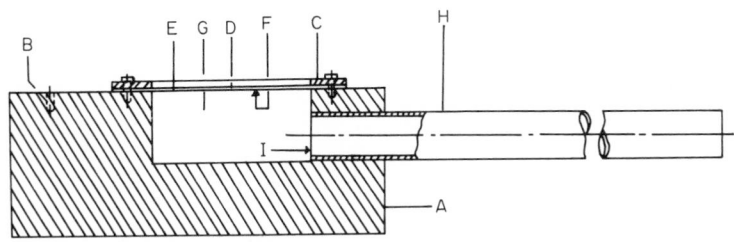

Fig. 3. Irradiation cell for high-temperature studies.

built of brass with glass insulated nichrome heater wire wrapped around the outside. B is a small hole for a thermocouple. A brass frame (C) holds a thin titanium or aluminum window (D) with silicone rubber gasket (E) over the top opening of the cell (F). (The cell is used in a horizontal position with the electron beam striking the window in a vertical direction.) G is the evacuated space into which the film, in its metal frame or melted onto a glass plate, is placed. The radiation zone is evacuated through the stainless steel tube (H). The thermal conductivity of the latter is sufficiently low to allow a rubber tube connected to a vacuum pump to be attached to its far end.

Williams and Dole (1959) used a cell similar to that shown in Fig. 2 for the irradiation of polyethylene films held between two aluminum plates around which glass-insulated heating wires were wrapped and to which a thermocouple was attached. After irradiations above the melting point the polymer film could be recovered, but it adhered so strongly to the aluminum plates that the assembly had to be soaked in boiling water for 0.5–1 hr before the plates could be separated from the film.

D. CELLS FOR LOW-TEMPERATURE IRRADIATIONS

Dole and Böhm (1968) described a combined irradiation and spectro-scopic cell which enabled them to irradiate polyethylene films at 77°K in a vacuum and then to examine them spectroscopically also at 77°K without an intervening heating. The windows of the cell could also be changed after the irradiation at 77°K without exposure of the films to air. A more recent modification of this cell (Waterman and Dole, 1970a) is illustrated in Fig. 4.

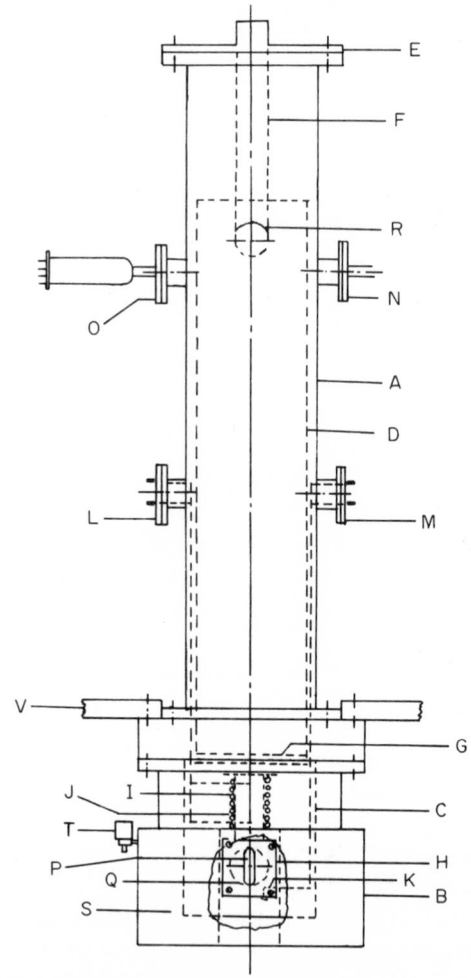

Fig. 4. Cell for combined radiation and spectroscopic studies.

The outer jacket (A) and the two demountable bottom sections (B and C) were made of brass. The vacuum jacketed liquid nitrogen reservoir (D) was made of copper and the top plate (E) of stainless steel. The stainless steel filling tube (F) had a right-angled bend in it at R which enabled the whole cell to be laid in a horizontal position for the electron beam irradiations without spillage of the liquid nitrogen. The Dewar section was a hollow stainless steel cylinder with a copper bottom plate (G). The sample holder (H) in two pieces was fastened to the bottom plate. Nichrome heating wire was wrapped around the solid cylindrical holder (J) of the cell so that the temperature of the cell could be varied from 77°K to higher temperatures. K is a hole for a thermocouple and ports L and M in the outer jacket were for the heating and thermocouple wires. The other two ports (N and O) were used for a pump-out and for the installation of a vacuum gauge.

In use, the sample to be irradiated was held mechanically over the opening indicated by P and between the two halves of the cell which were bolted together. A thin titanium window on one side and a solid brass plate on the other covered the openings of the inner demountable brass compartment (C). The whole interior section of the cell was continuously evacuated through N during and after the irradiation. During the irradiation the cell was laid on its side so that the electron beam passed in a vertical direction through the titanium window and polymer film. After the irradiation sodium chloride or quartz windows were affixed to the outer demountable compartment (B) and the space between B and C evacuated through the small valve (T). With the release of the air pressure which initially held on the titanium window and brass plate, the latter two objects fell off and could be rolled aside. A continuous light path through the cell and sample was thereby provided. The whole cell could then be placed in an infrared or ultraviolet spectrometer in a position which permitted the sample film to be analyzed spectroscopically.

Salovey *et al.* (1970) have recently described a cryostatic cell in which they irradiated films of poly(vinyl chloride) (PVC) and observed them spectroscopically at low temperatures. With the film holder at 80°K they found the film temperature to be 160°K, and using a dose rate of 600 Mrad hr^{-1}, they observed that the film rose in temperature 30° during the irradiation and cooled rapidly thereafter.

In working with liquid nitrogen, which may be exposed to high energy radiations, one should realize that the liquid nitrogen will dissolve molecular oxygen from the air. On exposure to the ionizing radiations ozone will form and may be produced in dangerously explosive amounts. This is true if the liquid nitrogen is replenished as it evaporates from time to time without pouring out the residual liquid nitrogen–oxygen mixture. In our work we routinely dump out the residual liquid nitrogen before adding new. The danger of ozone in liquid nitrogen has recently been investigated by Chen and Struss (1969).

Some irradiations at $77°K$ can be carried out by floating the object to be irradiated on the surface of liquid nitrogen using small Styrofoam floats attached to each end of the tube or vessel containing the sample. Such a system can then be placed under the exit nozzle of the electron beam generator and irradiated by electron beams impinging on the sample from above. The liquid nitrogen can be contained in a glass Dewar flask or simply in a vessel made of Styrofoam.

II. Radiation Dosimetry

A. The Fricke Dosimeter

As mentioned above it is most important to know the absolute dose of high-energy radiation that a macromolecule receives during an irradiation period so as to obtain quantitative and reproducible results of significance to other investigators in the field. The most convenient method of calibrating the radiation intensity is to use the Fricke dosimeter, whose preparation and properties have been ably summarized by Allen (1961). When the aqueous solution in the Fricke dosimeter is exposed to ionizing radiations, ferrous sulfate is oxidized to ferric sulfate at the rate of 15.6 molecules per 100 eV of energy absorbed. The solution should be prepared from redistilled distilled water; it should be 0.003 M in ferrous and chloride ions, 0.8 N in sulfuric acid, and the solution should be saturated with air or oxygen. Allen gives the following recipe for the dosimeter solution: 2 g $FeSO_4 \cdot 7\,H_2O$, 0.3 g NaCl, and 110 cm^3 concentrated (95–98%) H_2SO_4 dissolved in distilled water to make 5 liters of solution. The presence of the chloride ions is to inhibit the occurrence of chain reactions due to organic impurities which would interfere with the correct yield. If the dose is too great, the dissolved molecular oxygen is all consumed and the oxidative yield goes down. For solutions saturated with oxygen, the total dose should not exceed 0.2 Mrad. In general, the dose rate should not exceed 10^8 rad sec^{-1}. After the irradiation the amount of ferric ion produced should be measured by spectrophotometry at 305 nm where the absorption of the ferric ion is at a maximum and that of the ferrous ion is at a minimum. The extinction coefficient, ε, to use is 2200 liters $mole^{-1}$ cm^{-1} at 25°C. The temperature coefficient of ε is quite large, $+0.7\%$/deg. Hochanadel and Ghormley (Ghormley, 1961) recommend that the Fe^{3+} ion concentration be measured using light of 275-nm wavelength where the temperature coefficient is negligible. The Fe^{3+} extinction coefficient at 275 nm is 1830 liters $mole^{-1}$ cm^{-1}.

The absorbance due to Fe^{3+} should be determined for several different radiation exposures and the optical density plotted as a function of the time. A straight line extrapolating to zero optical density at zero time should be obtained. If the time taken to move the Fricke dosimeter into and out of the radiation zone is a significant fraction of the total time of exposure, a correction should be applied after determining the amount of ferric ion produced on inserting the cell into the radiation zone and instantly removing it. This difficulty does not exist when the dosimeter can be stationed in front of the electron beam and then the beam turned on. However, in the latter case if the maximum intensity of the beam is attained only slowly, again a correction might have to be applied.

Letting S equal the slope of the plot of the optical density versus the time, the dose rate, dr/dt, in units of electron volts per gram per hour, as measured in the Fricke dosimeter is given by the equation

$$\frac{dr}{dt} = \frac{S(3600)N_A}{1000\varepsilon l \rho_{H_2SO_4}(0.156)}$$

in which 3600 is the number of seconds per hour, N_A is Avogadro's number, ε is the extinction coefficient of the Fe^{3+} ions in liters per mole per centimeter, 1000 converts to milliliters, l is the length of the light absorption path in the spectrophotometer cell in centimeters, and $\rho_{H_2SO_4}$ is the density of the 0.8 N H_2SO_4 solution and converts the dose rate to grams^{-1} instead of milliliters^{-1}. The dose rate as calculated from the above equation for a polymer has to be modified by multiplying the right-hand side by the ratio

$$\frac{(\Sigma Z/\Sigma A)\text{polymer}}{(\Sigma Z/\Sigma A)0.8\,N\,H_2SO_4}$$

where $\Sigma Z/\Sigma A$ is the ratio of the sum of the atomic numbers to that of the atomic weights of all of the atoms in a polymer repeating unit or in the Fricke solution. In the case of polyethylene the above ratio is simply $(6 + 2)/(12 + 2)$ and taking the 0.8 N H_2SO_4 to be the same as water $(2 + 8)/(2 + 16)$ for the water. The ratio of $\Sigma Z/\Sigma A$ for polyethylene divided by that for the water is 1.03. In the case of polypropylene oxide the ratio is 0.993.

B. POLYETHYLENE FILMS IN DOSIMETRY

In the use of a solution in dosimetry such as that of the Fricke dosimeter, it is necessary, of course, to place the solution in exactly the same location with respect to height, width, and length that the sample to be irradiated will occupy. With the spectroscopic cells described above this is almost impossible to do. Instead, we have found it convenient to calibrate the high- and low-temperature cells at room temperature by measuring the rate of decay of the terminal unsaturation, the vinyl group, in Marlex-type (linear) polyethylene. The first-order decay constant in units of grams per electron volts had previously been determined at room temperature in Pyrex cells where the dose rate was well known from measurements with the Fricke dosimeter. Williams and Dole (1959) found 1.6×10^{-21} g eV^{-1} at 25° while Kang, Saito, and Dole (1967) found 1.3×10^{-21} for the first-order vinyl decay constant at 35°. Dole *et al.* (1958) stated that the first-order decay relation was valid up to a decay of about 50% of the initial vinyl concentration.

C. CALIBRATION OF HIGH-INTENSITY RADIATION SOURCES

Although the Fricke dosimeter is satisfactory for most radiation sources commonly used, sometimes the need exists for a high dose rate dosimeter. Frank and Benton (1970) have recently described a polycarbonate (Lexan) plastic dosimeter for measuring absorbed doses in the range of 10^7–10^9 rad. After exposure of strips of the plastic to γ-rays from ^{60}Co (apparently in air) they were etched in 6.25 N NaOH solution with 0.5% Benax surfactant at 70.4°C. By weighing the strips before irradiation and after the etching it was found that the etching rate was a nonlinear function of the dose. It was thought that by calibration a useful high-dosage dosimeter could be developed.

D. NEUTRON DOSIMETRY*

Although most radiation chemistry studies are conducted with radiation sources that produce either pure γ-ray fields or high-energy electron beams, the investigator frequently desires to study the effects of neutrons upon his system. There are several types of neutron sources available to the radiation chemist. These sources include, in order of their importance, reactors, accelerator produced neutrons, and radioisotopic neutron sources. All of these sources consist of a mixture of γ-rays and neutrons.

A reactor produces neutrons ranging in energy from the maximum in the fission spectrum down to the thermal distribution of energies. The fission neutron energy spectrum for ^{235}U, measured by Frye *et al.* (1954), shows a peak energy just below 1 MeV and an average energy of about 2 MeV. The neutron energy spectrum is, however, greatly affected by any moderating,

* By K. C. Humpherys, Nuclear Science and Instrumentation Labs, EG & G, Goleta, California.

scattering, or shielding material, making neutron dosimetry a necessary part of every experiment performed in the reactor.

Both steady state and pulsed reactors have been widely used for radiation chemistry studies. Unmoderated fast burst reactors such as the Godiva type have the capability of yielding pulses of over 10^{17} fissions (Wimett, 1965), with pulse halfwidths of 50–100 μsec. Water moderated fast burst reactors of the Triga type yield pulses of 10^{18} fissions (Mikesell, 1966), with pulse halfwidths of about 10 msec.

Accelerator-produced neutrons are generally obtained from one of the following reactions:

$$^2H_1 + {}^2H_1 \rightarrow {}^3He_2 + {}^1n_0$$

$$^3H_1 + {}^2H_1 \rightarrow {}^4He_2 + {}^1n_0$$

The bombardment of deuterium or tritium targets with deuteron beams gives rise to these reactions which produce nearly monoenergetic neutrons of 2.5 and 14 MeV, respectively. Neutron generators based on these reactions are commercially available and have yields of up to 2×10^{11} neutrons/sec. Another accelerator approach to neutron production is the bombardment of a uranium block with high-energy electrons. Variable pulsewidths are obtainable from less than 10^{-10} sec up to and including steady state operation. Neutron yields vary directly with the pulse width and are slightly less than that obtained from a reactor pulse of the same width.

Radioisotopic sources of neutrons are of low intensity and therefore of limited value to the radiation chemist. These sources generally consist of an alpha emitting radioisotope mixed with beryllium. Neutrons are produced according to the following reaction:

$$^9Be_4 + {}^4He_2 \rightarrow {}^{12}C_6 + {}^1n_0$$

The radioisotopes used for the alpha ($^4He_2$) emission include polonium (^{210}Po), plutonium (^{239}Pu), radium (^{226}Ra), and americium (^{241}Am). The typical neutron output from such sources is 10^5–10^7 neutrons/sec. The source half-life corresponds to the radioisotope used. The neutron energy distributions range from around 10 MeV down to thermal energy with the average being a few million electron volts.

The energy deposition, or dose, in materials from neutron irradiation is usually obtained by means of a calculation based upon the chemical composition of the material and the neutron energy spectrum and fluence. Fluence is defined as the total number of particles or photons of radiation traversing any part of a 1-cm-diam sphere during the entire period of exposure. It can be obtained from threshold foil measurements, which are general and can be adapted by the user to his particular neutron radiation source and material being studied. Generally calculation of the first collision dose (Kerma—NBS Handbook No. 84) is sufficient. If the irradiated sample is large compared to a neutron removal mean free path, corrections must be made to convert the first collision dose into a depth-dose distribution. These corrections can be made accurately by means of Monte Carlo calculations, or can often be measured in very large specimens by use of detectors placed throughout the material. If the sample thickness is less than 1 cm, the first collision dose will approximate the actual dose within experimental error of the neutron fluence measurements.

The basic relationship governing the dose to materials from neutron irradiation is

$$D(E) = KE_n \sum_i f_i Q_i \sigma_i \tag{1}$$

where $D(E)$ = dose in rads to material per unit neutron fluence of energy E, K = constant for conversion of million electron volts per gram to rads, 1.6×10^{-8} rad/MeV/g, E_n = neutron energy in million electron volts, f_i = average fraction of energy lost by the neutron to the ith atom in each interaction, Q_i = number of atoms of the ith element in 1 g of sample, and σ_i = elastic scattering cross section for the ith element for neutrons of energy E_n. The average energy lost by the neutron to the atom upon an interaction, f_i, is defined by

$$f_i = 2mM_i/(m + M_i)^2 \tag{2}$$

where m = atomic weight of a neutron, 1.009, and M_i = atomic weight of the ith atom of the sample. This calculation of f_i assumes that the neutron–sample atom interactions will be isotropic in the center of mass system and that the neutrons will make only elastic collisions. Both assumptions are reasonably accurate for neutrons up to several million electron volts. The number of atoms per gram of sample, Q, is calculated as follows:

$$Q_i = F_i N_A/M_i \tag{3}$$

where F_i = fraction of ith element of sample by weight and N_A = Avogadro number, 6.02×10^{23}. Combining Eqs. (1)–(3), the dose to a material from neutrons can be rewritten as follows:

$$D(E) = 1.0 \times 10^{16} E_n \sum_i \frac{F_i \sigma_i m}{(m + M_i)^2} \tag{4}$$

For an irradiation with neutrons of a distribution of energies, the does to a material can be written

$$D = \int_0^\infty D(E)\phi(E)\,dE \tag{5}$$

where D = total first collision dose to sample in rads and $\phi(E)$ = differential neutron fluence as a function of energy. If the neutron fluence can be measured above several threshold energies, the fluence within several energy increments can be obtained. The energy increments can be selected such that the variation in $D(E)$ is small within each increment, and the dose to the material can then be written

$$D = \sum_i \Phi_i \overline{D(E)_i} \tag{6}$$

where Φ_i = neutron fluence in the ith energy increment and $\overline{D(E)_i}$ = average value of $D(E)$ for the ith energy increment. For any given neutron energy increment, $\overline{D(E)}$ becomes a summation of the terms in Eq. (4) for the various elemental constituents making up the irradiation sample. Then, the total dose to the sample is a summation of Eq. (6) for the several energy increments used throughout the neutron energy spectrum.

Most neutron fluence measurements for radiation effects or radiation chemistry studies are made with neutron threshold foil techniques. Other passive techniques that may be applied in characterizing the neutron environment include the use of nuclear emulsions and condenser-type ionization chambers. In addition to these passive techniques, a variety of dynamic-readout systems can also be used, including proton-recoil counters, solid state spectrometers, intrinsic thermocouples, and proportional counters. All of these systems, with the exception of threshold foil detectors, require complex instrumentation and long experiment times. Threshold foil techniques, on the other hand, require very little time, are generally not expensive, and can be applied in as large a number as desired. Threshold foil detectors are not sensitive to gamma radiation. For these reasons the present discussion will be limited to this technique of neutron fluence measurement. The threshold foil detectors may be obtained from a number of companies such as Reactor Experiments, Inc.

Many nuclear reactions of the types (n, particle) or (n, fission) exhibit a threshold energy, E_t, below which the reaction does not occur and above which it does. Materials that exhibit such a threshold energy for some nuclear reaction represent a means of detecting the neutron fluence above that energy. Hence, by using a series of such materials as threshold detectors and by taking the differences in measured fluences above the various threshold energies, one can determine the fluences in several energy increments throughout the neutron energy distribution (Humpherys, 1969).

The cross section for the nuclear reaction rises from zero at the threshold energy and varies as a function of neutron energy above it. Furthermore, the threshold energy and cross section as a function of neutron energy are peculiar to each particular reaction. Many of the isotopes formed by these reactions are radioactive. In such cases, the quantity of radioisotope is obtained by measurement of the induced radioactivity of the material. The neutron fluence above the threshold energy is then related to this value by

$$\Phi_{E_t} = \int_{E_t}^{\infty} \phi(E)\, dE = \frac{A_0}{N}\left(\int_{E_t}^{\infty} \phi(E)\, dE \middle/ \int_0^{\infty} \sigma(E)\, \phi(E)\, dE \right) \tag{7}$$

where Φ_{E_t} = neutron fluence above threshold energy E_t in neutrons per square centimeter, $\phi(E)$ = differential neutron fluence as a function of energy, A_0 = radioactivity induced in material, disintegrations per second at zero decay time, N = number of atoms in the material subject to the reaction, and $\sigma(E)$ = cross section for the reaction as a function of neutron energy in square centimeters per atom. The activity of the detector material expressed in practical terms is

$$A_0 = CPM\varepsilon / e^{-\lambda t} \tag{8}$$

where CPM = counts per minute at time t as measured with a nuclear counter, ε = the nuclear counter efficiency, and $e^{-\lambda t}$ = decay correction for elapsed time t, from neutron exposure to time of counting. The number of atoms in the detector material subject to the reaction N, is

$$N = WN_A I / M \tag{9}$$

where W = detector material weight in grams, N_A = Avogadro's number, I = isotopic purity in fractional weight, and M = isotopic weight in grams per mole. Combining Eqs. (7)–(9) we have

$$\Phi_{E_t} = \left(\frac{CPM}{\varepsilon W e^{-\lambda t}} \frac{M}{N_A I} \right)\left(\int_{E_t}^{\infty} \phi(E)\, dE \middle/ \int_0^{\infty} \sigma(E)\phi(E)\, dE \right) \tag{10}$$

In practice, for most neutron fluence measurements, the threshold energy and the cross section as a function of neutron energy are replaced with an effective threshold energy $E_{t\,\text{eff}}$, and a constant effective cross section, σ_{eff}, weighted by the ^{235}U fission neutron energy spectrum. This relationship is expressed as follows:

$$\int_0^\infty \sigma(E)\phi(E)\,dE = \sigma_{\text{eff}} \int_{E_{t_{\text{eff}}}}^\infty \phi(E)\,dE \tag{11}$$

or

$$\frac{1}{\sigma_{\text{eff}}} = \left(\int_{E_{t_{\text{eff}}}}^\infty \phi(E)\,dE \middle/ \int_0^\infty \sigma(E)\phi(E)\,dE \right) \tag{12}$$

Therefore Eq. (10) can be expressed as

$$\Phi_{E_t} = \left(\frac{CPM}{\varepsilon W e^{-\lambda t}} \frac{M}{N_A I} \right) \frac{1}{\sigma_{\text{eff}}} \tag{13}$$

Table I lists several nuclear reactions used as threshold detectors, along with their effective threshold energies and effective cross sections.

TABLE I

NEUTRON THRESHOLD DETECTORS

Material	Reaction	Effective threshold energy (E_t) (MeV)	Effective cross section (10^{-24} cm²)
Plutonium	^{239}Pu(n,f) Fission products	0.010	1.7
Neptunium	^{237}Np(n,f) Fission Products	0.6	1.6
Rhodium	^{103}Rh$(n,n')^{103}$Rh	0.6	0.65
Uranium	^{238}U (n,f) Fission products	1.5	0.55
Indium	^{115}In$(n,n')^{115}$In	1.5	0.36
Sulfur	^{32}S $(n,P)^{32}$P	3.0	0.30
Nickel	^{58}Ni $(n,P)^{58}$Co	3.0	0.29
Silicon	^{28}Si $(n,P)^{28}$Al	5.5	0.05
Magnesium	^{24}Mg $(n,P)^{24}$Na	6.3	0.05
Aluminum	^{27}Al $(n,a)^{24}$Na	8.1	0.11
Iodine	^{127}I $(n,2n)^{126}$I	11.0	0.98
Zirconium	^{90}Zr $(n,2n)^{89}$Zr	14.0	1.60

III. Analytical Techniques

A. GAS ANALYSIS

It has already been explained above how the gas generated by the irradiation can be separated from the polymer, partitioned into fractions which can be condensed at 77°K and which can pass through a trap at that temperature, and measured in a Toepler pump. It should be pointed out that no gases diffuse with finite rates through polymers at 77°K, except possibly helium, with which we are not concerned. Hence, in irradiating a sample at 77°K, the sample has to be heated to room temperature before pumping off the gases.

In their investigation of the gaseous products of the radiolysis of polyoxymethylene (POM) Fischer and Langbein (1967) swept helium through their irradiation cell to carry the gaseous products into a commercial gas chromatographic column. The column was calibrated with known substances and in the case of formaldehyde by means of an internal standardization. They were able to identify H_2, CH_4, HCHO, CH_3CHO, CH_3OCHO, CH_3OCH_2OCHO, and H_2O and to prove the absence of O_2, N_2, CO_2, C_2H_6, C_2H_4, C_2H_2, CH_3OH, C_2H_5OH, and $C_2H_5OC_2H_5$. H_2, CH_4, CO, and formaldehyde were determined quantitatively. All other gases had G values less than 0.1.

Mass spectroscopy can also be used for gas analysis. This approach is especially convenient when only a few products have to be determined. In a radiolysis of linear PE, for example, the composition of the gas was found to be 99.7% H_2 as determined by mass spectral analysis (Williams and Dole, 1959). When a mixture of products results, the use of a gas chromatographic column is to be preferred. Mass spectroscopy has recently been used in the analysis of irradiated solid glycine (Clark *et al.*, 1970). They were able to prove that acetic acid was a direct product of the radiolysis and did not require the prior dissolving of the irradiated glycine in water to produce the acetic acid as had previously been thought. In this case the solid irradiated glycine was placed directly into the ion chamber of the mass spectrometer in an insertion probe at a source temperature of 220°.

B. INFRARED SPECTROSCOPY

An enormous amount of information concerning the chemical effects of radiations has been obtained by means of optical spectroscopy. Using infrared or ultraviolet techniques groups bound to the polymer can be quantitatively determined. In situations where the polymer has been rendered insoluble during the irradiation by three-dimensional cross-linking, the optical methods are the only reliable analytical methods that can be used. For infrared analysis films about 0.05–0.25 mm thick (Fischer and Langbein, 1967, cut films with a microtome from POM 0.005 to 0.070 mm thick) can be used. The thickness should be adjusted for the concentration and extinction coefficient of the group to be studied. Tryon and Horowitz (1962) have reviewed ir techniques for polymers.

When films of the polymer are not available commercially, they can be cast from a solution on a glass plate; the concentration of the solution should be adjusted to give the desired film thickness. Isotactic poly(propylene oxide) films could be made in this way by casting from a 1 % benzene solution. Self-supporting films of atactic poly(propylene oxide) could not be made, however, by this technique. In this case the atactic polymer was cast on a nickel screen (Roberts *et al.*, 1971) 100 mesh with a maximum transmission of 75 %. Another technique is to pour the solution of the polymer containing a volatile solvent onto a spinning platform according to the method of Kellö and Tkáč (1953). The nickel mesh is attached to the platform and by varying the spinning rate of the platform, the concentration of the polymer solution and the rate and frequency of pouring the solution, the thickness and uniformity of the resulting film can be controlled.

The relative thickness of the films can be monitored by measuring the absorbance of some standard band, after the absorption of that band has been calibrated with a film of known thickness. If the weight, area, and density of the film are known, the thickness can be calculated. However, in the calculation of *G* values, only the mass per unit area of the film needs to be known. In the study of poly(propylene oxide) mentioned above the infrared bands at 835, 935, and 2970 cm^{-1} had absorbances which increased linearly with the mass to area ratio of the film.

The polymer films have to be held in a brass frame that fits into the sample well of the infrared spectrometer. The polymer coated nickel mesh is similarly handled. By using a cryoscopic cell like that of Fig. 4, Dole and Böhm were able to obtain infrared spectra at 77°K (this was the temperature of the copper block which held the film) where the absorption bands are very much sharper than they are at room temperature.

Some absorption bands are sensitive to the crystallinity of the polymer. In fact, the crystallinity of polymers has often been determined by measuring the absorption of bands that either grow or decrease in intensity as the crystallinity is increased. An example of this is the work of Hendus and Schnell (1961), who measured the ratio of the absorbance due to the crystalline band at 1900 cm^{-1} to that of the amorphous band at 1300 cm^{-1} in polyethylene and from this ratio and from the absorbance at 1900 cm^{-1} extrapolated to 100% crystallinity calculated the crystallinity of different PE samples. It is important to determine whether the absorbance of the particular band being measured is sensitive to the crystallinity of the polymer or not. Comparative measurements should also be made at the same temperature.

De Kock *et al.* (1964) have made a study of unsaturation in polyethylene by the infrared method. They determined extinction coefficients using known oligomers containing the desired type of unsaturation dissolved in solid *n*-hexatriacontane. They pointed out that the extinction coefficient of *trans*-vinylene groups was less when measured in a liquid solution, than when measured in a solid. Curiously, the opposite effect was observed in the case of the vinyl group. In Table II are collected extinction coefficients for infrared radiation absorbing groups commonly encountered in high polymers.

TABLE II

EXTINCTION COEFFICIENTS FOR INFRARED ABSORPTION BANDS

Group	Wavelength (cm^{-1})	ϵ (liters mole^{-1} cm^{-1})	Reference
$-CH=CH_2$	910	153	Dole et al. (1958)
$-CH=CH_2$	910	122	de Kock et al. (1964)
$-CH=CH-(trans)$	964	169	de Kock et al. (1964)
$R_1R_2 C=CH_2$	885	129	de Kock et al. (1964)
$-CH=CH-CH=CH-$			
(trans–trans)	990	360	Ahlers et al. (1953)
ROH	3480	54	Cross et al. (1950)
ROH	3543	127	Ravens and Ward (1961)
Associated $-OH$	3450	260	Langbein (1964)
$R_1R_2 C=0$	1730	188	Cross et al. (1950)
$-CH_2OCHO$	1735	2170	Fischer and Langbein (1967)
$-\dot{C}HCH=CH-$	943	133	Waterman and Dole (1970a)
$-\dot{C}HCH=CH-CH=CH-$	943	222	Waterman and Dole (1970a)

It will be noted that there is a rather large discrepancy between the vinyl group extinction coefficient as determined by Dole et al. (1957) and by de Kock et al. (1964). In both investigations the value was checked by chemical measurement of the unsaturation; hence, it would appear that further work needs to be done to clarify the discrepancy.

C. ULTRAVIOLET SPECTROSCOPY

The use of ultraviolet spectroscopic methods in certain specific cases can yield valuable information concerning chemical effects of high-energy radiations. For groups that absorb strongly in the UV range of wavelengths, such as the dienes, polyenes, and allyl and polyenyl free radicals, UV spectroscopy is a valuable complement to IR or ESR studies.

Ultraviolet technique in principle is much like that discussed above for the infrared measurements. However, UV radiation is more strongly absorbed than infrared; hence, thinner films have to be used. In the case of PE films, they can be supported on a stainless steel frame in a quartz spectrophotometer cell. By filling the cell with hexane, which has a refractive index near to that of PE, scattering of the UV light at the surface of the film is reduced. Low-density PE is more transparent to UV light than the same thickness of a high-density PE because of more scattering of light by the spherulites present in the high-density sample. Partridge (1966) and, later, Budzol and Dole (1971) found that soaking the PE film in hexane before the UV measurements considerably reduced the absorption in the 200–300 nm wavelength range. Such soaking made it possible to observe far more clearly the conjugated diene peak in irradiated PE. Partridge attributed the reduced absorption to the removal of oxidation products.

By making the UV absorption measurements at liquid nitrogen temperature, Bodily and Dole (1966) were able to demonstrate a significant sharpening of the UV absorption bands. The sharpening was especially pronounced in the case of the absorption band of the allyl free radical at 258 nm.

TABLE III

EXTINCTION COEFFICIENTS FOR ULTRAVIOLET ABSORPTION BANDS

Group	Wavelength of maximum absorption (nm)	ϵ (liters mole^{-1} cm^{-1})	Reference
$-CH{=}CH{-}CH{=}CH-$			
(*trans, trans*)	236	25,000	Fallgatter and Dole (1964)
$-CH_2(CH{=}CH{-})_3CH_2-$	274	41,800	Fallgatter and Dole (1964)
$-CH_2(CH{=}CH{-})_4CH_2-$	310	58,900	Bohlmann (1953)
$-CH_2\dot{C}HCH_2-$	215	1,800	Waterman and Dole (1970a)
$-\dot{C}HCH{=}CH-$	258	7,300	Waterman and Dole (1970a)
$-\dot{C}HCH{=}CH{-}CH{=}CH-$	285	29,000	Waterman and Dole (1970a)

In Table III are collected extinction coefficients of ultraviolet absorption bands. The values of the diene and triene were estimated by Fallgatter and Dole from values measured by others. They are data obtained from measurements of low molecular weight dienes and trienes in solution, and, as the results varied with the solvent, there is no assurance that the values given in Table III for any of the polyenes are valid for polyenes in solid polyethylene. The order of magnitude is probably correct. The extinction coefficients of the free radicals determined by Waterman and Dole are more reliable as they were obtained on the basis of ESR measurements of the free radical concentrations in linear polyethylene.

D. ESR SPECTROSCOPY

The technique of preparing the samples and irradiation tubes for the ESR measurements has already been described above. Normally, the irradiations are carried out in evacuated tubes to eliminate the complications arising from the presence of peroxy- or oxy- free radicals. It should be noted, however, that gases are produced by the irradiation, and in the case of PE, for example, molecular hydrogen produced by the irradiation has a marked effect in catalyzing the alkyl to allyl free radical conversion (Waterman and Dole, 1970b). To avoid this further complication, the sample should be pumped out after the irradiation. In the case of irradiations at 77°K, the sample tubes should be pumped out as the sample warms up to room temperature.

In making ESR measurements it is important (1) that the modulation amplitude be adjusted so that the ESR spectrum will not be distorted and (2) that power saturation be avoided. The amplitude of sweep must be less than a tenth of the linewidth for a display of the true line shape (Ingram, 1958). The sensitivity decreases as the amplitude of sweep is reduced; hence, a compromise has to be made.

With respect to power saturation, this is especially significant in the case of ESR measurements of the allyl free radical (Waterman and Dole, 1970b). Using a Varian E-4 spectrometer the power has to be kept below 0.1 mW; i.e., in the power region where the signal intensity is linearly proportional to the square root of the power.

Timm and Willard (1969) have described equipment for temperature control for ESR measurements in the temperature range 10–80°K.

ESR measurements are used both for the qualitative identification of free radicals and for the quantitative estimation of their concentration. The theory of free radicals is treated in Chapters 2–6; here we shall be concerned only with the technique of quantitatively estimating free radical concentrations. If there is only one type of free radical present in the spectrum, the total area of the spectrum can be measured and related to the number of spins through a calibration with a known source of free radicals such as solid DPPH, 1,1-diphenyl-2-picryl hydrazil, or

in which there is one unpaired electron per molecule. In making measurements at 77°K a frozen solution of DPPH in benzene can be used in which case a singlet is observed. At room temperature a solution of 0.025 wt% DPPH in benzene gives a quintet spectrum with relative intensities 1:2:3:2:1.

When the pure spectrum has been calibrated with DPPH, relative spin concentrations can be estimated by comparing peak heights of separate spectra. In making this comparison it is necessary to choose a peak which does not include a signal from any other free radical. In the case of a mixture of allyl and alkyl free radicals, for example, the alkyl free radical concentration can be estimated from the height of the outermost or wing peak, because the allyl free radical spectrum is narrower than the alkyl and does not overlap with the wing peak of the alkyl free radical.

A more detailed description of ESR techniques is given in a review by Dole (1972).

E. NMR Spectroscopy

In the application of high-resolution nuclear magnetic resonance spectroscopy to the analysis of radiation products, it is necessary to dissolve the polymer in solution. In the case of polymers which cross-link, only the soluble component could be so studied, and as far as this author is aware no such application of narrow line NMR spectroscopy has been made. In the irradiation of polyoxymethylene, POM, the degradation is so extensive that the polymer is completely soluble in phenol if heated at 148° for 25 hr (Fischer and Langbein, 1967). NMR measurements were made by these authors at 110° using a Varian DP 60 NMR spectrometer and enabled them to determine the cross-link yield because the chemical shift of the proton resonance of the hydrogens attached to the bridged carbon atoms, e.g.,

$$-\text{CH}_2\text{OCHOCH}_2-$$
$$|$$
$$-\text{CH}_2\text{OCHOCH}_2-$$

was at 5.25 as compared 2.0 for the $-\text{CHO}$ end group, and 6.7 for the $-\text{OCH}_3$ end group. The value of 5.25 agreed well with that of 5.35 for the peak due to the CH–CH bridge in napthodioxan

Fischer and Langbein used a solution that contained 100 g of the irradiated POM per liter of solution.

REFERENCES

Ahlers, N. H. E., Brett, R. A., and McTaggart, N. G. (1953). *J. Appl. Chem.* **3**, 433.
Allen, A. O. (1961). "The Radiation Chemistry of Water and Aqueous Solutions," pp. 20–23. Van Nostrand–Reinhold, Princeton, New Jersey.
Bodily, D. M., and Dole, M. (1966). *J. Chem. Phys.* **45**, 1428.
Bohlmann, F. (1953). *Chem. Ber.* **86**, 63.
Budzol, M., and Dole, M. (1971). *J. Phys. Chem.* **75**, 1671.
Charlesby, A. (1952). *Proc. Roy. Soc. (London)* **215A**, 187.
Chen, C. W., and Struss, R. G. (1969) *Cryogenics* **9**, 131.
Clark, J., Kushelevsky, A. P., and Slifkin, M. A. (1970). *Radiat. Effects* **2**, 303.
Cross, L. H., Richards, R. B., and Willis, H. A. (1950). *Discuss. Faraday Soc.* **9**, 235.
Davidson, W. L., and Geib, G. (1948). *J. Appl. Phys.* **19**, 427.
de Kock, R. J., Hol. P. A. H. M., and Boz. H. (1964). *Z. Anal. Chem.* **205**, 371.
Dole, M. (1948). Paper presented before the Div. of Phys. and Inorg. Chem., Amer. Chem. Soc., Portland, Oregon. September.
Dole, M. (1972). "Advances in Radiation Chemistry" (M. Burton and J. L. Magee, eds.), Vol. 4. Wiley (Interscience), New York.
Dole, M., and Böhm, G. G. A. (1968). Advances in Chemistry Series, No. 82, p. 525, Amer. Chem. Soc. Washington, D.C.
Dole, M., Milner, D. C., and Williams, F. (1958). *J. Amer. Chem. Soc.* **80**, 1580.
Fallgatter, M. B., and Dole, M. (1964). *J. Phys. Chem.* **68**, 1988.
Fischer, H., and Langbein, W. (1967). *Kolloid Z.* **216–217**, 329.
Frank, A. L., and Benton, E. V. (1970). *Radiat. Effects* **2**, 269.
Frye, G. M., Gammel, J. H., and Rosen, L. (1954). USAEC Rep. LA-1670.
Ghormley, J. A. (1961). *In* "Radiation Chemistry of Gases" (S. C. Lind, ed.), p. 59. Reinhold, New York.
Hendus, H., and Schnell, G. (1961). *Kunststoffe* **51**, 69.
Humpherys, K. C. (1969). *In* "Fast Burst Reactors" (R. L. Long and P. D. O'Brien, eds.), pp. 497–518. Symposium Series 15, USAEC Div. of Tech. Information.
Ingram, D. J. E. (1958). "Free Radical," p. 80. Butterworths, London and Washington, D.C.
Kang, H. Y., Saito, O., and Dole, M. (1967). *J. Amer. Chem. Soc.* **89**, 1980.
Kellö, V. and Tkáč, A. (1953). *Chem. Zvest*, **7**, 129.
Langbein, G. (1964). *Kolloid Z.* **200**, 10.
Mikesell, R. E. (1966). General At. Rep. GA-1695 (Rev. 5).
Partridge, R. H. (1966). *J. Chem. Phys.* **45**, 1679.
Ravens, D. A. S., and Ward, I. M. (1961). *Trans. Faraday Soc.* **57**, 150 (1961).
Roberts, G. P., Budzol, M., and Dole, M. (1971). *J. Polym. Sci.* A-2 **9**, 1729.
Salovey, R., Albarino, R. V. and Luongo, J. P. (1970). *Macromolecules* **3**, 314.
Timm, D., and Willard, J. E. (1969). *Rev. Sci. Instrum.* **40**, 848.
Tryon, M., and Horowitz, E. (1962). *In* "Analytical Chemistry of Polymers" (G. M. Kline, ed.), Part II, p. 291. Wiley (Interscience), New York.
Waterman, D. C., and Dole, M. (1970a). *J. Phys. Chem.* **74**, 1906.
Waterman, D. C., and Dole, M. (1970b). *J. Phys. Chem.* **74**, 1913.
Williams, F., and Dole, M. (1959). *J. Amer. Chem. Soc.* **81**, 2919.
Wimett, T. F. (1965). USAEC Rep. SM 62/53.

13

Radiation Chemistry of Linear Polyethylene

Leo Mandelkern

Department of Chemistry and Institute of Molecular Biophysics,
Florida State University, Tallahassee, Florida

I. Introduction

In studying the radiation chemistry of polymers there are a variety of objectives depending primarily on the interests of a particular investigator. One main, general objective, which has engaged the attention of many, is to develop an understanding of the different chemical acts which occur when a polymer is subject to the action of high-energy ionizing radiation and to relate these to observed changes in physical properties. Starting with the pioneering studies of polyethylene, which began several decades ago (Dole, 1950; Charlesby, 1952), this road has been a rather long and arduous one. As should become clear to any objective observer, it has not as yet been successfully traversed. Based on the chemical formula, or repeating unit, linear polyethylene appears to be such a simple molecule that it might at first sight appear somewhat surprising that all the problems involved in its radiation chemistry have not been successfully resolved a long time ago.

The simplicity of its chemical formula is, of course, deceiving. It is the constitution and structure of a system comprised of a collection of such molecules that is important and which has a great bearing on the experimental observations and subsequent interpretations. Thus, before directly proceeding to a presentation and analysis of the experimental results dealing with the chemical events and the changes in properties, it is incumbent to set forth in some detail the conformational, structural, and morphological properties of linear polyethylene. This procedure will yield the set of independent thermodynamic and molecular variables that are involved. By necessity these must be specified and separated in order to set the proper basis for the carrying out of experiments and their analysis.

Linear polyethylene, as is characteristic of all chain molecules of regular structure, can exist in either the crystalline or amorphous state, depending upon the intensive variables that characterize the system. Above the melting temperature, in the completely liquid or amorphous state, both the isolated molecule and a collection of such molecules have been subject to what has become a classical statistical mechanical conformational analysis (Flory, 1969). In the amorphous state the molecule adopts a statistical conformation, which has been popularly termed a random coil. It has a characteristic ratio of 6.7 at 140° (Flory, 1969). With such a classical melt structure it would be natural to anticipate that the irradiation properties in this state would be delineated in a straightforward manner. However, as will become evident in the more detailed discussion given below, this simple expectation is not fulfilled. The main reason is that the linear polyethylene sample that has been studied most commonly (the Marlex-50 type) contains one vinyl end group per molecule. This end group becomes particularly reactive when the polymer is subject to ionizing radiation and thus plays a far more influential role than its relatively minimal concentration would indicate.

Below the melting point, in the crystalline state, the structure becomes more complex and more difficult to specify quantitatively. In the first effort in delineating this problem it is necessary to specify the mode of crystallization. In the absence of deformation processes there are two extreme types of crystallization.* One of these is crystallization from the pure melt or bulk crystallization. The other is crystallization from a very dilute solution. For bulk crystallization complete crystallinity is rarely if ever attained. Rather the crystallization from the pure melt results in a polycrystalline material (Mandelkern, 1964, 1967). On a completely theoretical equilibrium basis, molecular crystals (as are typified by the n-hydrocarbons) will not be formed

---

* In the context of the present work we are only concerned with the undeformed crystalline state. Crystallization under deformation, which leads to oriented structures, will not be considered here.

even for a well-fractionated sample. This type of crystalline structure requires that all the molecules be of exactly the same length which cannot be practically attained with long-chain molecules. It has become well recognized and is widely accepted that a characteristic structural feature of bulk crystallized homopolymers is a lamella-like crystallite, whose lateral dimensions are much greater than its thickness (Mandelkern, 1967). A wide array of experimental evidence, such as low-angle X-ray diffraction (Mandelkern *et al.*, 1961; Sella and Torillat, 1958), electron microscope examination of replicas of fracture surfaces (Keller, 1957; Geil, 1960; Eppe *et al.*, 1959; Anderson, 1964; Mandelkern *et al.*, 1966), examination of the internal structure of spherulites (Fujiwara, 1960), and the study of preferentially oxidized samples (Palmer and Cobbald, 1964; Keller and Sawada, 1964) strongly support the concept of a lamella-like crystallite. Unfortunately, the appearance of lamella-like crystallite structure has led to the concept that it must be accompanied by a regularly folded interface. Irrespective of how appealing a regularly structured interface may be, or how vociferously it has been argued, detailed physical–chemical studies have clearly indicated that it represents an untenable point of view (Mandelkern, 1966, 1967).

Deviations in the density, in the enthalpy of fusion, and in other thermodynamic properties from that of the unit cell are well recognized in crystalline polymers in general and in crystalline linear polyethylene in particular. Also observed in crystalline polymers are haloes in the wide-angle X-ray diffraction patterns and certain infrared absorption bands characteristic of bonds in nonordered orientations. Within the concept of a regularly folded interfacial structure, wherein the long-chain molecule is confined to a single crystallite, the thermodynamic deviations, the diffraction haloes, and the infrared spectra must by necessity result from contributions from the interfacial structure and from internal defects within the crystallite interior, since there are not any chain units connecting crystallites. Thus, in this view, crystalline polymers are treated as a complete set of crystals of small thickness, with a large concentration of internal defects. As a survey of some of the properties given below will indicate, although this argument could be made plausible by the analysis of isolated bits of experimental data, it cannot be substantiated when the wide range of properties that can be developed are examined.

As an example we consider the variation of the density and the enthalpy of fusion, at room temperature, as a function of the molecular weight and the crystallization temperature. These data, for molecular weight fractions, as compiled from various sources are given in Figs. 1 and 2 (Ergöz, 1970). As the most casual examination of these figures indicates, there is a wide range in the values of these thermodynamic quantities that are observed.

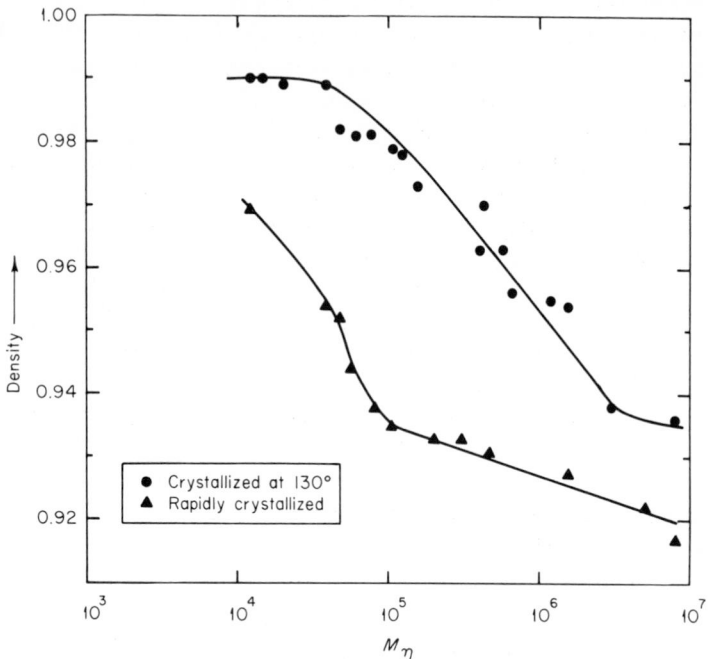

Fig. 1. Plot of density, at room temperature, of linear polyethylene as function of molecular weight for different modes of crystallization (Mandelkern, 1967).

The values are very dependent on the molecular weight and crystallization conditions. The measured densities of linear polyethylene vary from about 0.99 to 0.92 g/cm³. Hence this quantity is not restricted to very narrow limits as has been implied. Smaller values are obtained at the lower crystallization temperatures (more rapid crystallization), and the same trends are observed with molecular weight at the different crystallization temperatures. At the highest crystallization temperature studied (130°C), for molecular weights less than about 100,000, the densities are very high (0.98–0.99) and approach that of the unit cell (1.00). However, as the molecular weight is increased, the density decreases monotonically with increasing chain length and attains a value of about 0.95. For the very highest molecular weights a limiting value appears to be reached. Much lower values are observed, for comparable molecular weights, when the crystallization process is conducted more rapidly.

The measured enthalpy of fusion, ΔH^*, displays a very similar trend with molecular weight and crystallization temperature. Values of about 66 cal/g are obtained for the lower molecular weights, crystallized at high temperatures. This is very close to the value expected for the macroscopic perfect crystal (Quinn and Mandelkern, 1958). With increasing molecular weight and decreasing crystallization temperature the enthalpy of fusion decreases in a systematic way, in analogy to the changes observed in the density. Values as low as 24 cal/g are obtained for high molecular weight specimens which are allowed to crystallize very rapidly. The main conclusion to be drawn from these data, for the purposes at hand, is that a very wide range in properties is observed. Morphologically, it is shown by either direct electron microscopic examination of replicas of fracture surfaces (Mandelkern *et al.*, 1966) or by the examination of the residue of selective oxidation (Ergöz, 1970) that the lamella-like crystals are still the primary morphological entity over the complete range of molecular weights and crystallization temperatures.

Since a wide range of properties can be developed in molecular weight fractions of linear polyethylene, they represent an ideal set of data to examine the concept of the degree of crystallinity. In this concept the properties in the crystalline state are taken to result from the additive contribution of the crystalline and amorphous regions with contributions also being possible from the interfacial regions. The amorphous regions represent chain units

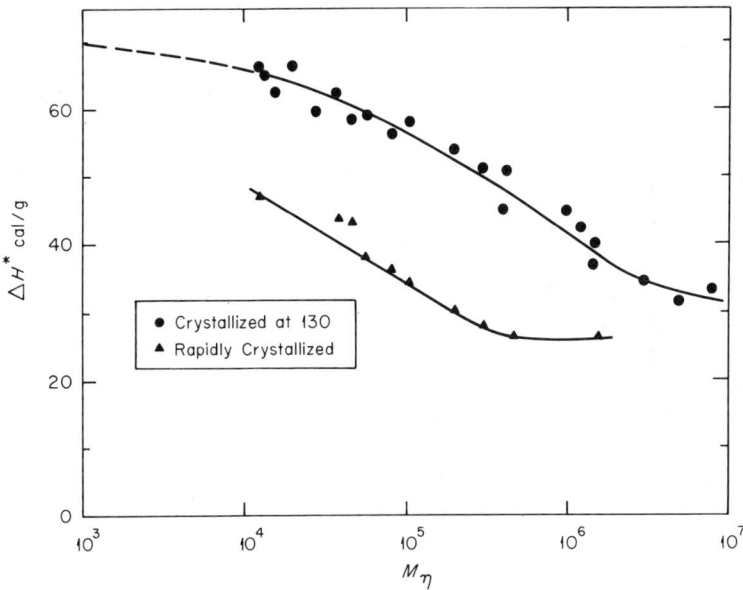

Fig. 2. Plot of enthalpy of fusion, ΔH^*, of linear polyethylene as function of molecular weight for different modes of crystallization (Mandelkern, 1967).

in nonordered conformations whose properties are characteristic of the pure melt. The degrees of crystallinity calculated from density and enthalpy of fusion (Mandelkern, 1967), infrared (Okada and Mandelkern, 1967; Hendus and Schnell, 1967), wide-angle X-ray diffraction (Hendus and Schnell, 1967; Gopalan and Mandelkern, 1967), and broad-line NMR (Bergmann and Nawatki, 1967) all give very concordant results.* These results give strong support to the presence of interzonal regions, i.e., chain units having the same conformational properties as the pure melt. For linear polyethylene, the levels of crystallinity that can be attained range from about 0.50 for high molecular weight fractions which are crystallized very rapidly, to about 0.95 for low molecular weight fractions crystallized slowly. Thus, at temperatures below its melting point, we have two major regions to consider whose relative proportions depend on the molecular weight and the conditions of crystallization. There is the strong possibility that there will be differences in chemical reactivity in these two states, as a consequence of the distinctly different chain conformations and molecular packing that are involved. This aspect of the morphology must obviously be clearly specified in some detail before a meaningful analysis of irradiation data can be made.

From the physical–chemical data that have been accumulated, and from the electron microscopy and other morphological types of observations, a very simple structural model of the crystalline state can be developed. This model, as is illustrated in schematic form in Fig. 3, consists of three major

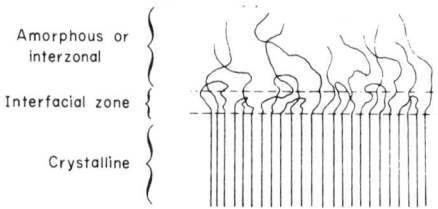

Fig. 3. Morphological schematic representation of crystallite.

regions or zones. One of these, the crystalline region is represented by the straight vertical lines and consists of chain units in the requisite ordered conformation. (For polyethylene all bonds are in the *trans* rotational state and the units are in a planar zigzag conformation.) In contrast, there is also the disordered, amorphous, or interzonal region, where the chain units are in nonordered conformation and connect one crystallite with another. The third region is the interfacial zone, which is many units thick, is not

---

* It should be noted that in this analysis there is a major contribution of the interfacial enthalpy to the measured enthalpy of fusion (Mandelkern *et al.*, 1968).

regular in structure, and is located at the diffuse boundary between the crystalline and disordered regions. Some of the chains emerging from the crystalline zone will return to the crystallite of origin, but not necessarily in an adjacent position. Other chain units will traverse the interfacial region and will become part of the interzonal space. Thus a diffuse, ill-defined interfacial region will result. The model depicted has been successful in explaining the thermodynamic data, the infrared spectra and the wide-angle diffraction results in a quantitative manner. It also can explain in a detailed manner the results from dynamic mechanical and broad-line NMR experiments (Stehling and Mandelkern, 1970).

The relative proportions of the different regions and the detailed nature of the interfacial structure depend on the molecular weight and crystallization conditions. These in turn are related to the ratio of the lamella thickness to the extended chain length. At the lower molecular weights, where the very high levels of crystallinity are attained, the crystallite thickness is found to be comparable to the extended chain length (Mandelkern, 1967; Mandelkern *et al.*, 1966). Thus, for this situation, there will be a negligible contribution to the interfacial properties from the interzonal region and less restraints on chain units in the interfacial regions when compared to higher molecular weight chains. For the higher molecular weights, the crystallite thickness reaches an asymptotic value at a degree of polymerization of about several thousand (Mandelkern, 1967; Mandelkern *et al.*, 1966). Thus, the crystallite thickness is no longer comparable to the extended chain length but decreases with increasing molecular weight. In this molecular weight range, which corresponds to the development of lower levels of crystallinity, there will be a significant contribution to properties from the interzonal region and also severe steric restraints on the chain units in the interfacial region. The experimental results on the crystallite thickness, together with the model of Fig. 3, explain the fact that the interfacial free energy associated with the basal plane has a relatively low value for samples of high levels of crystallinity and increases very markedly for the high molecular weight samples of low density (Mandelkern, 1967; Mandelkern *et al.*, 1966). For the study of the radiation chemistry of linear polyethylene we must recognize that chain length is a very important factor in delineating the level of crystallinity, morphology and crystallite thickness.

Another factor which is pertinent to the present considerations is the variation of the degree of crystallinity with temperature. This is particularly important when the result of high-energy ionizing radiation is examined as a function of temperature in the crystalline state. The molecular weight, crystallization temperature, and prior thermal history are found to exert an important influence. Results of the fusion process (plotted as the degree of

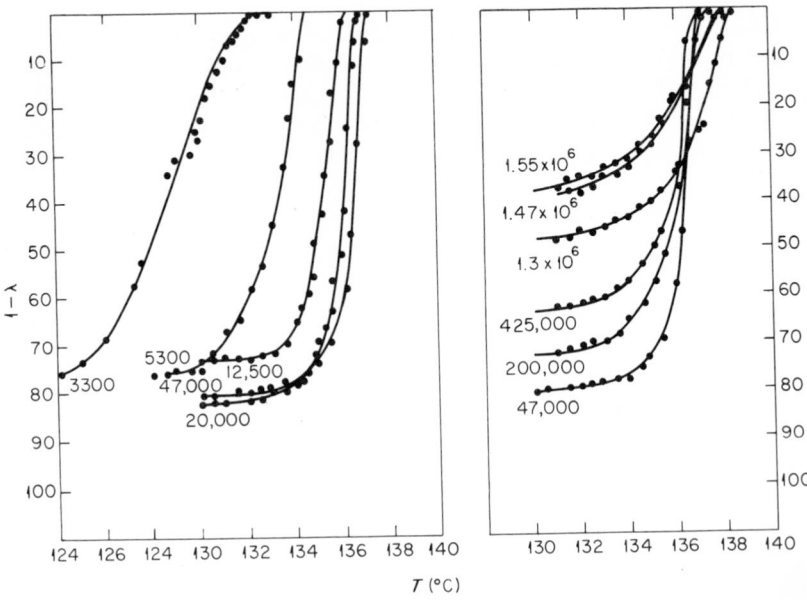

Fig. 4. Plot of degree of crystallinity, 1–λ (after crystallization at high temperature), as a function of temperature for indicated molecular weight fractions of linear polyethylene (Fatou and Mandelkern, 1965). [Reprinted from *J. Phys. Chem.* **69,** 417 (1965). Copyright (1965) by the American Chemical Society. Reprinted by permission of the copyright owner.]

crystallinity as a function of temperature) are illustrated in Fig. 4 (Fatou and Mandelkern, 1965) for molecular weight fractions. These results are for samples which have never been cooled below the isothermal crystallization temperature. For the two lowest molecular weight samples, the fusion curve is relatively broad and the changes in the level of crystallinity occur over the whole temperature range. In this case the relatively broad melting range can be attributed, in a straightforward manner, to the decreasing concentration of chain ends in the noncrystalline regions as fusion progresses. In contrast, for the molecular weight range 12,500–47,000 the melting curves become quite sharp. Here about 80–90% of the transformation occurs in a 2° interval. However, as the molecular weight is increased further, the fusion process broadens appreciably again. For example, in contrast to the values just cited, for $M = 425,000$ only 40% of the crystallinity disappears in the last 2° below the melting temperature. For molecular weights greater than about 10,000, there appears to be a correlation between the level of crystallinity and the breadth of the melting process.

In order to examine the change in the level of crystallinity over a much broader temperature range, representative molecular weights have been cooled to room temperature, after isothermal crystallization. The degree of crystallinity was then determined, as a function of temperature, on a subsequent heating cycle. These results, which may have more pertinency to irradiation over an extended temperature range, are shown in Fig. 5 (Fatou

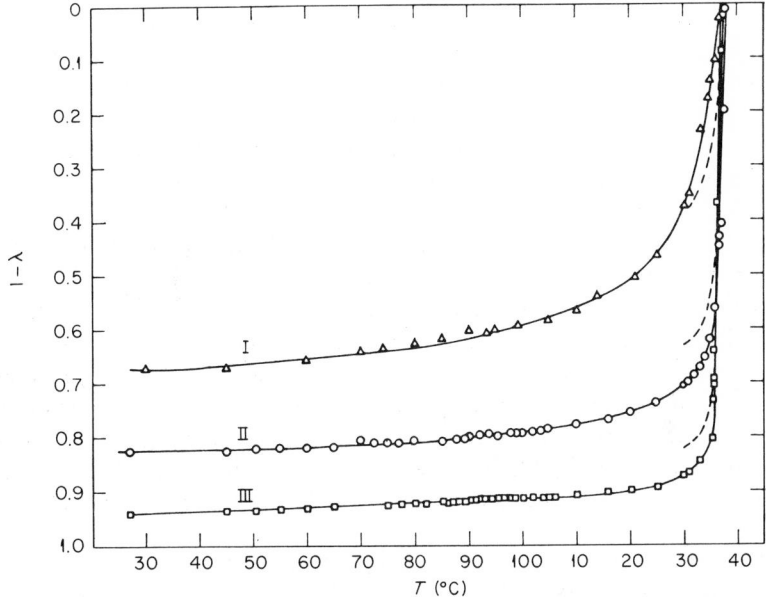

Fig. 5. Plot of degree of crystallinity, $1-\lambda$ (after cooling to room temperature), as a function of temperature for molecular weight fractions of linear polyethylene. Curve I, $M = 1.47 \times 10^6$; curve II, $M = 4.25 \times 10^5$; curve III, $M = 2 \times 10^4$ (Fatou and Mandelkern, 1965). [Reprinted from *J. Phys. Chem.* **69**, 417 (1965). Copyright (1965) by the American Chemical Society. Reprinted by permission of the copyright owner.]

and Mandelkern, 1965). The dashed lines in this figure represent the data from Fig. 4 for the corresponding fractions. For the two lowest molecular weight samples illustrated ($M = 20,000$ and $425,000$) the additional crystallization that has developed upon cooling has not affected in any major way the course of the fusion process. That is to say the level of crystallinity remains essentially constant up to the melting range, which is about 130° for the lowest molecular weight sample and just slightly less for the other sample. However, for the highest molecular weight sample, $M = 1.47 \times 10^6$, an appreciable amount of crystallinity develops upon cooling. This in turn manifests itself in a continuously changing level of crystallinity upon subsequent heating. Major changes in the level of crystallinity begin to occur, in this case, at about 100°C. If the temperature is reduced to still lower temperatures, it can be anticipated that only minor changes will occur in the level of crystallinity for the two lower molecular weight samples. By the same token, the trend displayed in Fig. 5 by the high molecular weight sample would be expected to continue.

Since the chain ends are usually structurally different from the main body of chain units, their spatial disposition within the crystalline polymer is a matter of interest. Within the context of the present problem, because of the presence of one vinyl end group per molecule, this becomes a matter of extreme importance. The chain ends can be considered to play the role of built-in impurities so that the central question is whether these units participate in the crystallization. The melting of the two lowest molecular weight samples, illustrated in Fig. 4 is very broad. The crystallinity content changes over the complete temperature range. For these molecular weights the end group concentration represents a significant fraction of the total material present. Hence, if the end groups are excluded from the crystalline regions, then their concentrations in the residual noncrystalline regions will decrease as fusion progresses. This will lead to the broad fusion process that is observed and provides a simple and clear explanation of the phenomena. For polymers possessing a most probable molecular weight distribution a quantitative comparison between theory and experiment demonstrates quite clearly that the end groups are excluded from the crystal lattice (Evans et al., 1950).

Also of importance for polyethylene, in the crystalline state, is the location and molecular nature of the structural transitions. For linear polyethylene there is a transition in the vicinity of 80°–90°C which has been termed the α transition (Boyer, 1963; McCrum et al., 1963; Illers, 1964). There is another transition, in the vicinity of $-120°$ to $-150°$, termed the γ transition (Stehling and Mandelkern, 1970; Boyer, 1963; McCrum et al., 1963; Illers, 1964). The β transition, which is quite pronounced in the vicinity of $-20°$ to $-30°$C for branched polyethylene (Stehling and Mandelkern, 1970) has not been observed in linear polyethylene even for samples with low levels of crystallinity (Stehling and Mandelkern, 1970). Dynamic mechanical experiments, on both bulk and solution crystallized samples, indicate that the α transition is characteristic of the crystalline regions and involves some type of molecular motion within these regions (McCrum et al., 1963; Takayanagi, 1965). It has also been shown that there are changes in the unit–cell lattice parameters in this temperature region (Takayanagi, 1960). The fact that this transition represents the onset of molecular motions within the crystalline region must be borne in mind when considering chemical reactivity as a function of temperature.

Dynamic mechanical experiments indicate that the γ transition is a complex one which probably consists of at least two unresolved transitions (Stehling and Mandelkern, 1970). The transitions at $-120°C$ can definitely be assigned to the glass temperature of linear polyethylene and hence is a property of the amorphous, noncrystalline regions. The molecular origin of the other transition, which is located in the vicinity of $-140°C$, is not as yet as well established. However, a consistent interpretation of dynamic mechanical (Stehling and Mandelkern, 1970) and broad-line NMR (Bergmann and Nawatki, 1967) measurements would assign it to the interfacial region (Stehling and Mandelkern, 1970).

The morphological structures that are formed subsequent to crystallization from dilute solution are much simpler and more easily defined than those formed in the bulk. It is well established that when crystallized from dilute solution homopolymers form platelets or lamella-like structures (Keller, 1957; Till, 1957; Fischer, 1957). Typically, the crystals formed from dilute solution are several microns in the lateral dimensions and are of the order of 100 Å thick. Selected area electron diffraction studies have demonstrated that the chain axes are preferentially oriented perpendicular to the wide faces of the crystallites. Since the platelet thicknesses are usually no more than the order of a few hundred angstroms thick and this crystal habit is observed for very high molecular weight chains, it is clear that a given molecule must traverse a crystallite many times in order to satisfy the molecular weight requirements, the crystallite thickness requirements and the required orientation.

Although the above facts are well agreed upon, the requirement that a chain must return to its crystallite of origin has led to a great deal of study and discussion with regard to the nature of the interfacial structure associated with the basal plane. Two extreme types of interfacial structure have been envisaged (Mandelkern, 1964), each of which is consistent with electron microscopy and low-angle X-ray diffraction studies. In one of these, which we shall designate as the regularly folded array, each molecule crystallizes to the fullest extent possible. Crystallization is complete, except for the number of chain elements required to make the fold or connection between successive crystalline sequences. The minimum number of chain atoms, consistent with the hindrances to bond rotation, form the hairpin-like connection between the crystalline sequences. Thus, in this model, crystalline sequences from the same molecule are adjacent to one another, and a regularly structured interface, comprised of a plane of folded chains, is formed.

Alternatively, it has been proposed that adjacent reentry of a chain does not occur. In this model, the crystalline sequences are connected more or less at random with one another, with the connecting loops also being of random length. A chain would thus traverse a nonordered or amorphous overlayer before rejoining the ordered crystal lattice. In this model, the system is clearly not completely crystalline, but deviates from the idealized situation according to the concentration of chain units involved in the disorganized interfacial region. Minor variants of these two extremes have also been proposed (Hoffman *et al.*, 1969; Peterlin, 1969; Zachmann and Peterlin, 1969).

The molecular nature of the interface becomes more than a matter of general interest when one is concerned with the chemical reactivity of such crystalline entities. Electron microscope techniques, and other measurements based on direct observations, cannot discriminate between these two extremes of structure. Physical chemical measurements and deductive reasoning power have to be brought to bear on the problem. It is not our intent, nor is it appropriate in the present context, to exhaustively and critically review all the kinds of measurements that have been reported on this problem. These have been reviewed elsewhere in great detail (Mandelkern, 1966, 1970). The evidence from a variety of thermodynamic, spectral, and other physical–chemical measurements conclusively demonstrates the presence of a disordered interfacial structure for crystals formed from dilute solution. About 15–20% of the chain units are in disordered conformation. This conclusion is given further strong support by the direct observation of glass formation in these systems by dynamic mechanical studies (Stehling and Mandelkern, 1969) and by the analysis of the temperature coefficient of the intensity maximum observed in low-angle X-ray diffraction (Fischer *et al.*, 1969; Fischer and Kloos, 1970).

Detailed studies of the crystallization kinetics from dilute solution (Mandelkern, 1954; Devoy *et al.*, 1970) and an analysis of the relation between the crystallite thickness and the crystallization temperature (Mandelkern, 1966; Jackson and Mandelkern, 1968; Sharma and Mandelkern, 1970) have shown that the crystallite thickness is controlled by a nucleation process. This conclusion can be reached from the most general considerations of nucleation theory. As a consequence the detailed structure of the nucleus and the mode of nucleation cannot be specified. A comparison of the interfacial free energy, associated with the basal plane, for nucleation with that characteristic of the mature crystallites indicates that these two quantities are significantly different (Jackson and Mandelkern, 1968). This conclusion is again reached from the most general considerations and must reflect differences in the interfacial structures in the two cases.

From the summary of properties that has been given above it is clear that for solution formed crystals, although we are dealing with samples which are highly crystalline they are not completely so. Thus in assessing the effects of chemical reactions and the effect of high-energy ionizing radiation on such systems the existence of units in disordered conformations must be recognized. Cognizance must also be taken of the fact that adjacent crystalline sequences are not required to be part of the same molecule, although there may be more nearest neighbor sequences from the same molecule as compared to crystallization in the bulk.

It should be noted that because of the restricted thickness of solution formed crystals the temperatures available for study in the crystalline state are reduced. The α transition is still observed in the vicinity of 80°–90° (Sinnott, 1966), and the same type of molecular motions associated with the crystalline regions of bulk crystallized polymers can be expected. When crystals prepared from dilute solution are separated from the mother liquor and subsequently heated no changes are observed in properties up to about 115°C. However, after heating or annealing at slightly higher temperatures major changes are observed in morphology, low-angle X-ray diffraction and thermodynamic stability (Statton and Geil, 1960; Statton, 1961; Fischer and Schmidt, 1962; Jackson *et al.*, 1963). In particular, the thickness of the crystals and their solubility temperature increase substantially as the temperature is raised. A detailed, quantitative study of the critical temperature at which these effects are first observed, as a function of crystallite size, has demonstrated that this temperature corresponds to the melting temperature of the solution formed crystals (Mandelkern *et al.*, 1969). This reduction in melting temperature, relative to bulk crystallized samples, is a consequence of the relatively small crystallite size of such systems. The major change in properties that are observed, including as much as a four- to fivefold increase in thickness, results from melting or partial melting and rapid recrystallization (Mandelkern *et al.*, 1969). The practical consequence of these observations is to restrict significantly the highest temperature at which experiments can be performed while still maintaining the same morphology and level of crystallinity. The temperature interval between the α transition and the melting temperature is significantly reduced for solution formed crystals.

The purpose of the brief review of the structure and morphology of linear polyethylene was to present the nature of the system whose radiation chemistry we shall be concerned with and to establish further the variables that need to be considered, and the interrelation between them. In addition to the usual specification of molecular weight as a variable in problems concerned with long-chain chain molecules there are some very important specific implications to the problem at hand. The viscosity of the melt is of course very dependent on molecular weight which could possibly have an effect on chemical reactivity. As has been indicated previously most of the linear polyethylene samples that have been studied contain one vinyl end group per molecule. Hence the concentration of this species present in the pure melt is dependent on the molecular weight. The molecular weight is also a very important variable in determining the constitution and morphology of the crystalline state. As we have seen, it has a rather important influence in establishing the crystallite size, the interfacial structure and the level of crystallinity. Since the chain ends are rejected from the crystal lattice their effective concentration in the noncrystalline region is also molecular weight dependent. The complex influence of molecular weight on the morphology and level of crystallinity, and the anticipated dependence of chemical reactivity on these factors make mandatory the use of molecular weight fractions in any quantitative studies. Samples, polydisperse with respect to molecular weight, can only help to establish the major chemical events that are involved in the radiation chemistry of linear polyethylene. Their use, for quantitative purposes, would not be appropriate at the present time for the reasons cited above. For example, Marlex-50, the most popular unfractionated sample that has been studied has a very broad molecular weight distribution, M_w/M_n, of about 15. It has a very low number average molecular weight and therefore a concomitantly high vinyl end-group concentration.

Another independent variable, which is of obvious importance with respect to chemical reactivity, is the temperature. As the temperature is varied, however, the possibility of changes in other structural features must also be considered. These include changing the level of crystallinity, complete melting, glass formation, or passage through one of the structural transitions. Thus, one must be concerned not only with the temperature coefficient of chemical reactions per se but must be cognizant of other changes that can be taking place concomitantly. The influence of the level of crystallinity is of obvious importance, but if it is to be studied properly the other variables must be held constant. Ideally, for example, one would like to compare the consequences of exposure to high-energy ionizing radiation of completely molten and completely crystalline samples of the same molecular weight as a function of temperature. Obviously this cannot be achieved since a completely crystalline sample cannot be attained. A compromise is then sought with high levels of crystallinity. This then restricts the molecular weight range that can be studied. Furthermore, there are restrictions with respect to the temperature range within which this comparison can be accomplished. One is concerned with maintaining a high level of crystallinity which sets the melting range as an upper temperature limit. At the same time, to maintain the sample completely molten, a lower temperature limit is set by the crystallization rate relative to the time of irradiation. As we shall see subsequently, there is a small temperature interval where the necessary conditions can be fulfilled.

II. Experimental Results and Discussion

Before proceeding with a detailed analysis it is convenient to examine first in a more general way the major chemical acts that result when linear polyethylene is exposed to the influence of high-energy irradiation. These have been primarily established from experiments involving unfractionated samples. For polymers in general the two possible major events that can occur are gel formation and chain scission. In addition with linear polyethylene chemical reactions are observed which are analogous to those which take place during the radiolysis of the normal hydrocarbons. These reactions include hydrogen evolution, intermolecular cross-linking through carbon–carbon bond formation, and the development of internal chain unsaturation. The disappearance of the vinyl end groups, when initially present, finds its analogy in the radiation chemistry of the ethylene-type hydrocarbons.

Intermolecular cross-linking is a major event that occurs during the radiation of linear polyethylene. This conclusion is made self-evident by the observation of gel formation and of studies of the partitioning of a sample between sol and gel past the gel point. As a consequence of gel formation there are changes in a variety of physical and physical–chemical properties. Particularly, the elastic and swelling properties above the melting temperature change (Charlesby and Hancock, 1953) as do the morphological and thermodynamic properties in the crystalline state below the melting temperature (Slichter and Mandell, 1958; Charlesby and Ross, 1953; Charlesby and Callaghan, 1958; Mandelkern *et al.*, 1960). A compilation of the literature, for different types of samples, as given by Chapiro (1962a) indicates that there is strong influence of temperature on the cross-linking efficacy. This compilation is highlighted by temperature invariance below the glass temperature and an enhancement to cross-linking, which increases with temperature, above this temperature. However, since the compiled data represent many different kinds of samples, with only the irradiation temperature being specified, one cannot develop a more detailed analysis at this point.

A survey of the early literature (Chapiro, 1962b; Charlesby, 1960) yields contradictory results as to whether main chain scission occurs during the radiation of linear polyethylene. One reason for this difficulty is the failure to distinguish, in many cases, between the branched and linear polymers.

Another reason is that virtually all the work reported has been with whole, unfractionated polymers. Unfractionated linear polyethylene samples are notorious in possessing very broad initial molecular weight distributions. All the theories, including the more sophisticated ones (Charlesby and Pinner, 1959; Inokuti 1963; Saito *et al.*, 1967), require a precise analytical specification of the molecular weight distribution in order to analyze the partitioning between sol and gel. This has been one of the major methods by which the ratio of chain scission to intermolecular cross-linking has been assessed. Theoretically this approach has the advantage that the experimental observations do not have to be carried out to high levels of cross-linking. However, unless molecular weight fractions are utilized, or the experiments are carried out to very high levels of cross-linking, unequivocal results cannot be obtained. Alternatively, if a careful examination is made of the limiting fraction sol that is obtained for high irradiation dosages (corresponding to high cross-linking levels) an assessment can also be made of the relative importance of chain scission. In particular, if the fraction sol approaches zero, then chain scission is obviously not an important factor. Alexander and Toms (1956) and Schumacher (1958) found that the sol fraction tended toward zero with high irradiation dose for both branched and linear polyethylene. On the other hand, Charlesby and Pinner (1959) concluded that, although both types of polymer behaved in a similar manner, the gel fraction did not approach unity as a limit and thus that chain scission played a significant role.

Radiolysis studies, on polymethylene (high molecular weight linear polyethylene) by Miller *et al.* (1956), showed that the gaseous yield is 99.9% hydrogen. This result finds an analogy in the radiolysis of *n*-alkanes, where the evolved gases are found to be mainly comprised of hydrogen, together with small amounts of hydrocarbons resulting from the cleavage of chain ends (Chapiro, 1962c). The compilation of a great deal of data by Chapiro (1962d) for the hydrogen evolution from polyethylene indicates a very marked temperature effect to this process. Below the glass temperature, the hydrogen evolution appears to be independent of temperature. Above the glass temperature the amount of hydrogen evolved is a strongly increasing function of temperature. However, since only the irradiation temperature is specified, and a variety of different types of polyethylene samples were studied, a more detailed quantitative analysis cannot be given at this point.

Another chemical event that is observed in linear polyethylene, as well as in the *n*-alkanes, is the formation of internal unsaturation. The formation of *trans*-vinylene in irradiated polyethylene has been extensively reported (Miller *et al.*, 1956; Dole *et al.*, 1966; Kang *et al.*, 1967; Black, 1958) while the formation of *cis*-vinylene unsaturation has not as yet been detected

(Chapiro, 1962e). For low radiation dosages, the growth of *trans*-vinylene double bonds is linear with the dosage. The formation of this species is reported to be independent of temperature, from the glass temperatures to the temperature of the pure melt (Dole *et al.*, 1958; Charlesby and Davison, 1957; Lawton *et al.*, 1958a). The yields of *trans*-vinylene appear to be independent of the level of crystallinity (Charlesby and Davison, 1957; Lawton *et al.*, 1958a). The formation of conjugated dienes and higher polyenes has also been reported (Fallgatter and Dole, 1964). This type of unsaturation has been observed after irradiation in the crystalline state but has not been detected after irradiation in the pure melt.

The importance of the vinyl end group in the irradiation chemistry of linear polyethylene has been brought to the fore and stressed by Dole and his collaborators (Dole *et al.*, 1958; Williams and Dole, 1959; Dole *et al.*, 1966). The most common linear polyethylene studied contains one vinyl end group per molecule as a consequence of its method of synthesis. These double bonds disappear rapidly under influence of irradiation and this reaction possesses a very definite temperature coefficient. At liquid nitrogen temperatures the disappearance of the vinyl group is virtually suppressed. The concentration of this group is solely dependent on the number average molecular weight in the pure melt, and its effective concentration in the crystalline state depends on the morphology and level of crystallinity.

Another process that has been postulated to occur, based on infrared studies, is a cyclization which could be either intra- or intermolecular in origin. The intramolecular process is the equivalent to the formation of intramolecular cross-links (Dole *et al.*, 1957, 1958).

With this general background we can now examine, in more detail, the influence of the independent variables, that have been specified, on the different physical and chemical processes that occur during the radiation of linear polyethylene. The first set of phenomena that we shall investigate involves gel formation, the partitioning of the sample between sol and gel with irradiation dose, and a critical examination of the possibilities of chain scission. For linear polyethylene, one of the questions of extreme interest is a comparison of properties between the completely molten or amorphous state and a state containing a high level of crystallinity. Since the temperature is also an independent variable, a true comparison cannot be attained by changing the level of crystallinity merely by varying the temperature over wide limits. However, for linear polyethylene there is a small temperature range, about 130°–133°C, wherein a true comparison can be made. An analysis of the isothermal crystallization kinetics of molecular weight fractions of linear polyethylene (Mandelkern, 1958, 1964; Ergöz, 1970) indicates that in this temperature range there is an appreciable time interval before crystallinity begins to develop. This time interval has proved to be more than adequate to perform the necessary irradiations. Hence a sample can be maintained in a completely molten state, below its melting temperature for the required time period (Kitamaru *et al.*, 1964; Kitamaru and Mandelkern, 1964). For the other type of experiment, if the molecular weight of the sample is not too large, high levels of crystallinity can be attained by isothermal crystallization. As has been indicated previously, such samples melt relatively sharply and the melting temperatures are in the vicinity of 138°C. Such samples can thus be irradiated in the range 130°–133°C, while a high level of crystallinity is maintained. Hence, the necessary comparative irradiation experiments can be performed at a constant, although a high temperature.

A typical result of such an experiment is illustrated in Fig. 6 (Kitamaru *et al.*, 1964). The sample is a molecular weight fraction, $M_n = 1.82 \times 10^5$, and the level of crystallinity is 0.83 for the crystalline sample. The fraction sol is plotted as a function of the radiation dose, for both the crystalline and completely amorphous polymer. It is evident that the gel point is achieved at a much lower dosage for the crystalline sample. For the modest dosages involved complete gel is formed for this sample, while in contrast 20% sol still remains in the amorphous sample. The relative efficiency of the cross-linking, as is indicated by the critical conditions for gelation, is 1.9 times greater in the crystalline state than in the amorphous state. These results, and conclusions based thereon, are valid of course only at 133°C.

It could be argued, based solely on the data of Fig. 6, that the main reason for the differences observed is that significant chain scission processes occur in the amorphous state (Dole and Katsuura, 1965). As has been indicated above, the most direct assessment of this possibility involves the determination of the asymptotic value of the fraction sol at high radiation dose. Obviously, the data in Fig. 6 are not sufficient to provide this information. Alternatively, the data for the early portions of the partitioning curve could, in principle, be analyzed (Inokuti, 1963; Dole and Katsuura, 1965). However, an analytic specification of the exact initial molecular weight distribution has to be made in order for this method to be of utility.* Despite this concern, such an analysis has been made of the data of Fig. 6. It was concluded that a major reason for the difference in the dosage required for gel formation was the difference in chain scission tendencies in the two states (Dole and Katsuura, 1965). As will be seen below, this conclusion is incorrect since complete gel is attained at the asymptotic limit in both states.

---

* It should be recognized that the partitioning between sol and gel is theoretically very sensitive to the molecular weight distribution and to even small amounts of polydispersity.

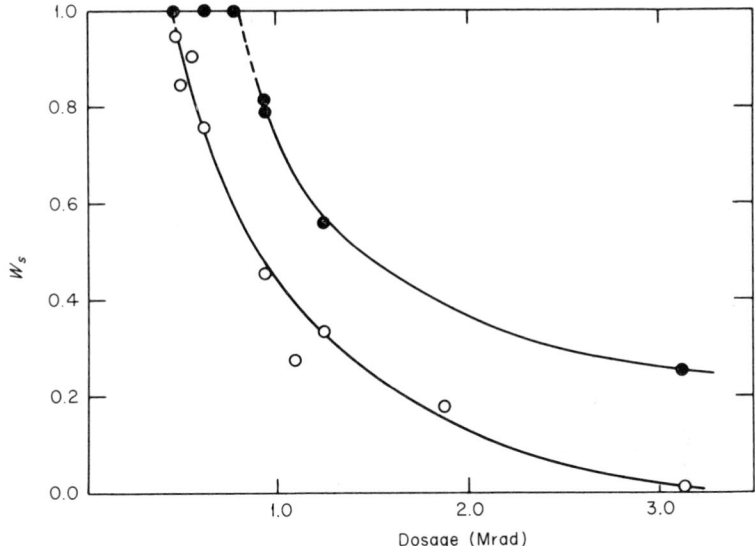

Fig. 6 Plot of weight fraction sol against radiation dose for a molecular weight fraction of linear polyethylene, $M = 1.82 \times 10^5$. Samples irradiated at 133°C. Completely amorphous polymer (●); highly crystalline polymer (○) (Kitamaru *et al.*, 1964). All irradiations of this and subsequent figures *in vacuo*.

Much higher radiation dosages are required, however, in order for the completely amorphous polymer to attain this condition.

The direct comparison between the highly crystalline and completely amorphous sample can, for the reasons cited above, only be made in this very narrow temperature range. The completely molten state cannot be maintained, for the necessary time, if the temperature is lowered below 130°C. Therefore, a direct comparison of the effect of temperature cannot be made. However, it is of interest to investigate the influence of temperature on the cross-linking efficiency in the highly crystalline state. When properly carried out, curves of the type illustrated in Fig. 6 should be obtained at a series of temperatures. However, since meaningful results require employing the same molecular weight fraction at all temperatures, in order to maintain constant the level of crystallinity and concentration of vinyl end groups, it becomes technically prohibitive to collect an adequate quantity of sample to perform the necessary experiments. Because of this limitation, the pertinent data presently available are limited to studying the partitioning between sol and gel as a function of temperature at a fixed irradiation dose.

Some representative data, resulting from this type of investigation, are

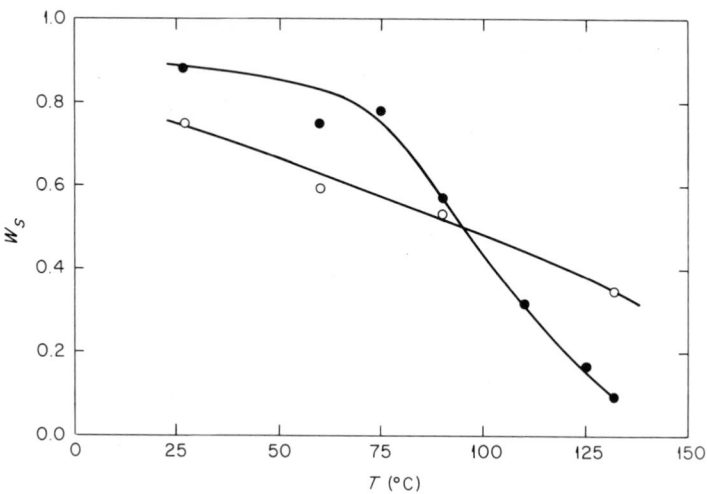

Fig. 7. Plot of weight fraction sol against temperature for molecular weight fraction $M_\eta = 2.07 \times 10^5$; radiation dose constant at 2.75 Mrad. Highly crystalline sample (●); rapidly crystallized sample (○) (Kitamaru and Mandelkern, 1964).

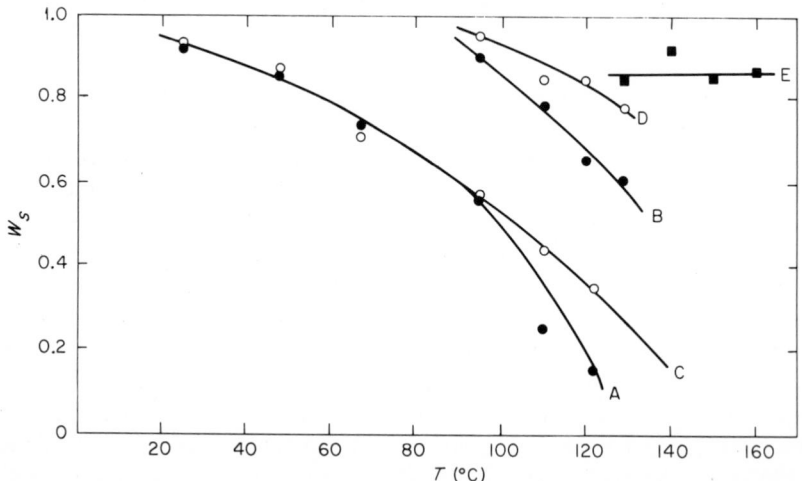

Fig. 8. Plot of weight fraction sol against temperature for molecular weight fraction $M_\eta = 4.25 \times 10^5$: Curves A and B, well-crystallized polymer, dosage 1.05 and 0.5 Mrad, respectively; curves C and D, rapidly crystallized sample, dosage 1.05 and 0.5 Mrad, respectively; curve E, completely amorphous polymer, dosage 0.46 Mrad (Kitamaru and Mandelkern, 1964).

illustrated in Figs. 7 and 8 (Kitamaru and Mandelkern, 1964). Here the weight fraction sol is plotted as a function of temperature for final dosages. The samples also represent different controlled crystallizations, as well as the completely amorphous polymer for the higher molecular weight fraction. At the lower temperatures, there is no significant difference in the amount of gel formed, at a constant dosage, for the different levels of crystallinity. However, as the temperature is increased, a larger proportion of gel is produced from the more highly crystalline specimens. This effect begins to manifest itself at about 80°–90°C. At 130°C it has become quite pronounced. A direct comparison can again be made between the highly crystalline polymer and the completely amorphous one. In the latter case, only about 15% gel has formed after 0.5 Mrad irradiation, while for the highly crystalline experiment the gel content has reached 40%. We can also note in Fig. 8, that for the limited range studied, the cross-linking efficiency is independent of temperature for the completely amorphous polymer.

The results in Figs. 7 and 8, at the higher temperatures, are in good accord with the data of Fig. 6, where the temperature was maintained constant and the radiation dose varied. We note, however, that the enhanced cross-linking efficiency, characteristic of the highly crystalline state, is not maintained as the temperature is lowered. Detailed studies of the specific volume–temperature relations (Kitamaru and Mandelkern, 1964) for the same samples, indicate that there is no major change in the level of crystallinity as the temperature is lowered. Hence the change in the amount of gel formed must be attributed to molecular changes that occur within the crystalline state as the temperature is lowered. The temperature region where these changes occur coincides with the α transition determined by dynamic mechanical methods. This transition is assigned to the crystalline regions of the polymer. The mechanical energy absorption can be attributed to some type of motion of the crystalline units within the lattice. Without having to delineate, in the present context, the detailed nature of the molecular motions involved it is clear, however, that at low temperatures (less than about 90°C) the chain units in the crystalline sequences are relatively immobile. Above this temperature some type of molecular motion develops. Although the latent possibility exists for the same cross-linking efficacy at the lower temperatures, the necessary contact between units, cannot be realized in a relatively immobile system. However, with the onset of molecular motions, above 90°, this cross-linking capability can be realized. It is unfortunate that, detailed sol–gel partitioning studies as a function of radiation dose and the establishment of the critical condition for gel formation, for highly crystalline molecular weight fractions at the lower temperatures are presently not available for analysis. An incomplete study

(Kitamaru and Mandelkern, 1964) indicates a very much retarded partitioning curve at low temperatures (25°C) relative to the completely amorphous
polymer at 133°C. The possibility that chain scission processes may be
occurring at the low temperatures, where the mobility is retarded, has been
suggested. In accord with this suggestion it has been noted that gel formation,
at several dosages, was lower by a factor of about two for a highly crystalline
polyethylene (density 0.98–0.99) when irradiated at 25° and −78°C as
compared to an ordinary bulk crystallized sample of much lower density
(Salovey, 1964). However, this aspect of the problem, as will be discussed
subsequently, is complicated by the fact that the partitioning between sol
and gel is very dependent on the details of the postirradiåtion treatment,
when the irradiation is carried out at temperatures below 90°C.

At this point, we digress briefly from the discussion of the bulk crystallized
system and consider the irradiation behavior of linear polyethylene crystallized from dilute solution. The highlights of the morphology that results
from this mode of crystallization have already been described above. Typical
data, for the partitioning between sol and gel for this system, as a function
of temperature at a fixed irradiation dose, are given in Fig. 9. The dashed
line in this figure, which is presented for comparative purposes, represents
the results for the well-crystallized bulk polymer. At comparable radiation
dosages, particularly at the lower temperatures, gel formation is greatly

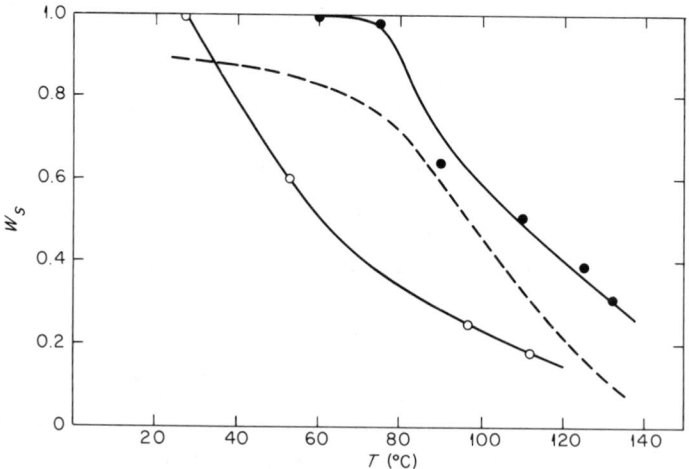

Fig. 9. Plot of weight fraction sol against temperature for crystals prepared from
dilute solution for molecular weight fraction $M_\eta = 2.07 \times 10^5$. Dosage 2.75 Mrad (●);
dosage 5.5 Mrad (○); dashed line for well-crystallized bulk sample at 2.75 Mrad
(Kitamaru and Mandelkern, 1964).

retarded in the solution formed crystals as compared with the bulk crystallized sample. However, gel is still formed at relatively modest dosages, indicative that intermolecular cross-linking can still take place. This difference appears to become diminished as the irradiation temperature is increased. As has been indicated previously, at temperatures greater than about 120°, well-defined morphological changes occur in the solution formed crystals. Therefore a proper comparison cannot be made between solution formed crystals and highly crystalline samples prepared in the bulk at elevated temperatures. This is the temperature region of most interest to the present problem, since it represents the condition wherein the cross-linking efficacy of the crystalline regions is greatly enhanced.

Qualitatively similar differences were observed by Salovey and Keller (1961; Salovey, 1962) in a more detailed study of an unfractionated linear polyethylene conducted at room temperature. A typical set of their results is shown in Fig. 10 (Salovey, 1962), where the gel fraction is plotted as a

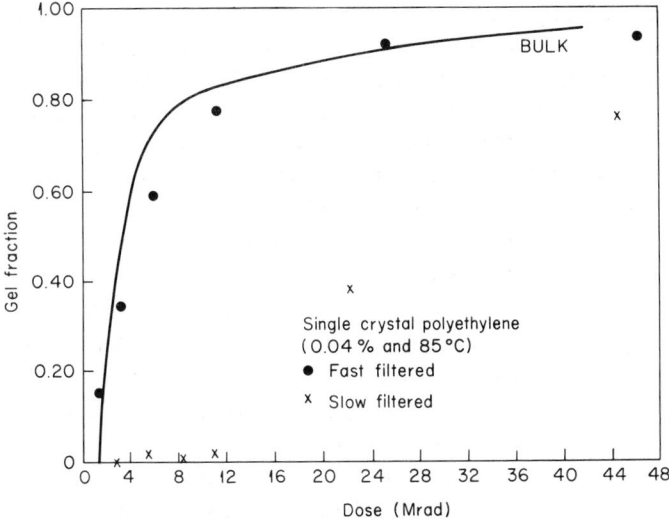

Fig. 10. Plot of gel fraction against radiation dose for bulk crystallized (solid line) and dilute solution crystallized linear polyethylene filtered by two different methods (Salovey, 1964).

function of the irradiation dose. The solid line represents the results for the bulk crystallized sample whose crystallization history and level of crystallinity are not described (Salovey, 1962). We note, according to this study, that the amount of gel formed at a given dosage, for the solution crystals, depends on the mode of their separation from the mother liquor (Salovey, 1962).

The rapidly filtered material yields identical results to that for the bulk specimens. However, for samples which were filtered slowly, over a period of several hours, the critical condition for gelation is much greater. Much less gel is formed at comparable dosages past the gel point. However, the same asymptotic value, of almost complete gel formation, is approached at high irradiation dose for all the cases studied.* If we assume that there is a real difference in the irradiation properties, as is illustrated in the figure, then it must reflect the packing or compacting of the lamella-like crystallites (Salovey, 1962). Thus the disordered, amorphous interfacial structure would be involved as being a major contributor to gel formation. The mechanical entanglement of the amorphous interfacial regions between different crystallites would enhance the possibility of intermolecular cross-linking. This concept is consistent with the conclusions arrived at above, of the inherent reduced intermolecular cross-linking efficiency of the crystallization regions at room temperature, which is the temperature of the experiments being described. This process would be aided by the reduced amount of the crystallizing solvent imbibed, as compared with the more isolated slowly filtered crystals. One conclusion, that clearly cannot be made from these results, as one has been prone to (Salovey, 1962), is that solution-formed crystals possess a regularly structured, regularly folded interface. If this model is retained, then in fact the comparative experimental results just described remain virtually uninterpretable.

Returning to the properties of bulk crystallized samples, we wish to consider further, from an experimental point of view, the validity and basis for the enhanced cross-linking and gelation efficiency in the highly crystalline state at elevated temperatures. We shall consider two main points here; the influence of the vinyl end group and a detailed quantitative assessment of the possibilities of chain scission in either state at the elevated temperatures of interest. In regard to the influence of the vinyl end group the most straightforward mode of attack is to remove it by hydrogenation, and compare the irradiation properties of such a system with another, identical in all other aspects except for the presence of the vinyl end group. Such experiments have been reported (Okada *et al.*, 1967). Two samples, of the same molecular weight fraction, $M_n = 6.1 \times 10^4$, $M_n = 4.6 \times 10^4$, were studied. The

---

* The interpretation of these results can become quite obscure and the significance of any fundamental difference in the partitioning between sol and gel due to the method of filtration questioned, by the observation (Salovey *et al.*, 1964) that for the "slow-filtered" material about 0.2% wt of xylene is retained within the crystals. For the more rapidly filtered samples only about a tenth of this amount of xylene is retained. Since xylene acts as a radiation sink, the difference in the partitioning between sol and gel, as it depends on filtration procedures, could be attributed solely to this source.

densities at room temperature were 0.980 and 0.981, respectively, which corresponds to a degree of crystallinity of 0.90. The enthalpies of fusion were 58.5 and 60.4 cal/g. The only difference between the two samples was the presence or absence of the vinyl end groups. The pertinent results are presented in Table I as the critical dosage, R_c, at 130°, required for gel formation.

TABLE I

RESULTS OF IRRADIATION OF HYDROGENATED AND NONHYDROGENATED
POLYETHYLENE AT 130°

Sample	State	R_c (Mrad)
Nonhydrogenated	Completely amorphous	3.1
	Highly crystalline	1.6
Hydrogenated	Completely amorphous	4.2
	Highly crystalline	4.5

The results for the nonhydrogenated control sample reconfirm the previous conclusion that the critical dosage required for gelation is about twice as great for the completely amorphous sample as compared to the highly crystalline one. However, a strikingly different set of results is obtained for the hydrogenated sample, which was prepared and treated in exactly the same manner as the nonhydrogenated one and which have the same physical properties. The critical dosages required for gelation are found to be comparable for the crystalline and amorphous hydrogenated sample. The critical dosage for the hydrogenated samples is still much greater than that required for the completely amorphous nonhydrogenated sample. It thus becomes abundantly clear that the greater efficiency of cross-linking in the nonhydrogenated highly crystalline sample results from a major contribution from chemical reactions involving the vinyl end group. Although the same number of vinyl end groups are present in both the highly crystalline and completely amorphous samples, their concentration, based upon the amount of amorphous material, is much greater in the crystalline state since the end groups are excluded from the lattice. This contribution to the enhancement of cross-linking could result solely from the concentration effect or arise from some particular arrangement of the end groups. Irrespective of the detailed mechanistic reason, it is clear that the enhanced cross-linking efficiency is a real effect in the usual (nonhydrogenated) crystalline linear polyethylene. It cannot be attributed to any differences in chain scission tendencies in the two states.

Further evidence for the influence of vinyl end groups on cross-linking is found in the two different types of preparations of solution formed crystals. The concentration of vinyl end groups decreased by a factor of 5 after the rapidly filtered single crystals were irradiated to 3.6 Mrad (Salovey, 1962). For the same dosage, the vinyl end group concentration only decreased by a factor of $1\frac{1}{2}$ for the slowly filtered crystals. These results parallel the cross-linking efficiency for the two types of crystals. It should also be noted that about 90% of the end groups (for this unfractionated polymer) have been located in the disordered interfacial layer (Keller and Priest, 1968).*

As has been indicated in the introductory discussion experiments directed to the question of chain scission relative to cross-linking have resulted in contradictory conclusions. Analysis of data utilizing unfractionated material presents theoretical difficulties unless the asymptotic limit for the fraction gel is established. Hence, it is highly desirable, if not mandatory, that experiments designed to investigate this question be performed with molecular weight fractions. According to the general theory for gel formation, as developed by Flory (1941) and by Stockmayer (1944) when intermolecular cross-links are randomly introduced into a collection of long-chain molecules, the critical condition for gel formation is achieved when

$$\rho_c = \frac{1}{(\bar{y}_w - 1)} \cong 1/\bar{y}_w \tag{1}$$

where ρ_c is the fraction of units cross-linked and $\bar{y}_w$ is the weight-average degree of polymerization. It has been generally accepted that the fraction of units cross-linked varies directly as the radiation dose. The critical condition for gelation thus depends only on the initial weight-average molecular weight. However, the partitioning between sol and gel is very sensitive to the initial molecular weight distribution. For an idealized system, perfectly homogeneous with respect to molecular weight, the weight fraction sol w_s varies with the fraction of units cross-linked according to (Flory, 1941)

$$w_s = 1 - \rho(1 - w_s)^y \tag{2}$$

which can be written to a good approximation as

$$-\ln w_s/(1 - w_s) \cong R/R_c \tag{3}$$

where the assumption of the proportionality between the fraction of units cross-linked and the radiation dose has been invoked. Equation (3) is only valid for a system where the primary molecules are of uniform molecular weight. It serves as a convenient reference point with which to examine

---

* In a set of experiments, which yield apparently contradictory results, it was reported that the same differences in gel formation were obtained from either a slowly filtered or rapidly filtered hydrogenated sample (Salovey and Hellman, 1968a).

experimental data on the partitioning between sol and gel for real fractions. It is also a convenient relation by which to compare the results in the different states, since the inherent cross-linking efficiency is contained in the normalization factor R_c.

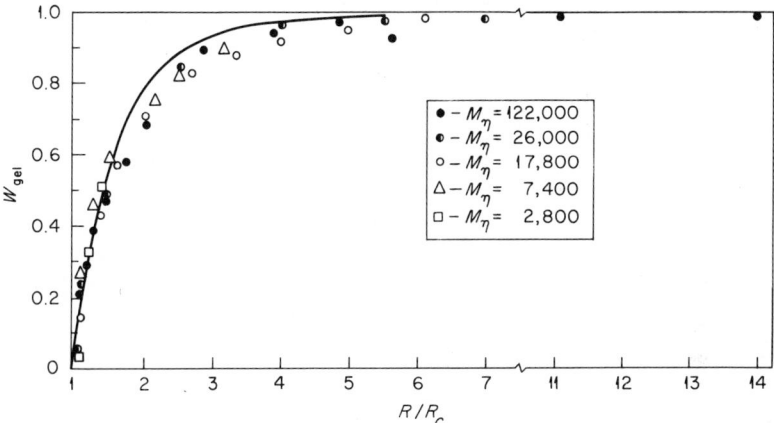

Fig. 11. Plot of gel fraction against R/R_c for indicated molecular weight fractions of linear polyethylene irradiated in melt at 133°C (Rijke and Mandelkern, 1971).

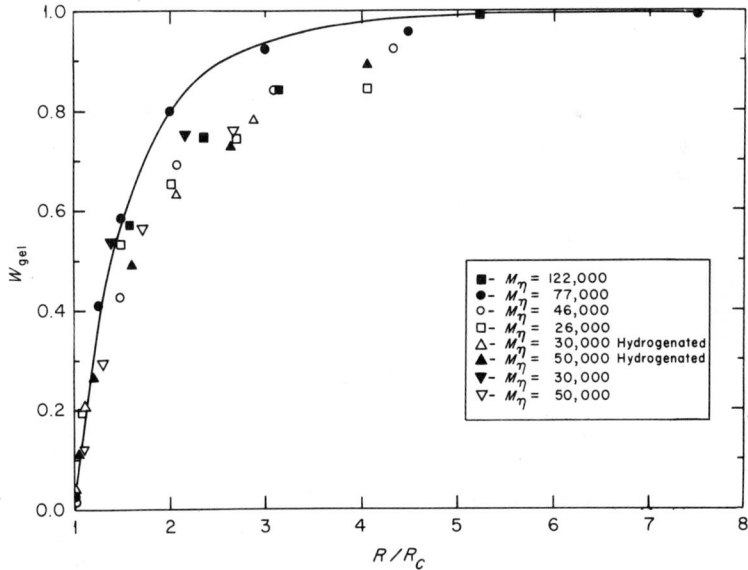

Fig. 12. Plot of gel fraction against R/R_c for indicated molecular weight fractions of linear polyethylene irradiated in crystalline state at 133°C (Rijke and Mandelkern, 1971).

In Figs. 11 and 12 (Rijke and Mandelkern, 1971) there is plotted an extensive amount of data, for irradiation at 133°, for a set of molecular weight fractions in the completely molten state and in the highly crystalline state. The latter set includes two pairs of hydrogenated and nonhydrogenated samples. The data are plotted according to the suggestion of Eq. (3) and the solid lines represent the theoretical curves. There are several major features that are common to both sets of data as are represented in the two curves. For the initial values of R/R_c both types of samples quantitatively adhere to the theoretical prediction for molecular weight fractions subject only to cross-linking. Deviations from the theoretical expectation set in when R/R_c is the order of 1.5. The deviations are such that less gel is produced than would be expected theoretically. However, for values of R/R_c of the order of 4 or greater, the theoretical line is rejoined. In this asymptotic region, the fraction gel produced is in the range of 0.98–1.00. Hence, there is no experimental evidence for the conventional type of random chain scission process. Admittedly, if the observations were limited to smaller values of R/R_c an argument could be developed, by incorrectly identifying the asymptotic radiation dose, that this type of chain scission could be playing an important role. However, even under this set of premises less chain scission would be assigned to the molten state as compared to the highly crystalline one.

The results in the molten state follow the theoretical expectations up to a weight fraction $w_g \cong 0.6$. For the highly crystalline state deviations for most of the fractions set in at a somewhat lower value of w_g. One fraction, $(M_\eta = 77,000 \ M_n = 75,000)$, which is obviously a very sharp one, adheres almost exactly to the theoretical expectation in the crystalline state. The deviations that are observed could be attributed to a variety of causes. We should note, however, that when compared at the same value of R/R_c there does not appear to be any major difference between the hydrogenated and unhydrogenated sample. Polydispersity could be a major factor as the partitioning is very sensitive to deviation from homogeneity. This factor can be demonstrated by plotting the data from Fig. 6 on the same basis as in Figs. 11 and 12. It is found that although this sample is a fraction deviations from the theoretical curve are observed at very low values of R/R_c, indicating that it is more polydisperse than would be usually anticipated. Another possibility, considering just the data presented in Figs. 11 and 12, would be for chain scission to occur with subsequent recombinations with the network. This latter possibility would allow for the experimental observation of complete gel formation. However, the major conclusion to be drawn from the data in Figs. 11 and 12 is that the random type of chain scission process (without any recombination) does not manifest itself in the irradiation of linear polyethylene in either state at elevated temperatures.

It has been well established (Lawton *et al.*, 1958a,b; Kang *et al.*, 1967) that the fraction gel obtained after low-temperature irradiation depends on the details of the postirradiation treatment. This phenomenon has been attributed to the different reaction pathways available to the trapped free radicals. Chain scission due to postirradiation oxidation could result, or further cross-linking due to radical reaction with the chain could occur. These distinctly different possibilities require that a very careful delineation and assessment be made of the postirradiation treatment. Such treatment is, of course, necessary to determine experimentally the fraction of gel, and it is important that it does not obscure the irradiation events. The results for two distinctly different modes of postirradiation treatment have been reported (Rijke and Mandelkern, 1971). In one of these, the sample was exposed to air for 3 days prior to extraction. In the other, the sample was heated in vacuum for 2 hr at 140° then cooled to room temperature and the extraction process initiated. For irradiation temperatures below 80°C major differences were observed. For a fixed irradiation dose a much higher fraction gel was obtained for the specimens heated to 140°C. The difference between the two treatments in the fraction gel obtained also depended on the initial level of crystallinity. The disparity between the methods became greater for higher levels of crystallinity. However, when the irradiation temperature was raised above 80°C the differences in the fraction of gel obtained diminishes. For irradiation at about 110°C the same fraction of gel is obtained irrespective of the details of the postirradiation treatment. This result is undoubtedly a consequence of the onset of motions within the crystal which would allow for a more effective cross-linking process. It is consistent with the previous discussion of the cross-linking efficiency as a function of temperature in the crystalline state. Even the postirradiation treatment of heating in vacuum at 140° for 2 hr, yields a very strong dependence of cross-linking efficiency on temperature in the crystalline state. We note further, that for irradiation at 133°C, in either state, the postirradiation treatment has no sensible effect on the results.

According to Eq. (1) the fraction of intermolecular cross-linked units required to achieve the gel point, and hence, the critical dosage necessary to achieve this condition is inversely proportional to the weight average molecular weight. This relation rests on a very firm theoretical foundation. It has, however, an underlying assumption that intermolecular cross-linking process is a random one. Thus chain units that are paired from different molecules must be selected at random from within the total system. For the particular problem at hand the additional assumption that ρ is directly proportional to R is made. The data that have been accumulated for molecular weight fractions of linear polyethylene, in the completely amorphous state and for samples of controlled crystallinity are well suited for analysis according to Eq. (1) (Kitamaru *et al.*, 1964; Okada *et al.*, 1967; Rijke and Mandelkern, 1971). A compilation of the data thus accumulated is presented in Fig. 13 (Rijke and Mandelkern, 1971) as a plot of R_c against M_η, the

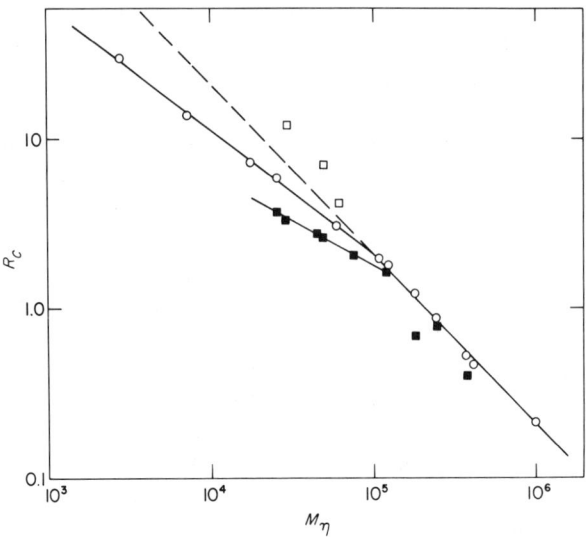

Fig. 13. Double logarithmic plot of critical condition for gelation against viscosity average molecular weight. Irradiation in molten state at 133°C (○ and □ hydrogenated); irradiation in crystalline state at 133°C (■) (Rijke and Mandelkern, 1971).

viscosity average molecular weight. The data encompass a molecular weight range of from 3×10^3 to 1×10^6 and the irradiations were conducted at 133°C. Two different types of samples are represented here. The vast majority of the fractions are the usual linear polyethylene and thus contain one vinyl end group per molecule. Three of the samples, designated by the open squares, have been hydrogenated so that the vinyl end group is absent.

In discussing the data in this figure it is most convenient that the different states of the polymer be analyzed separately before the complete problem is examined. We consider first, therefore, the simplest situation, namely, the irradiation of the completely molten polymer at 133°C. These data are represented by the open circles in Fig. 13. The seven samples in the molecular weight range 1×10^5–1×10^6 accurately fall on the solid straight line which is drawn and which has a slope of -1. Thus we conclude that in this molecular weight range the Flory–Stockmayer relation is rigidly adhered to. However, as the molecular weight is decreased below 1×10^5, in the completely molten state, systematic deviations are observed from the linear relation applicable to the higher molecular weights. The critical dosage for gelation is found to be less than that predicted. Thus, the intermolecular cross-linking process is enhanced as the concentration of vinyl end groups is increased. The influence of this end group in enhancing intermolecular cross-linking formation is thus manifested even in the completely molten state. Under these conditions the statistical considerations underlying the Flory–Stockmayer theory, and the assumption of the direct proportionality between cross-linking and radiation dosage are not fulfilled. On the other hand, in the higher molecular weight range, the concentration of vinyl end groups is sufficiently reduced so that its reactivity is not an important factor in the cross-linking process.

As has been indicated previously, within the experimental error, the critical condition for gelation (at 133°C) for the hydrogenated samples is the same in the completely molten and the highly crystalline samples. These results, based on the presently available experimental data, are restricted to the molecular weight range 3–6 × 10^4. If the presence of the vinyl end group is the only reason for the difference in cross-linking efficacy between the two states and for the deviation from the linear relation observed in the molten state at the lower molecular weights, then it is to be expected that the results for the hydrogenated samples (irrespective of state) should fall on the extension of the straight line of slope −1. As can be seen from Fig. 13, although the available data lie close to this line, they are slightly above it. A portion of this apparent discrepancy from our expectation, can be attributed to the usual experimental uncertainties inherent in the methods involved. In addition the possibility of slight oxidation taking place during the hydrogenation procedure would require a somewhat greater critical dose for gelation.* We note, however, that there is a much better adherence to the Flory–Stockmayer condition for the hydrogenated samples than for the completely amorphous or highly crystalline sample of the same molecular weight.

The conventional lower molecular weight, highly crystalline samples, designated by the solid rectangles in the figure display marked deviations from the extension of the straight line of slope −1. As has been discussed previously, in detail, the critical condition for gelation for these samples is substantially less than the corresponding completely molten fraction. These results are a consequence of the influence of the vinyl end group. For the higher molecular weight crystalline samples (> 1 × 10^5) the results are close to expectation but are not as yet completely unequivocal. There are

---

* Independent studies of slightly oxidized, but not hydrogenated, samples support this conclusion (Rijke and Mandelkern, 1971).

several inherent difficulties that are encountered in this molecular weight range, for the crystalline polymer, that hamper the interpretation. One of these, is that above a molecular weight of about 7×10^4 the level of crystallinity that can be attained monotonically decreases with increasing molecular weight. Thus, it is difficult to compare samples having exactly the same level of crystallinity. Furthermore, it is more difficult to prepare very sharp fractions as the molecular weight is increased so that the number average will be less than the weight average. Hence there will be a tendency for such samples to have a proportionately high concentration of end groups than would be characteristic of a given weight average molecular weight. Among the four crystalline samples that have been studied in this high molecular weight region, two of them adhere very closely to the straight line of slope -1. These results thus agree with the conclusions drawn from the study of the lowest molecular weight hydrogenated samples in that irrespective of the state of the polymer the vinyl end group concentration is the key factor in the observed differences in cross-linking efficiency. These two fractions are relatively sharp, the ratio of the weight to number average molecular weight is of the order of 1.3–1.4 (Rijke and Mandelkern, 1971). The data for the other two fractions fall below the reference line and thus indicate a more effective cross-linking reaction than would be predicted solely from the weight average molecular weights. The highest molecular weight fraction in this group has a ratio of weight to number average molecular weight of about 2.2 (Rijke and Mandelkern, 1971). Although the number average molecular weight is not available for the other sample, $M_\eta = 1.8 \times 10^5$, the nature of its partitioning between sol and gel (Kitamaru *et al.*, 1964; Dole and Katsuura, 1965) indicates that it is relatively more polydisperse. We tentatively conclude, therefore, that the deviations observed from the reference line, for the higher molecular weight crystalline samples, can be attributed to the vinyl end group concentration of these samples. If this conclusion is sustained by future work, then it is clear that the only reason for the difference in cross-linking efficiency in the different states of linear polyethylene, at temperatures above the α transition, can be attributed to the effective concentration of the vinyl end groups. Further experiments in this molecular weight range are clearly necessary in order to substantiate these conclusions. They will require, however, the preparation and characterization of high molecular weight samples which are very "monodisperse." Only when the results of these experiments are available can an unequivocal answer be obtained as to the influence of the drastically different molecular conformations in the two states on the cross-linking process. As has been indicated above, the results for the hydrogenated samples, which are in a lower molecular weight range, argue against any direct influence of conformation at these temperatures. This conclusion is in accord with the results obtained for the irradiation of poly-*trans*-1,4-isoprene in the crystalline state at 20°C and in the completely molten state at 100°C (Turner, 1966).

To complement the above discussion of gel formation and cross-linking it is necessary to consider in some detail the major chemical acts that occur as a consequence of exposure to high-energy ionizing irradiation. The main events that occur with linear polyethylene are hydrogen evolution, vinyl decay and the formation of *trans* unsaturation. These must of course be considered in terms of the independent variables that describe the system.

Some typical results for the dependence of the hydrogen evolution on radiation dose are given in Fig. 14 (Okada and Mandelkern, 1966) for the

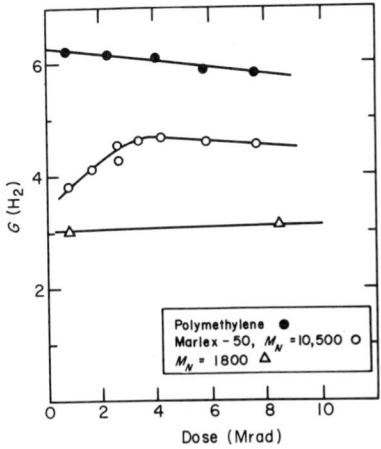

Fig. 14. Plot of $G(H_2)$ against radiation dose for indicated samples of linear polyethylene (Okada and Mandelkern, 1966).

completely molten samples. For the lowest molecular weight sample $G(H_2)$ is independent of the dosage. For very high molecular weight polymethylene, $M_\eta \cong 3 \times 10^6$, there is a slight decrease in $G(H_2)$ with increasing molecular weight. For unfractionated Marlex-50, a slight maximum in the hydrogen evolution is observed with increasing radiation dose. The important observation can be made from these data that at a fixed low dosage $G(H_2)$ is markedly dependent on the initial number average molecular weight. The significance of this point will be discussed in detail subsequently. Similar type data, for other conditions of irradiation, have not been reported. For a Marlex-type linear polyethylene, Dole and co-workers reported that for irradiation at 25°C, in the crystalline state $(1 - \lambda = 0.75-0.80)$, $G(H_2)$ decreases from about 3.7 to 3.2 as the radiation dose is increased to 15 Mrad. For the same sample, irradiated at 120°, virtually no change in $G(H_2)$ with radiation dose is reported (Kang *et al.*, 1967). No data are given for the dosage dependence of $G(H_2)$ at higher temperatures for this sample.

In order to establish the mechanism for the rate of vinyl decay the order of the reaction has to be determined following procedures of conventional kinetic theory. To accomplish this task it is necessary that the initial concentration of the species be varied. The order of the reaction cannot be established from experimental studies which are restricted to only one polymer sample and hence to one initial vinyl end-group concentration. However, the necessary data can be obtained by utilizing the range of samples that are available for study, i.e., by varying the number average molecular weight. It is then found that the rate of vinyl decay, in the molten state, cannot be represented by first-order reaction kinetics. However, as is indicated

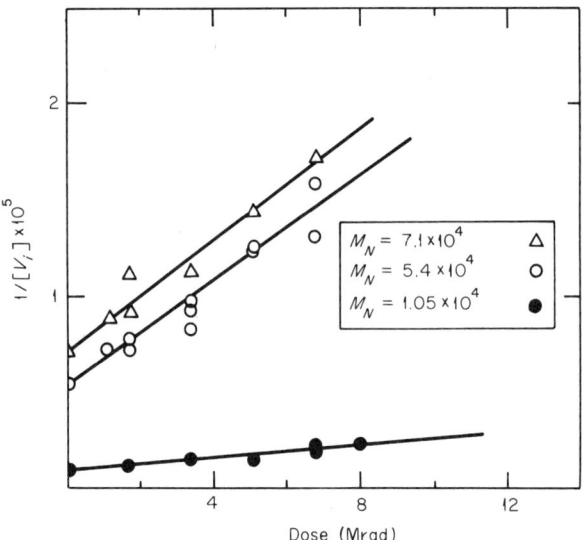

Fig. 15.   Plot of reciprocal of vinyl end-group concentration as a function of radiation dose for indicated molecular weight fractions of linear polyethylene (Okada and Mandelkern, 1966).

by the plots in Fig. 15 (Okada and Mandelkern, 1966) the data for each individual sample can be formally represented by second-order kinetics and obey the rate law

$$1/(V_i) - 1/(V_i)_0 = k_2 R \tag{4}$$

For the two highest molecular weight samples studied, $M_n = 7.1 \times 10^4$ and 5.4×10^4, the rate constant k_2 is essentially the same, being 1.35×10^4 and 1.56×10^4, respectively. However, the rate constants decrease for samples of lower number average molecular weight. For $M_n = 1.05 \times 10^4$, $k_2 = 1.24 \times 10^3$; for $M_n = 1.8 \times 10^3$, $k_2 = 1.00 \times 10^2$ g mole^{-1} Mrad^{-1}.

In contrast to the relative simple analysis that can be given to the kinetics in the completely molten state, the results for rate of vinyl decay in the crystalline state at 133° do not lend themselves to any clear-cut functional dependence (Okada and Mandelkern, 1966). Although the results for the unfractionated sample, $M_\eta = 1.7 \times 10^5$, $M_n = 1.05 \times 10^4$, give very good agreement to a second-order rate law, the other samples show significant deviations at relatively small dosages. Analysis of the data according to first-order kinetics does not show any better fit. The data for the rate of vinyl decay, in the crystalline state at 25°C, follow the first-order kinetic law for the initial dosages (Williams and Dole, 1959; Kang *et al.*, 1967). This first-order rate constant is essentially independent of the degree of crystallinity over the range $1 - \lambda = 0.60$–0.90. However, this rate constant depends on the initial number average molecular weight. It varies from 0.039 when V_i° is 55×10^{-5} to 0.16 when V_i° is 1.41×10^{-5} mole g^{-1}. The difficulties encountered in developing a formal analysis of the kinetics of vinyl decay in the crystalline state appear to be indicative of the complexities in mechanism that could be involved.

The formation of *trans*-vinylene is found to be linear with dosage at low dosage (Dole *et al.*, 1957, 1958; Sears, 1964; Charlesby *et al.*, 1964), and the rate is independent of the initial molecular weight (Okada and Mandelkern, 1972). The rate of formation of *trans*-vinylene varies in a relatively small, but perceptible manner with changes in temperature and dose. For higher doses, deviations from the linear relationships are observed, and the data display a tendency to level off at the higher dosages (Okada and Mandelkern, 1972). Dole and collaborators (Dole *et al.*, 1957, 1958, 1966; Dole, 1965) have shown that the kinetics of *trans*-vinylene growth can be quantitatively explained by the combination of zero-order growth combined with a first-order decay law.

The G values for the major chemical acts which result from the irradiation of linear polyethylene are summarized in Table II for a series of samples of varying molecular weight (Okada and Mandelkern, 1966, 1972). The results are listed for irradiation in the completely molten state at 133°C, in the crystalline state at 130°C and in the crystalline state at 25°C. Examining first the results in the completely molten state, we observe that $G(\text{H}_2)$ increases very markedly with increasing molecular weight (decreasing end-group concentration) and, as is seen in Fig. 16, reaches an asymptotic value

TABLE II

G Values for Major Chemical Events upon Irradiation of Linear Polyethylene[a]

M_η	M_n	$G(H_2)^b$	$G(-V_i)$	$G(V_t)$	$G_{app}(X)$
\multicolumn{6}{c}{Irradiated at 133°C in the completely molten state}					
3×10^{6c}	—	6.2	—	2.3	—
3×10^5	2×10^5	5.8	—	—	—
2.5×10^5	7.1×10^4	6.0	2.2	2.5	—
4.5×10^4	5.4×10^4	5.6	2.7	2.4	—
6.1×10^4	4.6×10^4	5.4	—	—	—
6.1×10^{4d}	4.6×10^4	6.2	—	2.7	1.9
1.7×10^{5c}	1.05×10^4	4.1	9.0	2.3	—
2×10^{3e}	1.8×10^3	3.1	25.5	—	—

M_η	M_n	$(1-\lambda)^f$	$G(H_2)^b$	$G(-V_i)$	$G(V_t)$	$G(X)$
\multicolumn{7}{c}{Irradiated at 130°C in the crystalline state}						
2.5×10^5	7.1×10^4	0.65	6.3	2.9	3.0	—
4.5×10^4	5.4×10^4	0.78	5.7	5.7	3.1	—
6.1×10^4	4.6×10^4	0.85	5.6	—	—	—
6.1×10^{4d}	4.6×10^4	0.85	5.7	—	2.9	1.8
1.7×10^{5c}	1.05×10^4	0.70	5.0	13.5	2.6	—
2×10^3	1.8×10^3	0.70	2.7	31.5	—	—

M_η	M_n	$(1-\lambda)$	$G(H_2)^g$	$G(-V_i)$	$G(V_t)$
\multicolumn{6}{c}{Irradiated at 25°C in the crystalline state}					
2.5×10^5	7.1×10^4	0.59	3.9	2.1	1.7
		0.86	3.4	1.8	2.1
4.5×10^4	5.4×10^4	0.66	3.6	2.1	2.0
		0.90	4.0	2.0	2.3
6.1×10^{4d}	4.6×10^4	0.90	3.9	—	—
1.75×10^{5c}	1.05×10^4	0.65	3.2	9.5	2.0
		0.89	3.4	7.3	2.4
2×10^3	1.8×10^3	0.86	2.8	19.5	1.9

[a] From Okada and Mandelkern (1966, 1972).
[b] Determined at 1.7 Mrad.
[c] Unfractionated sample.
[d] Hydrogenated sample.
[e] Irradiated at 124°.
[f] Degree of crystallinity.
[g] Calculated at 10 Mrad.

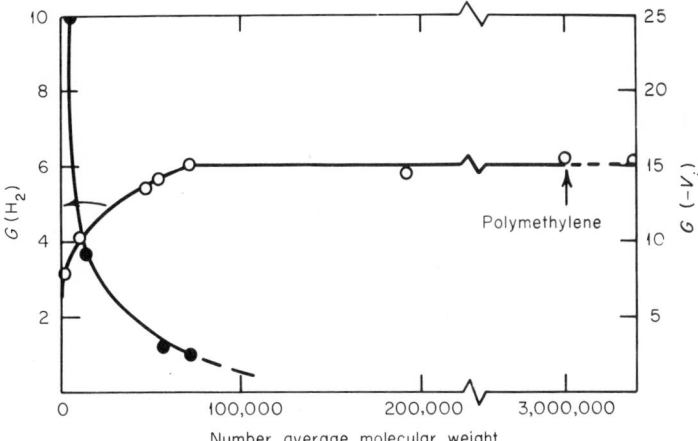

Fig. 16. Plot of $G(H_2)$ ($\bigcirc$) and $G(-V_i)$ ($\bullet$) as function of number average molecular weight for irradiation in molten state at 130°C (Okada and Mandelkern, 1972).

of 6.2 for the highest molecular weight. This asymptotic value for $G(H_2)$ agrees very well with that directly obtained from the hydrogenated sample of lower molecular weight. One can thus conclude that the experimentally determined value for $G(H_2)$ is dependent on the vinyl end-group concentration of the sample, a fact which accounts for the molecular weight effects that are observed. This conclusion is further supported by the observation that $G(-V_i)$ decreases with decreasing initial end-group concentration. There is a strong correlation, and almost one to one correspondence, between the decreasing value of $G(-V_i)$ and the increasing $G(H_2)$, with increasing molecular weight. This factor is also illustrated in Fig. 16. It becomes evident that the vinyl end group is implicated as being a sink for or "scavenger" of some of the H atoms being produced by the irradiation. At this point it should be recognized that if one is interested in studying the radiation chemistry of linear polyethylene for its own sake, then the presence of the vinyl end group introduces unnecessary complications which should be avoided. On the other hand, if one is interested in a particular item of commerce, then the reactions of the vinyl end group are of major importance. For example, for the unfractionated sample, $M_w = 1.7 \times 10^5$, $M_n = 1.05 \times 10^4$ (Marlex-50), $G(H_2) = 4.1$ while the asymptotic or inherent value for the linear polymer, devoid of end-group effects, is 6.2.

In contrast with the above rather dramatic results, the G values for the formation of *trans* unsaturation display no trend with molecular weight. All the results, for this temperature and state, are encompassed by $G(V_t) = 2.5 \pm 0.2$. The G value for cross-linking that is listed is an apparent one.

It is determined from the critical condition for gel formation and is thus based on the premise that randomly distributed, intermolecular cross-links are the only kind formed. As has been demonstrated in the previous discussion (see Figs. 11 and 12) chain scission plays no consequential role in these calculations. $G(X)$ is then found to be 1.9, based on the hydrogenated samples and those of very high molecular weight. Much higher values will of course be obtained for samples containing the vinyl end group. These latter values are not of any fundamental significance to the analysis of polyethylene as such but are of course very important to this particular type of polymer.

We note that there are no samples of very high molecular weight listed in the table for irradiation in the crystalline state at 130°C. Although admittedly this would be a very desirable set of data to have in hand, it cannot be obtained since one cannot produce as high a level of crystallinity in such samples as with the lower molecular weight species. For the highest molecular weight sample listed in this table a slight but perceptible decrease in the level of crystallinity that can be attained has already developed. We note that, under these conditions (130°C, highly crystalline), $G(H_2)$ also increases with increasing number average molecular weight. As is indicated

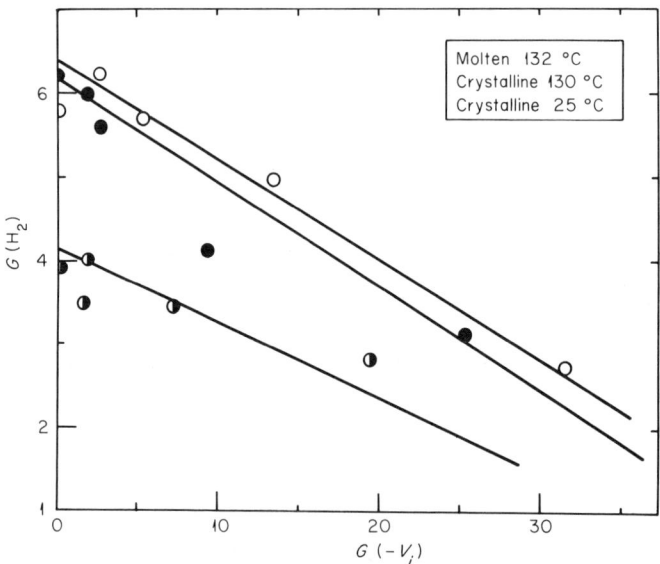

Fig. 17. Plot of $G(H_2)$ against $G(-V_i)$ for molecular weight fractions of linear polyethylene irradiated under conditions indicated (Okada and Mandelkern, 1972). (○) Molten; (●) 130°; (◐) crystalline at 25°.

in Fig. 17 (Okada and Mandelkern, 1972), where $G(H_2)$ is plotted as a function of $G(-V_i)$, there is a small experimental uncertainty in the asymptotic value that is attained. The hydrogenated sample directly yields $G(H_2) = 5.7$ while $G(H_2) = 6.3$ is obtained for the highest molecular weight sample studied. The plot in Fig. 17 indicates that, within the experimental error, $G(H_2)$ is about the same in the molten state and in the crystalline state at this temperature. Hence, at this elevated temperature, above the α transition, there is essentially no difference between the two states in the asymptotic limit of vanishing end-group concentration. There clearly is no marked increase in $G(H_2)$ in the molten state (Kang *et al.*, 1967).

Under these conditions $G(-V_i)$ again decreases with increasing molecular weight and follows a similar pattern as in the completely molten state. However, the magnitude of $G(-V_i)$ is much greater in this case. As can be seen in Fig. 17, when comparison is made at the same value of $G(H_2)$, $G(-V_i)$ for the crystalline state at 130°C is much greater than its value in the completely molten state at a comparable temperature. This difference correlates with the higher cross-linking efficacy in the crystalline state for samples containing vinyl end groups. Thus, the vinyl end group (or its H addition product) is again implicated in an additional type of cross-linking reaction.

The value for $G(V_t)$, under these irradiation conditions, is again found to be independent of molecular weight. $G(V_t)$ is 2.9 ± 0.2, slightly higher than the values in the completely molten state. The value for $G(X)$, calculated in the same basis as previously, is 1.8 and is thus virtually identical to that characteristic of the completely molten state. This result is merely a reflection of the virtually identical critical dosages required for gelation in the crystalline and amorphous states at this temperaure when the effect of the vinyl end group is discounted.

An examination of the data for irradiation at 25°C, for approximately the same level of crystallinity as at 130°C, indicates that $G(H_2)$ again increases with increasing molecular weight. However, the changes here are relatively small compared to the two other cases that have been studied in detail. An asymptotic value of $G(H_2) = 3.9$ is found for the high molecular weight samples which agrees very well with the directly measured value for the hydrogenated sample. The asymptotic value obtained at this low temperature, is significantly less than that obtained at the elevated temperature in either of the two states.

The dependence of $G(-V_i)$ on molecular weight follows a pattern similar to that of the other cases, as can be seen in Fig. 17. However, the magnitude of this quantity is the smallest of all the situations studied. This would appear to be related to the relatively low intermolecular cross-linking yield, based on gel formation at a fixed dosage, found at temperatures below the α transition. For the type of polyethylene containing a vinyl end group the differences in cross-linking efficiencies that are observed in the different states, at different temperatures, are thus directly related to the intensity at which the vinyl end group disappears upon irradiation.

The value of $G(V_t)$ is again found to be independent of molecular weight, but a small dependence on the level of crystallinity is indicated. For the higher level of crystallinity, comparable to that at 130°C, $G(V_t)$ has been reduced to 2.1 $\pm$ 0.2 and is further reduced by 0.3–0.4 units at the lower level of crystallinity studied. Unfortunately, systematic data with regard to the critical condition for gelation as a function of molecular weight and for hydrogenated samples are not presently available for irradiation in the crystalline state at 25°C. From studies of an unfractionated polymer, $M_w = 2.3 \times 10^5$, $M_n = 2 \times 10^4$, with $1 - \lambda = 0.75$–0.80 Dole and co-workers (Kang *et al.*, 1967) have deduced a value for $G(X)$ of about 1.0. If substantiated by further studies with samples having a larger number-average molecular weight, or with the vinyl end group absent, this result would indicate a substantial decrease in $G(X)$ in the crystalline state at this lower temperature of irradiation as compared with temperatures above the α transition.

It is of interest at this point to compare the G values that have now been assembled for the major chemical acts involving linear polyethylene with the corresponding values that have been obtained from the irradiation of the *n*-hydrocarbons. In studying the irradiation of *n*-hexane at liquid nitrogen temperature and at room temperature Dewhurst (1958) found that the unsaturation yield was independent of temperature. In another investigation (Dewhurst, 1957) a series of liquid *n*-hydrocarbons, from *n*-hexane to *n*-hexadecane were studied. The *trans*-vinylene yield increased with increasing chain lengths and approached a limiting value of $G(V_t) = 2.0$. This value is very close to that obtained for the long-chain polymer.

The products obtained from the irradiation of the *n*-alkanes can usually be divided into three main classes (Chapiro, 1962f). These are species whose molecular weights are less than that of the original hydrocarbon; species which are intermediate in molecular weight between the original hydro-carbon and a dimer, and the dimers. In some cases small amounts of higher molecular weight products have also been found (Chapiro, 1962f). The G value for dimer formation can thus be taken as a reliable lower limit for $G(X)$ in the *n*-hydrocarbons. For liquid *n*-hexane and *n*-hexadecane Dewhurst (1958) found G values of 2.0 and 1.8, respectively. For *n*-octacosane a G value of 2.8 was obtained, which has been attributed to the effect of the state (Miller *et al.*, 1956). More recently, G(dimer) for liquid and solid hexadecane were determined utilizing either γ or electron irradiation (Salovey and Falconer, 1965). Slightly different results were obtained from the two methods. In the liquid state at 19°C, γ irradiation yielded a G value of 1.59 while electron irradiation gave 2.60. The former value decreased to 1.50 and to 0.99 in the solid state at -80°C and -196°C, respectively. The latter value decreased to 2.30 in the solid at 4°C and was further reduced to 2.01 and 1.23 for the two lower temperatures cited.* The G value for dimerization of solid n-heptadecane at 4°C was 1.13 while for the liquid at 24° it was 1.63. The difference in the G values between n-hexadecane and n-heptadecane was attributed to the known difference in crystal structure (Salovey and Falconer, 1966). By utilizing gel permeation chromatographic techniques, Salovey and Hellman (1968b) determined the G value for cross-linking for the normal hydrocarbons, n-hexatriacontane (C-36) and n-pentatriacontane (C-35). At the irradiation temperature of 48°C the crystallographic structure is monoclinic for the former and orthorhombic (as is polyethylene) for the latter. For the C-36 hydrocarbon the G value of 1.1 ± 0.1 is the same in the crystalline state and in the liquid state at 78°C. For the C-35 polymer $G = 1.5$ in the crystalline state and increases slightly to 1.8 in the liquid state at 78°C. From the summary given above it is found that exact, concordant results have not as yet been attained between all investigators for the G cross-linking value for the n-hydrocarbons. A complete study of the effect of chain length, state, crystallographic structure, and temperature is not yet in hand. However, it appears clear that the G value for cross-linking of the n-hydrocarbons is comparable to that for linear polyethylene in the absence of any influence of the vinyl end group.

Dewhurst (1958) has reported that there is a very strong influence of temperature and state on the hydrogen yield of n-hexane. In the solid state, below -100°C, $G(H_2)$ is about 3. In the liquid phase, between -78° and $+30$°C, there is a linear increase of the hydrogen yield with temperature. In this state $G(H_2)$ is found to be 5.0 at 30°C for n-hexane. This represents the actual hydrogen that evolved. However, if there is any scavenging of the H atoms by the unsaturation that is concomitantly being formed (particularly at the ends of the molecules), then the quantity which is of theoretical interest, to the problem at hand, is not the actual hydrogen evolved but that which is actually formed by the interaction with the irradiation. Hardwick (1960, 1961) has attacked this problem from both a theoretical and experimental point of view. By a straightforward kinetic analysis he has shown that the asymptotic value for $G(H_2)$, the value in the absence of any scavenging, is 6.35 for n-pentane, 5.28 for n-hexane, 6.06 for n-heptane,

---

* Falconer and Salovey (1966) also report G values for the dimerization of liquid and solid n-hexadecane of 1.7 and 1.6, respectively.

6.18 for *n*-octane, and 6.05 for *n*-nonane. These values are very close to the asymptotic $G(H_2)$ values listed in Table II for the irradiation of linear polyethylene in the completely molten state and in the highly crystalline state at 130°C and clearly indicate a similar mechanistic process. We note here that the experimental values of $G(H_2)$ as a function of end-group concentration, for the completely molten polymer, follow Hardwick's analysis (Okada and Mandelkern, 1972). The G value for scavengable H atom production is the order of 4.0 which is very similar to that obtained by Hardwick for the liquid *n*-paraffins.

There does not exist at present such extensive data for hydrogen production for the *n*-hydrocarbons in the crystalline state. Falconer and Salovey (1966) found that at 4 Mrad, $G(H_2) = 4.0$ for solid *n*-hexadecane. For solid *n*-octacosane, at a dose of 56.5 Mrad, Chapiro (1962g) has calculated from the data of Miller *et al.* (1956) a $G(H_2)$ of 3.8. Although a coincidence may be involved we note that these values are virtually identical to the asymptotic value obtained for linear polyethylene, $1 - \lambda = 0.90$, when irradiated at 25°C in the crystalline state.

Based on the data that are presently available and, admittedly, there are still some serious gaps, a comparison of the G values for the *n*-hydrocarbons and linear polyethylene indicate that there is a very close similarity between the major chemical events. We should recall, however, that this conclusion can only be reached when the complications involving the vinyl end group are removed from the polymer. Experimental data based solely on studies utilizing the usual unfractionated polymer, with its high end group concentration, do not lead to the same conclusion. With the close similarity in G values it is reasonable to conclude that the mechanisms involved for the two cases must be the same. Thus, the burden in developing the details of the reaction mechanism can be placed on the *n*-hydrocarbons. There is, however, one important exception to this generalization. Upon the irradiation of the liquid *n*-hydrocarbons degradation products can be directly observed. However, for the polymer, the only sensitive physical chemical method of analysis has not as yet demonstrated any chain scission processes.

In order to analyze the results to ascertain if a material balance is obtained a distinction must be made between linear polyethylene chains which contain a vinyl end group and those which do not. Quite obviously the assumption that none of the evolved hydrogen back reacts with vinyl groups cannot be made (Kang *et al.*, 1967). Neither can the contribution of vinyl end groups, or its addition product to the formation of intermolecular cross-links be ignored (Dole *et al.*, 1966). The important influence of these factors, although well demonstrated experimentally, cannot at present be quantitatively separated. Therefore, it is prudent to focus one's attention, with regard to the material balance problem, on samples and experiments in which the vinyl end group has been removed. The distinction between the states and the temperature must still be maintained. Sufficient data exist to examine this question in the completely molten state at 133°C and in the highly crystalline state at this elevated temperature.

In the material balance analysis the initial assumption that is usually made is that the evolution of hydrogen occurs only through the production of cross-links and *trans*-vinylene formation. Therefore, the following relation should be obeyed

$$G(H_2) = G(X) + G(V_t) \qquad (5)$$

The pertinent data with which to assess this relation are summarized in Table III. We should recall that the $G(X)$ listed is calculated from the Flory–

TABLE III

G VALUES FOR MATERIAL BALANCE

	$G(H_2)$	$G(X)$	$G(V_t)$
Melt 133°C	6.2	1.9	2.5
Crystalline state 130°C	6.2	1.8	2.9

Stockmayer gelation condition so that only randomly distributed intermolecular cross-links are counted. It is apparent from the data in the table that even when the complexities introduced by the vinyl end group are removed from consideration and even when an allowance is made for a reasonable experimental error the requirement of Eq. (5) is not fulfilled. More hydrogen is evolved at these elevated temperatures, in either state, than can be accounted for by the simple postulate made. Other processes have been suggested to account for this discrepancy (Chapiro, 1962h). Chief among these are the formation of intramolecular cross-links and the possibility of nonrandom cross-linking, particularly in neighboring positions. These processes will clearly contribute to the hydrogen evolution but are not included in the quantity $G(X)$. Unfortunately, no reliable quantitative methods have as yet been developed to assess these other cross-linking processes. Intramolecular cross-links can naturally be expected to occur to some extent in the molten state where the chains adopt a statistical or random coil conformation. However, this process would be severely suppressed by the highly ordered chain conformation characteristic of the crystalline state. On the other hand, a nonrandom cross-linking process could be favored by an ordered structure.

Another possible source of hydrogen evolution could be the formation of other types of unsaturation besides *trans*-vinylene. *cis*-Vinylene has not however been detected following the irradiation of polyethylene. However, Fallgatter and Dole (1964) have demonstrated diene formation, when the sample is maintained in the crystalline state at room temperature and 120°C. However, diene formation is not detected following irradiation in the molten state. Therefore, no aid is received from this source in achieving material balance in the molten state. There is no direct evidence of diene formation at 130°C in the crystalline state. If the trend reported at lower temperatures is maintained, at 130°C, then about half of the deficiency in the material balance would be removed. However, the major discrepancy in the molten state is unaffected.

Analysis of the material balance in the crystalline state at temperatures below the α transition is severely hampered by the lack of a direct determination of $G(X)$ under these conditions. If a contribution of 0.5 is allowed for diene formation (Kang *et al.*, 1967), then at 25°C $G(X)$ would have to be 1.4 to achieve a material balance. As has been discussed previously, at a fixed radiation dose, the amount of gel formation is significantly reduced at the lower temperatures. This observation then requires a higher radiation dose for the critical condition of gel formation and thus a lower value for $G(X)$ than at the higher temperature. Although it appears that a qualitative material balance can be obtained under these conditions, no quantitative comparison can be made. Moreover, the possibilities of chain scission, under these conditions of irradiation, still remains to be experimentally examined in sufficient detail to establish its significance. One must conclude from the above that the total problem of demonstrating a material balance, under any fixed irradiation condition, still remains to be resolved. The claims of success, in this endeavor (Kang *et al.*, 1967) utilizing an unfractionated sample, containing a high concentration of vinyl end group, must be considered to be fortuitous, particularly since the very special and unique effects of this end group were not taken into account.

In summary, we can conclude that the major chemical acts accompanying the irradiation of linear polyethylene give quantitative yields which are very similar to those that have been reported for the lower molecular weight *n*-paraffins. One would therefore anticipate that the reaction mechanism for the polymer and the monomeric analogue should be the same. The major exception is the lack of detection to date of any chain scission processes in the polymer after irradiation in the completely molten state or in the crystalline state at temperatures above the α transition. The question of whether chain scission occurs upon irradiation in the crystalline state, at temperatures below that of the α transition, is still unresolved. The distinct possibility exists, however, for degradation to occur under these circumstances. It has also become clear that one of the major reasons for the complexities that have arisen in studying the radiation chemistry of linear polyethylene is the presence of one vinyl end group per molecule in the usual type of sample studied. Although the absolute concentration of this species is small, it exerts a profound influence acting as both a hydrogen sink and participating in an additional type of intermolecular cross-linking reaction. A very careful distinction must then be made when this type of polyethylene is studied. It clearly represents a very special situation which can not be taken to be representative of the species linear polyethylene.

An examination of the available experimental data indicates that the independent variables, some of which are of a subtle nature, must be clearly specified. These include the molecular weight, the state of the system, the level of crystallinity, the morphology and mode of crystallization, and the temperature. The latter quantity needs to be specified not only in the usual absolute sense but also in relation to the location of the different structural transitions typical of linear polyethylene.

The problem of achieving a material balance is still unresolved for all of the major conditions of irradiations that have been studied. In general, it is found that more hydrogen is evolved than can be accounted for by attributing its yield to random intermolecular cross-links and to the formation of unsaturation. A search for other reactions, in which hydrogen is evolved, is suggested, as is the development of experimental methods to detect intramolecular cross-links as well as spatially nonrandom intermolecular cross-links.

ACKNOWLEDGMENTS

Support from the National Aeronautics and Space Administration under Research Grant NSG-247-62 and from the Division of Biology and Medicine, Atomic Energy Commission is gratefully acknowledged.

REFERENCES

Alexander, P., and Toms, D. (1956). *J. Polym. Sci.* **22**, 343.
Anderson, F. R. (1964). *J. Appl. Phys.* **35**, 64; (1963). *J. Polym. Sci.* **3C**, 123.
Bergmann, K., and Nawatki, K. (1967). *Kolloid-Z.* **219**, 132.
Black, R. M., (1958). *J. Appl. Chem. (London)* **8**, 159.
Boyer, R. F. (1963). *Rubber Chem. Tech.* **36**, 1303.
Chapiro, A. (1962a). "Radiation Chemistry of Polymeric Systems," p. 433. Wiley (Interscience), New York.
Chapiro, A. (1962b). "Radiation Chemistry of Polymeric Systems," pp. 419 ff. Wiley (Interscience), New York.

Chapiro, A. (1962c). "Radiation Chemistry of Polymeric Systems," p. 407. Wiley (Inter-science), New York.

Chapiro, A. (1962d). "Radiation Chemistry of Polymeric Systems", p. 409. Wiley (Inter-science), New York

Chapiro, A. (1962e). "Radiation Chemistry of Polymeric Systems," p. 412. Wiley (Inter-science), New York.

Chapiro, A. (1962f). "Radiation Chemistry of Polymeric Systems," p. 79. Wiley (Interscience), New York.

Chapiro, A. (1962g). "Radiation Chemistry of Polymeric Systems," p. 75. Wiley (Inter-science), New York.

Chapiro, A. (1962h). "Radiation Chemistry of Polymeric Systems," p. 439. Wiley (Inter-science), New York.

Charlesby, A. (1952). *Proc. Roy. Soc. (London)* **A215**, 187.

Charlesby, A. (1960). "Atomic Radiation and Polymers," p. 211. Pergamon Press, New York.

Charlesby, A. and Callaghan, L. (1958). *J. Phys. Chem. Solids* **4**, 227, 306.

Charlesby, A., and Davison, W. H. T. (1957). *Chem. Ind. (London)* 232.

Charlesby, A., and Hancock, N. H. (1953). *Proc. Roy. Soc. (London)* **A218**, 245.

Charlesby, A., and Pinner, S. H. (1959). *Proc. Roy. Soc. (London)* **A249**, 367.

Charlesby, A., and Ross, M. (1953). *Proc. Roy. Soc. (London)* **A217**, 212.

Charlesby, A., Gould, A. R., and Ledbury, K. J. (1964). *Proc. Roy. Soc. (London)* **A227**, 348.

Devoy, C., Mandelkern, L., and Bourland, L. (1970). *J. Polym. Sci. A-2*, **8**, 869.

Dewhurst, H. A. (1957). *J. Phys. Chem.* **61**, 1466.

Dewhurst, H. A. (1958). *J. Phys. Chem.* **62**, 15.

Dole, M. (1950), *Rep. Symp. 9th Chem. Phys. Radiat. Dosimet.* Army Chem. Center.

Dole, M. (1965), *In* "Crystalline Olefin Polymers Part I" (R. A. V. Raff and K. W. Koak, eds.), Chapter 16. Wiley (Interscience), New York.

Dole, M., and Katsuura, K. (1965). *J. Polym. Sci.* **3B**, 467.

Dole, M., Keeling, C. D., and Rose, D. G. (1954). *J. Amer. Chem. Soc.* **76**, 4304.

Dole, M., Milner, D. C., and Williams, T. F. (1957). *J. Amer. Chem. Soc.* **79**, 4809.

Dole, M., Milner, D. C., and Williams, T. F. (1958). *J. Amer. Chem. Soc.* **80**, 1580.

Dole, M., Fallgatter, M. B., and Katsuura, K. (1966). *J. Phys. Chem.* **70**, 62.

Eppe, R., Fischer, E. W., and Stuart, H. A. (1959). *J. Polym. Sci.* **34**, 721.

Ergöz, E. (1970). Ph. D. Dissertation, Florida State Univ.

Evans, R. D., Mighton, H. R., and Flory, P. J. (1950). *J. Amer. Chem. Soc.* **72**, 2018.

Falconer, W. E., and Salovey, R. (1966). *J. Chem. Phys.* **44**, 3151.

Fallgatter, M. B., and Dole, M. (1964). *J. Phys. Chem.* **68**, 1988.

Fatou, J. G., and Mandelkern, L. (1965). *J. Phys. Chem.* **69**, 417.

Fischer, E. W. (1957). *Z. Naturforsch.* **129**, 753.

Fischer, E. W., and Kloos, F. (1970). *J. Polym. Sci.* **8B**, 685.

Fischer, E. W., and Schmidt, G. F. (1962). *Angew. Chem.* **1**, 488.

Fischer, E. W., Kloos, P. and Lieser, G. (1969). *J. Polym. Sci.* **7B**, 845.

Flory, P. J. (1941). *J. Amer. Chem. Soc.* **63**, 3097.

Flory, P. J. (1969). "Statistical Mechanics of Chain Molecules." Wiley (Interscience), New York.

Fujiwara, Y. (1960). *J. Appl. Poly. Sci.* **4**, 10.

Geil, P. H. (1960). *J. Polym. Sci.* **44**, 449; **47**, 65.

Gopalan, M. R., and Mandelkern, L. (1967). *J. Polym. Sci.* **5B**, 925.

Hardwick, T. J. (1960). *J. Phys. Chem.* **64**, 1623.

Hardwick, T. J. (1961). *J. Phys. Chem.* **65**, 101.

Hendus, H., and Schnell, G. (1967). *Kunststoffe* **57**, 193.

Hoffman, J. D., Lauritzen, J. I., Jr., Passaglia, E., Ross, G. S., Frolen, L. J. and Weeks, J. J. (1969). *Kolloid-Z.* **231**, 564.

Illers, K. H. (1964). *Rheol. Acta* **3**, 202.

Inokuti, M. (1963). *J. Chem. Phys.* **38**, 2999.

Jackson, J. F., and Mandelkern, L. (1968). *Macromolecules* **1**, 546.

Jackson, J. B., Flory, P. J. and Chiang, R. (1963). *Trans. Faraday Soc.* **59**, 1906.

Kang, H. Y., Saito, O., and Dole, M. (1967). *J. Amer. Chem. Soc.* **89**, 1980.

Keller, A. (1957). *Phil. Mag.* **2**, 1171; (1959). *Die Makromol. Chem.* **34**, 1.

Keller, A., and Priest, D. (1968). *J. Macromol. Sci.* **B2**, 479.

Keller, A., and Sawada, S. (1964). *Die Makromol. Chem.* **74**, 190.

Kitamaru, R., and Mandelkern, L. (1964). *J. Amer. Chem. Soc.* **86**, 3529.

Kitamaru, R., Mandelkern, L., and Fatou, J. G. (1964). *J. Polym. Sci.* **2B**, 511.

Lawton, E. J., Balwit, J. S., and Powell, R. S. (1958a). *J. Poly. Sci.* **32**, 257; (1960). *J. Chem. Phys.* **33**, 405.

Lawton, E. J., Balwit, J. S., and Powell, R. S. (1958b). *J. Polym. Sci.* **32**, 277.

Mandelkern, L. (1958). *In* "Growth and Perfection of Crystals" (R. H. Doremus, B. W. Roberts, and D. Turnbull, eds.). Wiley, New York.

Mandelkern, L. (1954). *Polymer* **5**, 637.

Mandelkern, L. (1964). "Crystallization of Polymers." McGraw-Hill, New York.

Mandelkern, L. (1966). *J. Polym. Sci.* **C15**, 129.

Mandelkern, L. (1967). *Polym. Eng. Sci.* **7**, 232.

Mandelkern, L. (1969). *Polym. Eng. Sci.* **9**, 255.

Mandelkern, L. (1970). *In* "Progress in Polymer Science" (A. D. Jenkins, ed.), Vol. 2. Pergamon, Oxford.

Mandelkern, L., Roberts, D. E., Halpin, J. C., and Price, F. P. (1960). *J. Amer. Chem. Soc.* **82**, 46.

Mandelkern, L., Posner, A. S., Dioro A. F., and Roberts, D. E. (1961). *J. Appl. Phys.* **32**, 1509.

Mandelkern, L., Price, J. M., Gopalan, M. R., and Fatou, J. G. (1966). *J. Polym. Sci.* **4A**, 385.

Mandelkern, L., Allou, A. L., Jr., and Gopalan, M. R. (1968). *J. Phys. Chem.* **72**, 309.

Mandelkern, L., Sharma, R. K., and Jackson, J. F. (1969). *Macromolecules* **2**, 644.

Matsuo, H., and Dole, M. (1959). *J. Phys. Chem.* **63**, 837.

McCrum, N. G., Read, B. E., and Williams, G. (1963). "Anelastic and Dielectric Effects in Polymeric Solids," pp. 353ff. Wiley, New York.

Miller, A. A., Lawton, E. J., and Balwit, J. S. (1956). *J. Phys. Chem.* **60**, 599.

Okada, T. and Mandelkern, L. (1966). *Abstr. Int. Symp. Macromol. Chem. Tokyo.*

Okada, T., and Mandelkern, L. (1967). *J. Polym. Sci. A-2* **5**, 239.

Okada, T., and Mandelkern, L. (1972). To be published.

Okada, T., Mandelkern, L., and Glick, R. (1967). *J. Amer. Chem. Soc.* **89**, 4790.

Palmer, R. P., and Cobbald, A. J. (1964). *Die Makromol. Chem.* **74**, 174.

Peterlin, A. (1969). *J. Macromol. Sci. Phys.* **B3**(1), 19.

Quinn, F. A., Jr., and Mandelkern, L. (1958). *J. Amer. Chem. Soc.* **80**, 3178; (1959). *ibid.* **81**, 6355.

Rijke, A. J., and Mandelkern, L. (1971). *Macromolecules* **4**, 594.

Saito, O., Kang, H. Y., and Dole, M. (1967). *J. Chem. Phys.* **46**, 3607.

Salovey, R. (1962). *J. Polym. Sci.* **61**, 463.

Salovey, R. (1964). *J. Polym. Sci.* **2B**, 833.

Salovey, R., and Falconer, W. E. (1965). *J. Phys. Chem.* **69**, 2345.

Salovey, R., and Falconer, W. E. (1966). *J. Phys. Chem.* **70**, 3203.

Salovey, R., and Hellman, M. Y. (1968a). *J. Polym. Sci.* **6B**, 527.

Salovey, R., and Hellman, M. Y. (1968b). *Macromolecules* **1**, 456.

Salovey, R., and Keller, A. (1961). *Bell Syst. Tech. J.* **15**, 1397, 1409.

Salovey, R., Malm, D. L., Beach, A. L., and Luongo, J. P. (1964). *J. Polym. Sci.* **A-2**, 3067.

Schumacher, K. (1958). *Kolloid-Z.* **157**, 16.

Sears, W. C. (1964). *J. Polym. Sci.* **A2**, 2455.

Sella, C., and Torillat, J. J. (1958). *C. R. Acad. Sci. Paris* **246**, 3246; (1959). **248**, 410, 1819, 2348.

Sharma, R. K., and Mandelkern, L. (1970). *Macromolecules* **3**, 758.

Sinnott, K. M. (1966). *J. Appl. Phys.* **37**, 3385.

Slichter, W. P., and Mandell, E. R. (1958). *J. Phys. Chem.* **62**, 334.

Statton, W. O. (1961). *J. Appl. Phys.* **32**, 2332.

Statton, W. O., and Geil, P. H. (1960). *J. Appl. Polym. Sci.* **3**, 357.

Stehling, F. C., and Mandelkern, L. (1969). *J. Polym. Sci.* **B7**, 255.

Stehling, F. C., and Mandelkern, L. (1970). *Macromolecules* **3**, 242.

Stockmayer, W. H. (1944). *J. Chem. Phys.* **12**, 125.

Takayanagi, M. (1960). *J. Polym. Sci.* **46**, 531.

Takayanagi, M. (1965). *Proc. Int. Congr. Rheolog. 4th Kyoto. Japan, 1963* p.161.

Till, P. H. (1957). *J. Polym. Sci.* **24**, 301.

Turner, D. T. (1966). *J. Polym. Sci.* **4B**, 717.

Williams, T. F., and Dole, M. (1959). *J. Amer. Chem. Soc.* **81**, 2919.

Zachmann, H. G., and Peterlin, A. (1969). *J. Macromol. Sci. Phys.* **B3**(1), 495.

14

Free Radicals in Irradiated Polyethylene

Malcolm Dole

Department of Chemistry, Baylor University, Waco, Texas

I. Introduction

Free radicals are most readily observed at 77°K (l.n.t.) after an irradiation at that temperature. Free radicals have also been observed through their spin signals by EPR during the irradiation, thanks to the development of equipment for this purpose in the laboratory of V. V. Voevodsky in the Institute of Chemical Physics of Moscow in 1958 [see the memorial lecture by Kondratiev (1967)]. However, most EPR measurements have been made subsequent to the irradiation. In a sense the technique of producing and trapping free radicals in polyethylene at l.n.t. is somewhat similar to the matrix isolation method of studying free radicals in that at l.n.t. the free radicals are frozen in and do not react. One irradiation of linear polyethylene at 4°K has been reported (Waterman and Dole, 1971) with interesting results as described below.

II. Type and Identification of Free Radicals

The free radicals of importance in the radiation chemistry of polyethylene are the alkyl radical, $-CH_2\dot{C}HCH_2-$, the allyl, $-\dot{C}-HCH=CH-$, and the polyenyl free radicals, $-\dot{C}-H(CH=CH)_n-$ although the latter are produced in significant yields only at high doses. As already pointed out in Chapter 12, relative concentrations of the alkyl free radical can be estimated from the height of the wing peak of its EPR spectrum. See Fig. 1 where the spectra of mixed alkyl and allyl free radicals and purely allyl free radicals are compared.

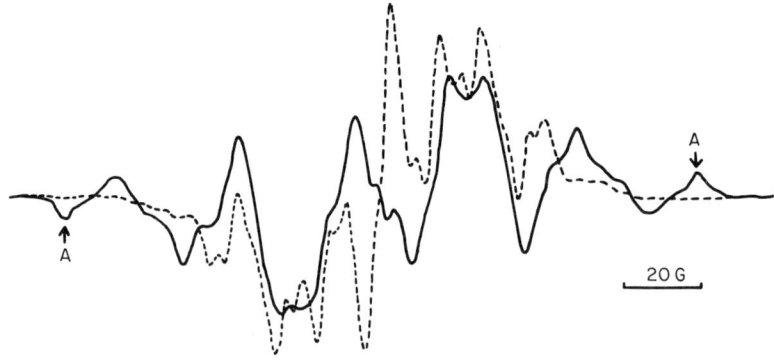

Fig. 1. Comparison of ESR spectra of mixed alkyl and allyl free radicals, solid line, and of the allyl free radical, dotted line. From Waterman and Dole (1970b). [Copyright (1970) by the American Chemical Society. Reprinted by permission of the copyright owner.]

The assignment of the EPR spectra to the alkyl and allyl free radicals is due to a number of workers such as Libby *et al.* (1960), Lawton *et al.* (1960), Ohnishi *et al.* (1961, 1962), and Charlesby *et al.* (1961). Libby *et al.* (1960) confirmed the assignment of the six-line spectrum to the $-CH_2\dot{C}HCH_2-$ radical by observing the spectrum at 77°K of a stretched polyethylene sample. This spectrum consisted of five doublets when the stretching axis was perpendicular to the applied magnetic field. A theoretically derived spectrum assuming an absorption peak ratio of 1:4:6:4:1, a main component separation of 28 G, a doublet splitting of 13 G, and a linewidth of 10.5 G agreed well with the observed spectrum. This treatment yields an unpaired spin density on the α-hydrogen only 13/28 of that on the four β-hydrogen atoms. It is the latter group of atoms that yields the five basic lines each of which is split into a doublet by the α-hydrogen atom. The six-line spectrum ordinarily seen in irradiated polyethylene is explained on the assumption of equal unpaired spin densities on all five α- and β-hydrogen atoms.

Charlesby *et al.* (1961) determined the spectrum due to the allyl free radical by subtracting the alkyl radical spectrum from the EPR spectrum of 17-pentatriacontene irradiated at l.n.t. but observed after heating to room temperature. However, the most complete analysis of the EPR allyl spectrum is that of Ohnishi *et al.* (1962). In considering the allyl free radical it is necessary to distinguish between the various carbon and hydrogen atoms as indicated in Fig. 2 where the nomenclature of Ohnishi *et al.* is used. Ohnishi

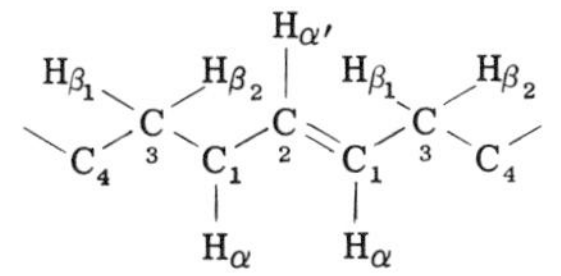

Fig. 2. Structure of the allyl free radical (after Ohnishi *et al.*, 1962).

used oriented linear polyethylene which was electron beam irradiated and studied the EPR spectrum with the long fiber axis parallel to the magnetic field. With this orientation the C_1–H_α bonds are at right angles to the field. Let θ be the angle between the C_3–H_β bond and the π-electron axis as projected on a plane vertical to the C_1–C_3 bond. At $-180°$ Ohnishi's results indicated that the C_3–$H_{\beta,1}$ angle θ_1 is different from that of the C_3–$H_{\beta,2}$ angle θ_2, but at 148°C chain oscillations average these two angles to 33°13′. At $-180°$ they are 27°30′ and 41°39′, respectively; in other words, the radical is twisted at $-180°$, due probably to the fact that the angle between the C_1–C_3 and the C_3–C_4 bonds is greater than the tetrahedral angle of 109°28′. Ohnishi *et al.* were able to calculate an activation energy of 410 cal mole^{-1} for the twisting motion. Taking the unpaired electron spin densities on carbon atoms C_1 and C_2 to be 0.622 and –0.231 as calculated by Lefkowitz *et al.* (1955) and the coupling constants to be $32.5 \times 0.622 = 20.2$ G and $32.5 \times 0.231 = 7.5$ G for H_α and H_α', respectively, at l.n.t. and 21.3 G for the H_β protons, Ohnishi *et al.* were able to get good agreement with the observed EPR spectra. For example, at 142° neglecting any splitting due to H_α', the two H_α and the four H_β protons should give rise to a seven-line EPR spectrum with intensities 1:6:15:20:15:6:1 inasmuch as the coupling constants for the two H_α protons can be taken to be equal to 21.3 G at that temperature. The interaction of the α' proton with the unpaired electron should cause an additional splitting of each of the above lines into a closely spaced doublet.

At low temperatures Ohnishi *et al.* (1962) observed that the EPR spectrum of their oriented PE fibers changed reversibly into a very complicated pattern which could be interpreted on the basis of a change of the $H_{\beta 1}$ and $H_{\beta 2}$ coupling constants with temperature, due to a change in the θ_1 and θ_2 angles. Figure 3 illustrates the temperature dependence of the different coupling constants.

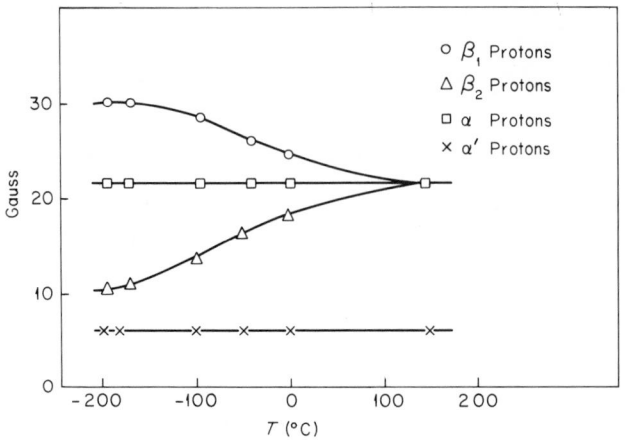

Fig. 3. Hyperfine coupling constants of the allyl free radical (Ohnishi *et al.*, 1962).

III. *G* Values

The abundance of the allyl free radical can only be determined from the EPR spectra after the alkyl free radical has decayed. In the work of Waterman and Dole (1970b) the decay of the alkyl free radical at room temperature (r.t.) was accelerated by the presence of 20 or 40 cm of hydrogen pressure (see below). The use of UV spectroscopy, however, makes it possible to detect and to measure the relative concentrations of the allyl free radical even in the presence of the alkyl (Waterman and Dole, 1970a, b). In this way the latter authors were able to demonstrate the negligibly small formation of allyl free radicals either at 77° or 4°K (Waterman and Dole, 1971).

In Table I are collected data of various investigators for the *G* values of the alkyl and allyl free radicals. Waterman and Dole (1970a) found that the alkyl free radical grew linearly with dose at least up to 80 Mrad when PE was irradiated at l.n.t. They also found that the UV absorption band at 215 nm which they attributed to the alkyl radical grew linearly with dose, but in these experiments the observations were only extended up to a dose of 5 Mrad.

TABLE I

G-VALUES FOR THE FORMATION OF FREE RADICALS IN IRRADIATED LINEAR POLYETHYLENE

Sample	Irradiation conditions		Dose (Mrad)	G(alkyl) Measured at 77°K	G(allyl) Measured at r.t.	Ref.
	Press.	Temp. (°K)				
Rigidex 15	0	77	50	2.4	0.2	a
Hostalen GF	0	77	50	2.8	0.3	a
Marlex 6002	0	77	80	3.3	0.23	b
Stretched						
Rigidex 15	0	77	50	1.3	0.4	a
Hostalen GF	0	77	50	1.4	0.3	a
Marlex 50	N_2	77	40	3.0	(Total radicals)	c
Marlex 50	0	77	20	6.1		d

[a] Charlesby *et al.* (1961).
[b] Waterman and Dole (1970a).
[c] Lawton *et al.* (1960).
[d] Ohnishi (1961).

The allyl G values were calculated from the allyl EPR spectra at r.t. after the alkyl radicals had decayed.

The agreement between the results of the different investigators is quite good except for the one result of Ohnishi (1961) for G(alkyl). This value of 6.1 is claimed by Ohnishi to include free radicals formed at l.n.t. and decaying during the irradiation. An interesting observation by Charlesby *et al.* (1961) is the rather low values of G(alkyl), but not G(allyl), for the stretched linear polyethylene samples. The reason for this is not known at the present time; it would be interesting to have $G(H_2)$, G(unsat.) and G(cross-link) data for these samples at l.n.t. Unfortunately, there is no method presently available for the measurement of either cross-links or hydrogen evolution at l.n.t. The sample has to be heated above l.n.t. to measure either of these quantities. One might guess that the lowered G(alkyl) value in the stretched linear polyethylene is related to the greater ease of forming the vinylene double bond in such stretched samples and to the correspondingly smaller number of irradiation produced excited states available for free radical formation. Such an explanation is in line with the statement of Charlesby *et al.* (1961) that there were fewer cross-links in the irradiated stretched samples. Inasmuch as cross-links are formed by the recombination of free radicals, with fewer of the latter, there would be fewer cross-links. This explanation is also in line with the allyl free radical yields in the stretched samples which are equal to or greater than those in the unstretched samples. With an increased G(unsat.) there would be a greater chance for allyl groups to form.

The dienyl free radical, $-\dot{C}H(-CH=CH)_2-$, is produced at a rather low G value of 0.015 ± 0.005 as determined by Waterman and Dole (1970a) from a combination of EPR, IR and UV measurements. Knowing G(allyl) and G(dienyl) the latter authors were able to determine the extinction coefficients in both the IR at 943 nm and in the UV at 258 nm (allyl) and 285 (dienyl) for the allyl and dienyl free radicals given in Tables I and II of Chapter 12.

IV. Kinetics of Decay

Theoretically, if the alkyl free radicals decay by recombination or disproportionation, one might expect the free radicals to decay by a second-order process, and a second-order decay has indeed been observed in a number of cases, especially by Charlesby *et al.* (1961) (see Fig. 4). These

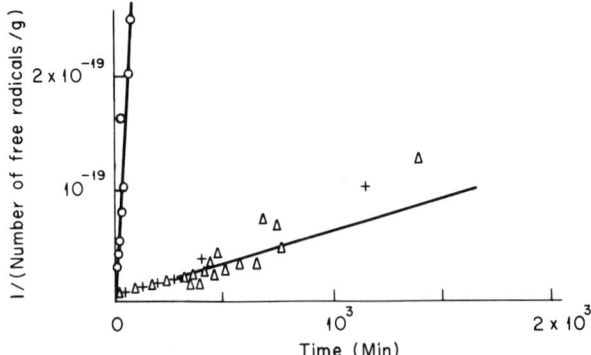

Fig. 4. Second-order decay plot of alkyl radicals (Charlesby *et al.*, 1961). Open circles, stretched linear polyethylene; crosses, the same unstretched; triangles, a very high molecular weight polyethylene.

authors irradiated their samples at 77°K, then heated them to 20°C where the decay rate was measured. Most of the free radicals decayed on heating to r.t. [in a later work Waterman and Dole (1970b) found that only 3.6% of the alkyl radicals survived the heating to r.t.], and Charlesby *et al.* obtained the alkyl free radical concentration after subtracting from the total radical concentration their estimated concentrations of the allyl free radicals. Some of their second-order decay constants, k_2, are given in Table II. Apparently, the less the crystallinity the greater the decay rate; the stretched Rigidex 15 also had a higher decay rate than the unstretched Rigidex 15. One difficulty not forseen by these authors was the effect of ambient hydrogen gas on the free radical decay rate.

TABLE II

Second-Order Alkyl Radical Decay Constants at 20°C[a]

Type of PE	Dose (Mrad)	k_2 (free radicals $g^{-1})^{-1}sec^{-1}$
Low-density alkathene Q 53	50	$>2 \times 10^{-21}$
High-density, Rigidex 15	50	9×10^{-24}
High-density, Hostalen GF	50	$\sim2 \times 10^{-21}$
Stretched alkathene Q 53	50	$>2 \times 10^{-21}$
Stretched Rigidex 15	50	5×10^{-22}
Stretched Hostalen GF	50	$>2 \times 10^{-21}$

[a] Date of Charlesby *et al.* (1961).

In 1962 Cracco *et al.* observed that the decay of the free radicals at r.t. in irradiated linear PE was somewhat greater in the presence of hydrogen than in its absence. Ormerod (1963) found that his second-order decay constants for the alkyl radical decay increased linearly with the pressure of hydrogen gas. Waterman and Dole (1970b) confirmed these observations, and demonstrated that both the first- and second-order alkyl radical decay constants increased linearly with hydrogen gas pressure. Deuterium gas was also used, but the effect of hydrogen was 4.0 and 5.0 times as great as deuterium on the first- and second-order free radical decay constants, respectively. The effect of hydrogen is quite pronounced: Doubling the hydrogen pressure almost doubles the decay rate constant. Thus, for accurate interpretation of the data, the ambient hydrogen pressure should be recorded along with other pertinent data such as temperature, type of PE, etc.

Waterman and Dole (1970b) found that in their experiments the 3.6% of the alkyl radicals which survived the heating to room temperature after an irradiation at l.n.t. decayed accurately according to the first-order law whether in the presence or absence of hydrogen gas (see Fig. 5). The second-order decay observed by them was exhibited by the alkyl radicals which had been regenerated by UV photolysis of the allyl free radicals at l.n.t. This ultraviolet regeneration process was first observed by Ohnishi *et al.* (1963). However, the second-order decay was observed only in the presence of hydrogen gas; the decay rate of the regenerated alkyl radicals at zero ambient pressure was immeasurably slow.

It was mentioned above that theoretically one would expect a second-order alkyl decay. This would be true for a free radical recombination reaction in a homogeneous medium. In a paper entitled "Conditions for First or Second-Order Kinetics during Multiple Zone Reactions," Dole and Inokuti (1963) examined mathematically the overall kinetics that would be observed if free radical recombination reactions were occurring simultaneously in a number of

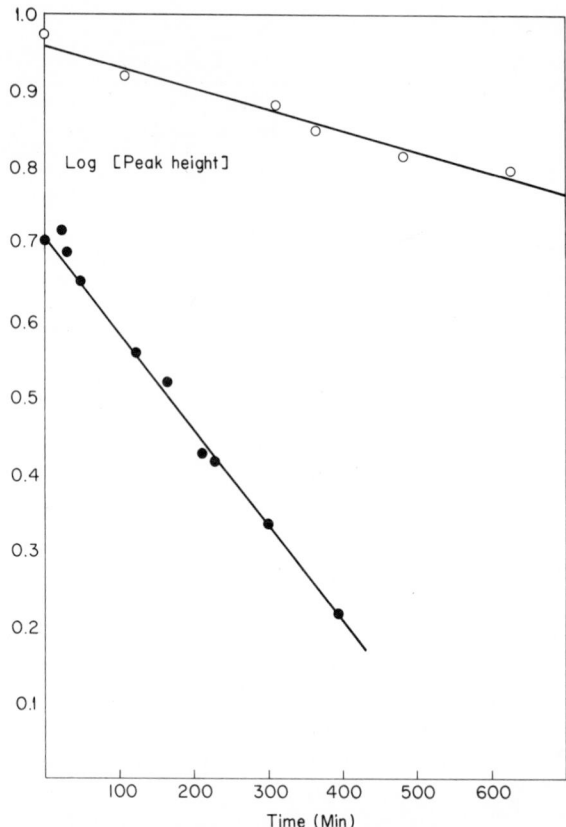

Fig. 5. First-order decay of alkyl free radicals at room temperature; open circles, *in vacuo*; closed circles, in the presence of 10 cm of H_2 pressure. Irradiation to 60 Mrad at l.n.t. Data of Waterman and Dole (1970b). [Reprinted from *J. Phys. Chem.* **74**, 1913 (1970). Copyright (1970) by the American Chemical Society. Reprinted by permission of the copyright owner.]

isolated zones. Assuming that the free radicals are produced in pairs in each zone and that there was no overlap between zones, it turns out that the recombination kinetics is first order when there is only one free radical pair in each zone. Figure 6 illustrates the way in which the kinetics changes from first to second order as the number of pairs in each isolated zone increases. By the time that there are four free radical pairs in each zone the observed overall kinetics would be almost 80% pure second order. Figure 6 is based on the assumption of an equal number of radical pairs in each zone. If, however, the radical distribution is a multinomial (random) distribution, then the overall observed reaction rate constant would be strictly second order. In

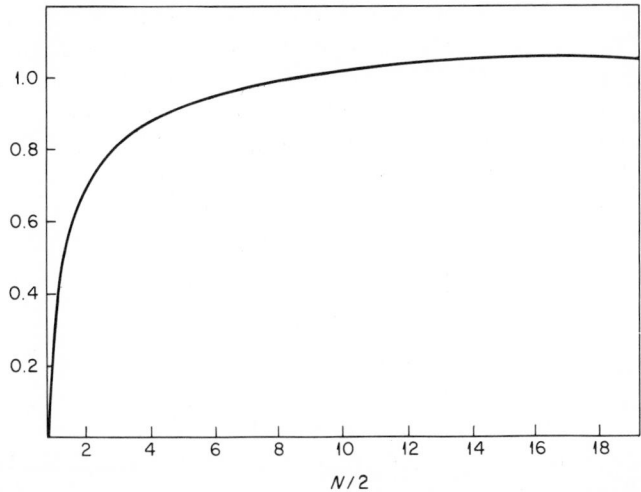

Fig. 6. Fraction of pure second-order kinetics in one reaction zone plotted as a function of the number of pairs of free radicals per zone. From Dole and Inokuti (1963).

this connection it is interesting to point out that Iwasaki *et al.* (1969) measured the concentration of free radical pairs in irradiated PE (type of PE not specified) in which the two free radicals were close enough together to give rise to a triplet type ESR signal, i.e., spectra in which $\Delta M_s = 2$. However, the ratio of the integrated $\Delta M_s = 2$ spectrum intensity to that of the $M_s = 1$ spectrum was only 6×10^{-5} in the case of PE.

As is rather obvious, the decay rate of free radicals in PE depends to a very great extent on the ease of molecular motions in the solid. A very convincing demonstration of this was obtained in a study of isotactic polypropylene by Nara *et al.* (1967a). These authors observed a second-order decay reaction at a number of temperatures in two temperature regions, namely, in the neighborhood of 170° and 260°K where the free radicals rapidly decayed. The activation energies, $\Delta E\ddagger$, as computed from the variation of the second order decay constant with temperature according to the Arrhenius relation,

$$k_2 = A \exp(-\Delta E\ddagger/RT) \qquad (1)$$

for the two regions were 11 and 48 kcal mole^{-1}, respectively. These values agreed rather closely with the activation energies for the γ and β dispersion of the mechanical motions in polypropylene, namely, 13 and 58 kcal mole^{-1}, respectively.

In the case of polyethylene Cracco *et al.* (1962) observed that the free radicals produced in Marlex-50 polyethylene by irradiation at l.n.t. began to decay at about 150°K with the most rapid rate of decay occurring at about 170°–180°K. Nara *et al.* (1967b) state that in their studies the most rapid rate of decay of free radicals in an irradiated linear PE occurred near 120, 200, and 250°K. Cracco *et al.* (1962) also found that the evolution of hydrogen, which had been trapped in the solid PE at l.n.t., paralleled the decay of free radicals, starting at about 160°K and reaching its greatest rate of evolution at about 190°K. Thus, the molecular motions which made possible the escape of hydrogen from the solid, also very probably were the segmental chain motions which permitted the free radical recombination reactions to occur. Stehling and Mandelkern (1970) have recently determined the glass-transition temperature T_g for linear PE to be about 143°K. This is the temperature, within the experimental limits of uncertainty, that the decay of free radicals and the evolution of hydrogen begin as an irradiated PE sample is heated from 77°K to room temperature.

In studying the temperature region between 4° and 77°K, Waterman and Dole (1971) observed two overlapping absorption bands in the IR at 973 and 966 cm^{-1} after an irradiation at 4°K; on heating to 77°K these two bands merged into a single sharp band characteristic of the vinylene group at 966 cm^{-1}. Its area was equal to the sum of the areas of the two overlapping bands, and a tentative assignment of the new peak to the $-CH^+=CH-$ group was made.

V. Mechanisms

Alkyl radicals produced by irradiation of PE in a vacuum at l.n.t. can only decay as the temperature is raised to room temperature by one of three processes: These are (a) recombination to form cross-links, (b) disproportionation to form a double bond and a saturated group, and (c) migration of the alkyl group to a double bond to form an allyl free radical. It should be realized that molecular hydrogen is formed during the irradiation and provides an atmosphere of hydrogen gas as the temperature is raised above 150°K. Although Dole and Cracco (1962) were able to demonstrate that the following reaction occurred at room temperature subsequent to the irradiation and after replacement of H_2 by D_2,

$$R'\cdot + D_2 + RH \rightarrow R'D + HD + R\cdot$$
$$(I)$$

there is no evidence that atomic hydrogen is produced during the reaction and reacts with free radicals.

Process (b), disproportionation, occurs only to a very small extent, if at all, because no increase in the vinyl or vinylene concentration is observed on heating to room temperature after the irradiation at l.n.t. In fact, the vinyl concentration was observed to decrease, while the vinylene concentration remained unchanged (Dole *et al.*, 1958). Some of the free radicals equal to a *G* value of about 0.14 form the allyl free radical (Waterman and Dole, 1970a, b) on the heating from l.n.t. to room temperature, but the great majority of them must recombine to form cross-links (process a). Waterman and Dole (1970b) found that the alkyl free radicals that survived the heating to r.t. practically quantitatively converted to allyl free radicals by migration to a double bond (process c).

It will be noted that reaction (I) either with D_2 or H_2 provides a chemical mechanism for the migration of the free radical through the solid PE. But it is difficult to decide whether free radicals decay either in the amorphous or crystalline regions, or both, or chiefly at the interfacial surfaces. Direct evidence that oxygen of the air reacts at room temperature with dienyl free radicals at two different rates is shown in Fig. 7 (Böhm, 1967). Note the

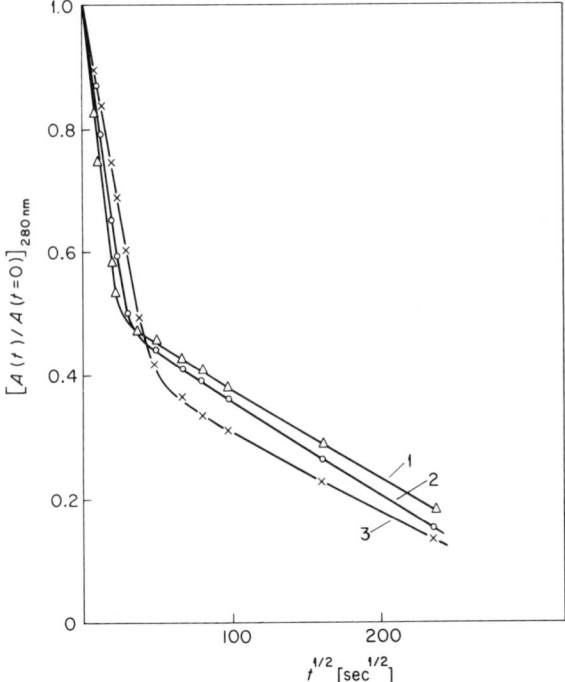

Fig. 7. Decay of dienyl free radicals on exposure to air at room temperature as observed by the UV absorption band at 285 nm. Data of Böhm (1967). Irradiations at l.n.t. to 20 Mrad, curve 1; 40 Mrad, curve 2; and 80 Mrad, curve 3. Curves taken after an annealing at room temperature for 5 hr. [By permission of John Wiley & Sons, Inc.]

initial sharp drop with time in the UV absorption band at 285 nm, and then when about 50–60% of the dienyl free radicals had decayed, the rate of oxidation became much slower. Böhm assumed that oxygen could penetrate only into the amorphous regions where initially the oxidation rate was about three-fold faster than in the later stages. The latter slow rate was assumed to be controlled by the rate of diffusion of the radicals out of the crystalline regions into the amorphous regions.

One problem of interest is whether free radical migration in the PE is along or across chains. If the migration occurs by means of reaction (I), it would presumably be intermolecular. An abstraction type of free radical transfer, as represented by the reaction

$$R\cdot + R'H \rightarrow RH + R'\cdot$$
(II)

would also be between chains, but a hydrogen atom "hopping" mechanism of the type

$$-CH_2\dot{C}HCH_2- \rightarrow -\dot{C}HCH_2CH_2-$$
(III)

would be intramolecular. Dole *et al.* (1969) estimated the activation energy of process (III) and concluded that it was too high for (III) to be able to compete significantly with (II).

The allyl and polyenyl free radicals are too stable, relative to the alkyl radical, to undergo hydrogen exchange reactions of the type illustrated by reactions (I) or (II). In fact, it should be pointed out that reactions (I) and (II) probably can occur only when the C–H bond energy is close to that of the H–H bond energy. In the case of the polyoxymethylene free radical, $-CH_2O\dot{C}HOCH_2-$, Lee and Dole (1971) found no acceleration in its decay rate due to ambient hydrogen. This is explained by them as due to the stability of the radical because of resonance between electrons of the lone electron pairs on the two neighboring oxygen atoms and the unpaired electron of the free radical.

REFERENCES

Böhm, G. G. A. (1967). *J. Polym. Sci. A-2* **5**, 639.
Charlesby, A., Libby, D., and Ormerod, M. G. (1961). *Proc. Roy. Soc. (London)* **A262**, 207.
Cracco, F., Arvia, A. J., and Dole, M. (1962). *J. Chem. Phys.* **37**, 2449
Dole, M., and Cracco, F. (1962). *J. Phys. Chem.* **66**, 193.
Dole, M., and Inokuti, M. (1963). *J. Chem. Phys.* **39**, 310.
Dole, M., Milner, D. C., and Williams, F. (1958). *J. Amer. Chem. Soc.* **80**, 1580.
Dole, M., Böhm, G. G. A., and Waterman, D. C. (1969). *Eur. Polym. J. Suppl.* 93.
Iwasaki, M., Ichikawa, T., and Ohmori, T. (1969). *J. Chem. Phys.* **50**, 1984.

Kontratiev, V. N. (1967). *Int. Symp. Free Radicals, 8th Novosibirsk*, U.S.S.R.
Lawton, E. J., Balwit, J. S., and Powell, R. S. (1960). *J. Chem. Phys.* **33**, 395.
Lee, P. N., and Dole, M. (1971). Unpublished results.
Lefkowitz, H. C., Fain, J., and Matsen, F. A. (1955). *J. Chem. Phys.* **23**, 1690.
Libby, D., Ormerod, M. G., and Charlesby, A. (1960). *Polymer* **1**, 212.
Nara, S., Kashiwabara, H., and Sohma, J. (1967a). *Rep. Progr. Polym. Phys. Japan* **10**, 479.
Nara, S., Shimada, S., Kashiwabara, H., and Sohma, J. (1967b). *Rep. Prog. Polym. Phys. Japan* **10**, 483.
Ohnishi, S. (1961). *Bull. Chem. Soc. Japan* **35**, 254.
Ohnishi, S., Ikeda, Y., Kashiwagi, M., and Nitta, I. (1961). *Polymer* **2**, 119.
Ohnishi, S., Sugimoto, S., and Nitta, I. (1962). *J. Chem. Phys.* **37**, 1283.
Ohnishi, S., Sugimoto, S., and Nitta, I. (1963). *J. Chem. Phys.* **39**, 2647.
Ormerod, M. G. (1963). *Polymer* **4**, 451.
Stehling, F. C., and Mandelkern, L. (1970). *Macromolecules* **3**, 242.
Waterman, D. C., and Dole, M. (1970a). *J. Phys. Chem.* **74**, 1906.
Waterman, D. C., and Dole, M. (1970b). *J. Phys. Chem.* **74**, 1913.
Waterman, D. C., and Dole, M. (1971). Abstracts, Phys. Chem. Div., Amer. Chem. Soc. Meeting, Los Angeles, Calif.

Author Index

Subject Index

A

Absorption bands
  of aliphatic polymers, 42
  of alkyl free radicals, 338
  of allyl free radicals, 340
  of biphenyl radical anion, 149
  of dienyl free radicals, 340
  of γ-irradiated hydrocarbon glasses, 149
  of radicals in 3-methyl pentane, 175
  of trapped electrons, 148
Acceptor molecule, 29, 32
Acrylate polymers electron attachment, 19
Activation energy, 81, 82, 84
  apparent, 82
  of dark conductivity, 135
  of decay of PTFE electret, 140
  of depolarization currents, 139
  of electron decay in PE, 183–185
  of fluorescence, 212
  from glow peak temperature, 200
  of molecular motion, 199, 343
  negative, 105
  optical, 197, 213
  of phosphorescence, 212
  of thermoluminescence in PE, 211, 212
    in PMMA, 218
    in PS, 218
    in PTFE, 219
  of the twisting motion in the allyl free radical, 337
Additives
  amines, 50
  aromatic, 45, 50
  benzene in c-hexane, 47
  2,5-diphenyl oxazole, 47
  pyrene, 46
  triphenylamine, 45

Alkane polymers
  C–C bond exciton migration in, 46
  excitation transfer in, 39
Alkanes
  excitation-transfer in, 39
  excited states in, 39
  ground states, 28
  ion pair yield in, 28
  positive ions in, 27
  protection of, 25
  radiolysis of compared to PE, 5, 302, 326–328
  thermoluminescence in, 215, 216
  trapped electron in, 178
Alkenes, mono, 49
Alkylbenzenes, 48
Alkyl free radical
  analysis of, 285
  EPR spectrum of, 163, 164, 336, 337
  mechanism of decay, 344, 345
  migration in PE, 345
  second order decay constants, 341 (table)
  splitting constants in, 68
  structure of, 68
  UV spectrum of, 338
Allyl free radical, 282, 283, 336
  EPR spectrum of, 337
  mechanism of formation, 345
  power saturation in EPR studies of, 284
  structure of, 337
  twisting motion in, 337
  UV absorption bands in, 337
Analytical techniques
  EPR spectroscopy, 283
  gas, 279, 280
  IR spectroscopy, 280
  NMR spectroscopy, 285
  UV spectroscopy, 282

2-3753